Geometry-Driven Diffusion in Computer Vision

Computational Imaging and Vision

Volume 1

Table of Contents

5 BAYESIAN RATIONALE FOR THE VARIATIONAL FORMULATION 135

David Mumford

10 MORPHOLOGICAL APPROACH TO MULTISCALE ANALYSIS: FROM PRINCIPLES TO EQUATIONS 229

Luis Alvarez and Jean-Michel Morel

13 NONLINEAR SCALE-SPACE   339

Luc Florack, Alsons Salden, Bart ter Haar Romeny, Jan Koenderink and Max Viergever

14 A DIFFERENTIAL GEOMETRIC APPROACH TO
ANISOTROPIC DIFFUSION   371

David Eberly

15 NUMERICAL ANALYSIS OF GEOMETRY-DRIVEN DIFFUSION EQUATIONS   393

Wiro Niessen, Bart ter Haar Romeny and Max Viergever

Preface

Scale is a concept the antiquity of which can hardly be traced. Certainly the familiar phenomena that accompany scale changes in optical patterns are mentioned in the earliest written records. The most obvious topological changes such as the *creation* or *annihilation* of details have been a topic of fascination to philosophers, artists and later scientists. This appears to be the case for all cultures from which extensive written records exist. For instance, chinese 17^{th}c artist manuals remark that "distant faces have no eyes". The *merging* of details is also obvious to many authors, *e.g.*, Lucretius mentions the fact that distant islands look like a single one. The one topological event that is (to the best of my knowledge) mentioned only late (by John Ruskin in his "Elements of drawing" of the mid 19^{th}c) is the *splitting* of a blob on blurring. The change of images on a gradual increase of resolution has been a recurring theme in the arts (*e.g.*, the poetic description of the distant armada in Calderón's *The Constant Prince*) and this "mystery" (as Ruskin calls it) is constantly exploited by painters.

It was facts as these that induced me to attempt a mathematical study of scale, especially when my empirical researches in the psychophysics of spatial contrast in the peripheral visual field of human observers convinced me that the brain samples and represents the optic array at many scales simultaneously. Events accelerated around 1980. I remember that I met Andy Witkin at the 1st David Marr memorial workshop thrown by Whitman Richards at Cold Spring Harbor: I believe that there I heard the term "scale-space" for the 1st time. By then I had set up the diffusion equation formulation that connects the scales formally though I didn't publish because the whole thing still had too much ad hockery to it. However, shortly thereafter the "causality constraint" occurred to me (triggered by a fascination with cartographic "generalization") and I saw the pieces click together except for a few frayed ends. (Although the notion of "causality" in the scale domain remains indeed a corner stone of the theory in my opinion, its original formulation—though correct—was perhaps unfortunate because often misunderstood even to the point were people believe it to be false.) Since then much has happened and I feel that today "scale-space" can stand on its own.

"Pure scale-space" can be constructed from a few basic assumptions that essentially express *total a priori ignorance*. Thus no scale, place or orientation is potentially special in any way. The resulting pristine structure has the beauty and fascination of something inevitable, a *discovery* rather than a mere *construction*. However, it tends to leave practical people unsatisfied because in any *real* application one knows many things that might advantageously be exploited in the analysis. This might be any form of prior knowledge, including rather abstract notions of what structure to expect, so that the actual data can be used to control the operations performed on themselves. It seems natural to try to use the freedom left in pure scale-space to attach such control handles and thereby turn the nymph into a handmaiden. It is perhaps fair to say that no one today knows how to proceed in an apparently *necessary* way here, thus the theory sprouts into a multitude of complementary and concurrent approaches.

I have read through the manuscript of this book in fascination. Most of the approaches that have been explored to tweak scale-space into practical tools are represented here. Many of the contributions have a certain attraction of their own and several appear to be promising in future applications. It is easy to appreciate how both the purist and the engineer find problems of great interest in this area. The book is certainly unique in its scope and has appeared at a time where this field is booming and newcomers can still potentially leave their imprint on the core corpus of scale related methods that will slowly emerge. As such the book is a very timely one. It is quite evident that it would be out of the question to compile anything like a textbook at this stage: This book is a snapshot of the field that manages to capture its current state very well and in a most lively fashion. I can heartily recommend its reading to anyone interested in the issues of image structure, scale and resolution.

One thing that the reader will notice is the nature of the pictures that go to illustrate many of the contributions: They are quite distinct in character from the type of illustration one meets in the pure diffusion literature. The difference lies in the very sharpness of the "blurred" images. Here we meet with the same fascination one finds in *e.g.*, Canaletto's figures in his paintings of Venice's squares: As the viewer looks at the painted figures farther and farther away in the perspective of the pavement details are lost though the figures are always made up of sharply delimited blobs of paint. In the near foreground a single blob may stand for the button of a waistcoat, in the far distance a whole head, yet everything appears mysteriously "sharp" at *any* distance. It is as with cartographic generalization where the general shape of a city may suddenly give way to a circular disc.

For those readers sensitive to such matters the mathematics also has a quite different flavor from the original diffusion literature. We meet with an *embarras de richesse* perhaps typical of a field in rapid development. It seems almost inevitable that many of the various strands will eventually come together in the weave of something novel.

I compliment the editor on this timely and fascinating book and promise the reader pleasure and profit.

Jan J. Koenderink, Utrecht University

Foreword

This book is the fruit of a European-American transatlantic collaboration, sponsored by an ESPRIT–NSF grant (travel and subsistence) awarded for the period 1993–1996 [96]. The book brings together the important groups active in this field, many of which have an outstanding record of achievements. The group formed in 1992, with a collaborating effort to write the grant proposal. Since then a number of successful meetings were held: In May 1993 the kick-off meeting was held in Berlin. Here the concept of the book was born. The next meeting was a joint meeting with the Newton Mathematical Institute, end November 1993. The workshop in Stockholm, May 1994, in connection with the 3rd European Conference on Computer Vision saw the final draft presentation of this book as first tangible result of the collaboration.

A short synopsis of the book:

Chps. 1 and 2 by Lindeberg and ter Haar Romeny give a tutorial introduction to linear scale-space. It includes the history, its (axiomatic) foundations, basic theory and an overview of the current research topics, related kernels such as wavelets and Gabor functions, and it introduces the notion of scaled differential operators and their applications. Directions into the study of the 'deep structure' of scale-space are indicated.

From Chp. 3 onwards the book focuses on non-linear or geometry-driven diffusion. Coarsely, two major approaches can be discriminated, each with their variations: the variational approach, were the energy of a functional is minimised, the functional being some cost function that can be manipulated, and on the other hand the nonlinear PDE approach, where the evolution of the image is expressed as some function of invariant properties of the image. Historically geometry-driven diffusion was introduced as being some local function of edge-strength. This is discussed in Chp. 3 by Perona, Shiota and Malik, together with a critical analysis of the resulting non-linear PDE and the discrete maximum principle. In Chp. 4 by Whitaker and Gerig not only edges are involved, but a complete 'feature space' gives rise to multi-valued or vector-valued diffusion, using both (higher order) features in a geometry-limited diffusion scheme, as local frequency decomposition in a spectra-limited diffusion scheme.

The variational approach is given a firm Bayesian basis by Mumford, one of its original authors, in Chp. 5. Four probabilistic models are presented from reasoning and axioms, and shown on examples.

The variational approach is a prototypical example of a 'free-discontinuity problem'. Critical questions like uniqueness of the solution and existence of an optimal segmentation are raised by Leaci and Solimini in Chp. 6.

Edges get displaced when they are blurred. Nordström in Chp. 7 studies edges as finite unions of smooth curves (line drawings) and demonstrates the existence of a solution in this particular case.

In the next two chapter a link is being established between the variational approach and the PDE-based evolutions. In Chp. 8 Richardson and Mitter extend the variational approach with the theory of Γ-convergence, i.e. the approximation of one functional by another. Replacing the edge set with a smooth function then leads to edge detection by *coupled* partial differential equations. These are further discussed and elaborated in Chp. 9 by Proesmans, Pauwels and van Gool, extending the applications to second order smoothing, multispectral images, optic flow and stereo.

An attempt to some unification is presented in Chp. 10 by Alvarez and Morel, who summarize the morphological approach to geometry-driven diffusion in the light of many other 'scale-space' or multi-scale approaches. The authors establish a number of basic principles for the visual pyramid and derive the 'fundamental equation' of shape analysis.

The general form of the nonlinear diffusion equations are always some function of a particular invariant. In Chp. 11 Olver, Sapiro and Tannenbaum first give a tutorial on basic invariant theory, starting from Lie groups and prolongations. Images can be considered embedded sets of curves. The authors then describe in detail the invariant *curve evolution* flows under Euclidean, similarity, affine and projective transformations, as well as the impact of certain conservation laws.

Kimia, Tannenbaum and Zucker take methods from optimal control theory in Chp. 12. Dynamic programming and the application of the Hamilton-Jacobi equation are applied to shape theory, morphology, optic flow, nonlinear scale-space and shape-from-shading.

Two chapters introduce and apply the notion of the covariant formalism for coordinate free reasoning. Florack et al. study in Chp. 13 a special nonlinear scale-space, which can be mapped onto a linear scale-space. The Perona and Malik equation is studied in this context, and a log-polar foveal scale-space is presented as one of the examples. Eberly in Chp. 14 gives detailed examples of nonlinear scale-space properties once a proper *metric* is chosen. The appendix of Chp. 13 is a tutorial for the covariant formalism.

The volume ends with a chapter on implementations. They can be found scattered in the book, but here Niessen et al. show the feasibility of many nonlinear PDE approaches by implementing them in a forward Euler fashion with Gaussian derivatives on a variety of (medical) images.

Ackowledgements

The atmosphere of enthusiasm and the flexible cooperation of all authors made editing this book a pleasure.
On behalf of all authors I would like to express gratitude to ESPRIT and NSF. Their grant for transatlantic collaboration enabled the organization of workshops and mutual visits, and stimulated the collaboration necessary for the endeavour of this volume.
I like to thank the members of the 3D Computer Vision Research Group, and my wife Hetty for their support, especially at busy times, and Utrecht University and the Utrecht University Hospital for the time and facilities to create this book.

Bart M. ter Haar Romeny, Utrecht, June 1994

Contributors

Luis Alvarez – Universidad de Las Palmas, Departamento de Informatica, Facultad de Informatica, Campus Universitario de tafira, 35017 Las Palmas, Spain. luis@amihp710.dis.ulpgc.es

David Eberly – University of North Carolina, Department of Computer Science, Sitterson Hall 083A, Chapel Hill, NC 27514, USA. eberly@cs.unc.edu

Luc M. J. Florack – 3D Computer Vision Research Group, Utrecht University, Heidelberglaan 100, E.02.222, 3584 CX Utrecht, the Netherlands. luc@cv.ruu.nl

Guido Gerig – Communication Technology Laboratory, Image Science Division, ETH-Zentrum, Gloriastr.35, CH-8092 Zürich, Switzerland. gerig@vision.ethz.ch

Luc van Gool – Catholic University Leuven ESAT/MI2, Kardinaal Mercierlaan 94, B-3001 Heverlee, Belgium. Luc.Vangool@esat.kuleuven.ac.be

Bart M. ter Haar Romeny – Utrecht University, 3D Computer Vision Research Group, Heidelberglaan 100, E.02.222, 3584 CX Utrecht, the Netherlands. bart@cv.ruu.nl

Benjamin B. Kimia – Brown University, Laboratory for Man/Machine Systems, Box D, 182 Hope Street, Providence, RI 02912, USA. kimia@lems.brown.edu

Jan J. Koenderink – Utrecht University, Department of Biophysics, Princetonplein 5, 3508 TA Utrecht, the Netherlands. koenderink@fysbb.fys.ruu.nl

Antonio Leaci – Universitá di Lecce, Dipartimento di Matematica, 73100 Lecce, Italy. dipmat@vaxle.le.infn.it

Tony Lindeberg – Royal Institute of Technology (KTH), Department of Numerical Analysis and Computing Science, S-100 44 Stockholm, Sweden. tony@bion.kth.se

Jitendra Malik – University of California Berkeley, Department of Electrical Engineering and Computer Science, 555 Evans Hall, Berkely, CA 94720, USA. malik@robotics.eecs.berkeley.edu

Sanjoy K. Mitter – Massachusetts Institute of Technology, Department of Electrical Engineering, Room 35-308, Cambridge, Massachusetts 02139, USA. mitter@lids.mit.edu

Jean-Michel Morel – Universite Paris Dauphine, CEREMADE, Room C-612, Place du Marechal du Lattre de Tassigny. 75775 Paris CEDEX 16, France. morel@paris9.dauphine.fr

David Mumford – Harvard University, Department of Mathematics, Science Center 519, Cambridge, MA 02138, USA. mumford@math.harvard.edu

Wiro J. Niessen – Utrecht University, 3D Computer Vision Research Group, Heidelberglaan 100, E.02.222, 3584 CX Utrecht, the Netherlands. wiro@cv.ruu.nl

Niklas Nordstrom – Royal Institute of Technology (KTH), Department of Numerical Analysis and Computing Science, S-100 44 Stockholm, Sweden. niklas@bion.kth.se

Peter J. Olver – University of Minnesota, Department of Mathematics, 206 Church Street SE, Minneapolis, MN 55455-0488, USA. olver@ima.umn.edu

Eric Pauwels – Catholic University Leuven ESAT/MI2, Kardinaal Mercierlaan 94, B-3001 Heverlee, Belgium. Eric.Pauwels@esat.kuleuven.ac.be

Pietro Perona – California Institute of Technology, Department of Electrical Engineering – MS 116-81, Pasadena, CA 91125, USA. perona@verona.caltech.edu

Marc Proesmans – Catholic University Leuven ESAT/MI2, Kardinaal Mercierlaan 94, B-3001 Heverlee, Belgium. Marc.Proesmans@esat.kuleuven.ac.be

Thomas J. Richardson – AT&T Bell Laboratories, 600 Mountain Av., Murray Hill, NJ 07974, USA. tjr@research.att.com

Alfons H. Salden – 3D Computer Vision Research Group, Utrecht University, Heidelberglaan 100, E.02.222, 3584 CX Utrecht, the Netherlands. alfons@cv.ruu.nl

Guillermo Sapiro – Massachusetts Institute of Technology, LIDS, MIT - 35 - 431, Cambridge MA 02139, USA. guille@lids.mit.edu

Takahiro Shiota – Kyoto University, Math Department, School of Science, Sakyo-ku, Kyoto 606-01, Japan. shiota@math.harvard.edu

Sergio Solimini – Universitá di Lecce, Dipartimento di Matematica, Via Lecce Arnesano 73100, Lecce, Italy. dipmat@vaxle.le.infn.it

Alan R. Tannenbaum – University of Minnesota, Department of Electrical Engineering, Minneapolis MN 55455, USA. tannenba@ee.umn.edu

Max A. Viergever – 3D Computer Vision Research Group, Utrecht University, Heidelberglaan 100, E.02.222, 3584 CX Utrecht, the Netherlands. max@cv.ruu.nl

Ross T. Whitaker – European Computer-Industry Research Center GmbH, Arabellastrasse 17, D-81925 Munchen, Germany. ross@ecrc.de

Steven W. Zucker – McGill University, Department of Electrical Engineering, 3480 University Street, Montreal, Quebec, Canada, H3A 2A7. zucker@lightning.mcrcim.mcgill.edu

LINEAR SCALE-SPACE I: BASIC THEORY

Tony Lindeberg

Royal Institute of Technology (KTH)
Computational Vision and Active Perception Laboratory (CVAP)
Department of Numerical Analysis and Computing Science
S-100 44 Stockholm, Sweden

and

Bart M. ter Haar Romeny

Utrecht University, Computer Vision Research Group,
Heidelberglaan 100 E.02.222,
NL-3584 CX Utrecht, The Netherlands

1. Introduction

Vision deals with the problem of deriving information about the world from the light reflected from it. Although the active and task-oriented nature of vision is only implicit in this formulation, this view captures several of the essential aspects of vision. As Marr (1982) phrased it in his book *Vision*, vision is an information processing task, in which an *internal representation* of information is of out-most importance. Only by representation information can be captured and made available to decision processes. The purpose of a representation is to make certain aspects of the information content *explicit*, that is, immediately accessible without any need for additional processing.

This introductory chapter deals with a fundamental aspect of early image representation—the notion of *scale*. As Koenderink (1984) emphasizes, the problem of scale must be faced in any imaging situation. An inherent property of objects in the world and details in images is that they only exist as meaningful entities over certain ranges of scale. A simple example of this is the concept of a branch of a tree, which makes sense only at a scale from, say, a few centimeters to at most a few meters. It is meaningless to discuss the tree concept at the nanometer or the kilometer level. At those scales it is more relevant to talk about the molecules that form the leaves

1

of the tree, or the forest in which the tree grows. Consequently, a multi-scale representation is of crucial importance if one aims at describing the structure of the world, or more specifically the structure of projections of the three-dimensional world onto two-dimensional images.

The need for multi-scale representation is well understood, for example, in cartography; maps are produced at different degrees of abstraction. A map of the world contains the largest countries and islands, and possibly, some of the major cities, whereas towns and smaller islands appear at first in a map of a country. In a city guide, the level of abstraction is changed considerably to include streets and buildings etc. In other words, maps constitute symbolic multi-scale representations of the world around us, although constructed manually and with very specific purposes in mind. To compute any type of representation from image data, it is necessary to extract information, and hence interact with the data using certain *operators*. Some of the most fundamental problems in low-level vision and image analysis concern: *what* operators to use, *where* to apply them, and *how large* they should be. If these problems are not appropriately addressed, the task of interpreting the output results can be very hard. Ultimately, the task of extracting information from real image data is severely influenced by the inherent *measurement* problem that real-world structures, in contrast to certain ideal mathematical entities, such as "points" or "lines", appear in different ways depending upon the scale of observation.

Phrasing the problem in this way shows the intimate relation to physics. Any *physical observation* by necessity has to be done through some *finite aperture*, and the result will, in general, depend on the aperture of observation. This holds for any device that registers physical entities from the real world including a vision system based on brightness data. Whereas constant size aperture functions may be sufficient in many (controlled) physical applications, e.g., fixed measurement devices, and also the aperture functions of the basic sensors in a camera (or retina) may have to determined *a priori* because of practical design constraints, it is far from clear that registering data at a fixed level of resolution is sufficient. A vision system for handling objects of different sizes and at difference distances needs a way to control the scale(s) at which the world is observed.

The goal of this chapter is to review some fundamental results concerning a framework known as *scale-space* that has been developed by the computer vision community for controlling the scale of observation and representing the multi-scale nature of image data. Starting from a set of basic constraints (axioms) on the first stages of visual processing it will be shown that under reasonable conditions it is possible to substantially restrict the class of possible operations and to derive a (unique) set of weighting profiles for the aperture functions. In fact, the operators that are obtained bear qualitative similarities to receptive fields at the very earliest stages of (human) visual processing (Koenderink 1992). We shall mainly be concerned with the operations that are performed directly on raw image data by the processing modules are collectively termed the *visual front-end*. The purpose of this processing is to register the information on the retina, and to make important aspects of it explicit that are to be used in later stage processes. If the operations are to be local, they have to preserve the topology at the retina; for this reason the processing can be termed *retinotopic processing*.

1.1. Early visual operations

An obvious problem concerns what information should be extracted and what computations should be performed at these levels. Is *any* type of operation feasible? An axiomatic approach that has been adopted in order to restrict the space of possibilities is to assume that the very first stages of visual processing should be able to function without any direct knowledge about what can be expected to be in the scene. As a consequence, the first stages of visual processing should be as uncommitted and make as few irreversible decisions or choices as possible.

The Euclidean nature of the world around us and the perspective mapping onto images impose natural constraints on a visual system. Objects move rigidly, the illumination varies, the size of objects at the retina changes with the depth from the eye, view directions may change etc. Hence, it is natural to require early visual operations to be unaffected by certain primitive transformations (e.g. translations, rotations, and grey-scale transformations). In other words, the visual system should extract properties that are *invariant* with respect to these transformations.

As we shall see below, these constraints leads to operations that correspond to spatio-temporal derivatives which are then used for computing (differential) *geometric* descriptions of the incoming data flow. Based on the output of these operations, in turn, a large number of feature detectors can be expressed as well as modules for computing surface shape.

The subject of this chapter is to present a tutorial overview on the historical and current insights of linear scale-space theories as a paradigm for describing the structure of scalar images and as a basis for early vision. For other introductory texts on scale-space; see the monographs by Lindeberg (1991, 1994) and Florack (1993) as well as the overview articles by ter Haar Romeny and Florack (1993) and Lindeberg (1994).

2. Multi-scale representation of image data

Performing a physical observation (e.g. *looking* at an image) means that some physical quantity is measured using some set (array) of measuring devices with certain apertures. A basic tradeoff problem that arises in this context is that if we are interested in resolving small *details* then the apertures should be narrow which means that less of the physical entity will be registered. A larger aperture on the other hand gives a stronger response and coarser details. Since we, in general, cannot know in advance what aperture sizes are appropriate, we would like to be able to treat the scale of observation as a *free* parameter so as to be able to handle all scales simultaneously. This concept of having a range of measurements using apertures of different physical sizes corresponding to observations at a range of scales is called a *multi-scale* measurement of data.

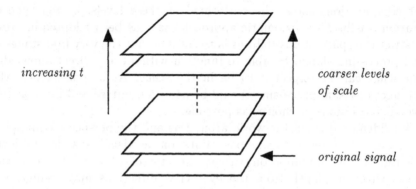

increasing t   *coarser levels of scale*

original signal

Figure 1.1.   A multi-scale representation of a signal is an ordered set of derived signals intended to represent the original signal at various levels of scale. (From [223].)

In case a set of measurement data is already given (as is the case when an image is registered at a certain scale using a camera) this process can be simulated by the vision system. The basic idea behind a multi-scale representation is to embed the original signal into such a one-parameter family of derived signals. How should such a representation be constructed? A crucial requirement is that structures at coarse scales in the multi-scale representation should constitute *simplifications* of corresponding structures at finer scales—they should not be accidental phenomena created by the smoothing method.

This property has been formalized in a variety of ways by different authors. A noteworthy coincidence is that the same conclusion can be reached from several different starting points. The main result we shall arrive at is that if rather general conditions are imposed on the types of computations that are to be performed at the first stages of visual processing, then the Gaussian kernel and its derivatives are singled out as the only possible smoothing kernels. The requirements, or axioms, that specify the Gaussian kernel are basically linearity and spatial shift invariance combined with different ways of formalizing the notion that structures at coarse scales should be related to structures at finer scales in a well-behaved manner; new structures should not be created by the smoothing method.

An simple example where structure is created in a "multi-scale representation" is when an image is enlarged by pixel replication (see figure 2). The sharp boundaries at regular distances are not present in the original data; they are just artifacts of the scale-changing (zooming) process.

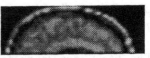

Figure 1.2. Example of what may be called creation of spurious structure; here by generating coarser-scale image representations by subsampling followed by pixel replication. (left) Magnetic resonance image of the cortex at resolution 240 * 80 pixels. (middle) Subsampled to a resolution of 48 × 16 pixels and illustrated by pixel replication. (right) Subsampled to 48 × 16 pixels and illustrated by Gaussian interpolation.

Why should one represent a signal at multiple scales when all information is present in the original data anyway? The major reason for this is to explicitly represent the multi-scale aspect of real-world images[1]. Another aim is to simplify further processing by removing unnecessary and disturbing details. More technically, the latter motivation reflects the common need for smoothing as a pre-processing step to many numerical algorithms as a means of noise suppression.

Of course, there exists a large number of possible ways to construct a one-parameter family of derived signals from a given signal. The terminology that will be adopted[2] here is to refer to as a "multi-scale representation" any one-parameter family for which the parameter has a clear interpretation in terms of spatial scale.

3. Early multi-scale representations

The general idea of representing a signal at multiple scales is not entirely new. Early work in this direction was performed by e.g. Rosenfeld and Thurston (1971), who observed the advantage of using operators of different

---

[1] At the first stages there should be no preference for any certain scale or range of scales; all scales should be measured and represented equivalently. In later stages the *task* may influence the selection, for example, do we want to see the tree or the leaves?

[2] In some literature the term "scale-space" is used for denoting any type of multi-scale representation. Using that terminology, the scale-space concept developed here should be called "(linear) diffusion scale-space".

size in edge detection. Related approaches were considered by Klinger
(1971), Uhr (1972), Hanson and Riseman (1974), and Tanimoto and Pavlidis
(1975) concerning image representations using different levels of spatial
resolution, i.e., different amounts of subsampling.

These ideas have then been furthered, mainly by Burt and by Crowley, to
one of the types of multi-scale representations most widely used today, the
pyramid. A brief overview of this concept is given below.

3.1. Quad tree

One of the earliest types of multi-scale representations of image data is the
quad tree[3] introduced by Klinger (1971). It is a tree-like representation of
image data in which the image is recursively divided into smaller regions.

The basic idea is as follows: Consider, for simplicity, a discrete two-
dimensional image I of size $2^K \times 2^K$ for some $K \in \mathbb{N}$, and define a measure
Σ of the grey-level variation in any region. This measure may e.g. be the
standard deviation of the grey-level values.

Let $I^{(K)} = I$. If $\Sigma(I^{(K)})$ is greater than some pre-specified threshold α,
then split $I^{(K)}$ into sub-images $I_j^{(K-1)}$ $(j = 1..p)$ according to a specified
rule. Apply this procedure recursively to all sub-images until convergence
is obtained. A tree of degree p is generated, in which each leaf $I_j^{(k)}$ is a
homogeneous block with $\Sigma(I_j^{(k)}) < \alpha$. (One example is given in figure 1.3.)
In the worst case, each pixel may correspond to an individual leaf. On the
other hand, if the image contains a small number of regions with relatively
uniform grey-levels, then a substantial data reduction can be obtained.

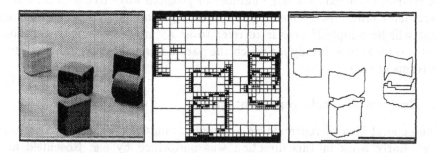

Figure 1.3.   Illustration of quad-tree and the split-and-merge segmentation algorithm;
(left) grey-level image, (middle) the leaves of the quad-tree, i.e., the regions after the
split step that have a standard deviation below the given threshold, (right) regions after
the merge step. (From [223].)

---

[3]For three-dimensional data sets, the corresponding representation is usually referred
to as *octtree*.

This representation has been used in simple segmentation algorithms for image processing of grey-level data. In the "split-and-merge" algorithm, a splitting step is first performed according to the above scheme. Then, adjacent regions are merged if the variation measure of the union of the two regions is below the threshold. Another application (when typically $\alpha = 0$) concerns objects defined by uniform grey-levels, e.g. binary objects; see e.g. the book by Tanimoto and Klinger (1980) for more references on this type of representation.

3.2. Pyramids

Pyramids are representations that combine the subsampling operation with a smoothing step. Historically they have yielded important steps towards current scale-space theories and we shall therefore consider them in more detail. To illustrate the idea, assume again, for simplicity, that the size of the input image I is $2^K \times 2^K$, and let $I^{(K)} = I$. The representation of $I^{(K)}$ at a coarser level $I^{(K-1)}$ is defined by a reduction operator. Moreover, assume that the smoothing filter is separable, and that the number of filter coefficients along one dimension is odd. Then, it is sufficient to study the following one-dimensional situation:

$$
\begin{aligned}
f^{(k-1)} &= \text{REDUCE}(f^{(k)}) \\
f^{(k-1)}(x) &= \sum_{n=-N}^{N} c(n)\, f^{(k)}(2x - n),
\end{aligned}
\tag{1.1}
$$

where $c\colon \mathbb{Z} \to \mathbb{R}$ denotes a set of filter coefficients. This type of *low-pass pyramid* representation (see figures 1.4–1.5) was proposed almost simultaneously by Burt (1981) and in a thesis by Crowley (1981).

Low-pass and band-pass pyramids: Basic structure. A main advantage of this representation is that the image size decreases exponentially with the scale level, and hence also the amount of computations required to process the data. If the filter coefficients $c(n)$ are chosen properly, the representations at coarser scale levels (smaller k) will correspond to coarser scale structures in the image data. How can the filter be determined? Some of the most obvious design criteria that have been proposed for determining the filter coefficients are:

- positivity: $c(n) \geq 0$,
- unimodality: $c(|n|) \geq c(|n+1|)$,
- symmetry: $c(-n) = c(n)$, and
- normalization: $\sum_{n=-N}^{N} c(n) = 1$.

Another natural condition is that all pixels should contribute equally to all levels. In other words, any point that has an odd coordinate index should

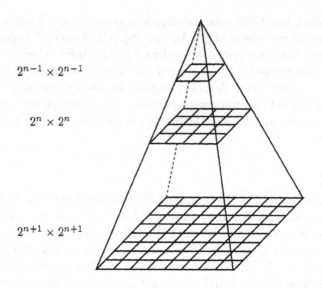

$2^{n-1} \times 2^{n-1}$

$2^n \times 2^n$

$2^{n+1} \times 2^{n+1}$

Figure 1.4. A pyramid representation is obtained by successively reducing the image size by combined smoothing and subsampling. (From [223].)

contribute equally to the next coarser level as any point having an even coordinate value. Formally this can be expressed as

– equal contribution: $\sum_{n=-N}^{N} c(2n) = \sum_{n=-N}^{N} c(2n+1)$.

Equivalently, this condition means that the kernel $(1/2, 1/2)$ of width two should occur as at least one factor in the smoothing kernel.

The choice of the filter size N gives rise to a trade-off problem. A larger value of N increases the number of degrees of freedom in the design at the cost of increased computational work. A natural choice when $N = 1$ is the binomial filter

$$\left(\frac{1}{4}, \frac{1}{2}, \frac{1}{4}\right) \tag{1.2}$$

which is the unique filter of width three that satisfies the equal contribution condition. It is also the unique filter of width three for which the Fourier transform $\psi(\theta) = \sum_{n=-N}^{N} c(n) \exp(-in\theta)$ is zero at $\theta = \pm\pi$. A negative property of this kernel, however, is that when applied repeatedly, the equivalent convolution kernel (which corresponds to the combined effect of iterated smoothing and subsampling) tends to a triangular function.

Of course there is a large class of other possibilities. Concerning kernels of width five $(N = 2)$, the previously stated conditions in the spatial domain imply that the kernel has to be of the form

$$\left(\frac{1}{4} - \frac{a}{2}, \frac{1}{4}, a, \frac{1}{4}, \frac{1}{4} - \frac{a}{2}\right). \tag{1.3}$$

Gaussian pyramid

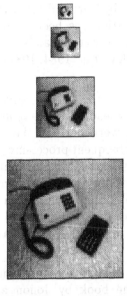

Figure 1.5. A Gaussian (low-pass) pyramid is obtained by successive smoothing and subsampling. This pyramid has been generated by the general reduction operator in equation (1.1) using the binomial filter from equation (1.2). (From [223].)

Burt and Adelson (1983) argued that a should be selected such that the equivalent smoothing function should be as similar to a Gaussian as possible. Empirically, they selected the value $a = 0.4$.

By considering a representation defined as the difference between two adjacent levels in a low-pass pyramid, one obtains a *bandpass pyramid*, termed "Laplacian pyramid" by Burt and "DOLP" (Difference Of Low Pass) by Crowley. This representation has been used for feature detection and data compression. Among features that can be detected are blobs (maxima), peaks and ridges etc (Crowley *et al.* 1984, 1987).

Properties. To summarize, the main advantages of the pyramid representations are that they lead to a *rapidly decreasing image size*, which reduces the computational work both in the actual computation of the representation and in the subsequent processing. The memory requirements are small, and there exist commercially available implementations of pyramids in hardware. The main disadvantage concerning pyramids is that they correspond to quite a coarse quantization along the scale direction, which makes it algorithmically complicated to relate (match) structures across scales. Furthermore, pyramids are not translationally invariant.

Further reading. There is a large literature on further work of pyramid representations; see e.g. the book by Jolion and Rosenfeld (1994), the books edited by Rosenfeld (1984), Cantoni and Levialdi (1986) as well as the special issue of *IEEE-TPAMI* edited by Tanimoto (1989). A selection of recent developments can also be found in the articles by Chehikian and Crowley (1991), Knudsen and Christensen (1991), and Wilson and Bhalerao (1992). An interesting approach is the introduction of "oversampled pyramids", in which not every smoothing step is followed by a subsampling operation, and hence, a denser sampling along the scale direction can be obtained. Pyramids can, of course, also be expressed for three-dimensional datasets such as medical tomographic data (see e.g. Vincken *et al.* 1992).

It is worth noting that pyramid representations show a high degree of similarity with a type of numerical methods called multi-grid methods; see the book by Hackbusch (1985) for an extensive treatment of the subject.

4. Linear scale-space

In the quad tree and pyramid representations rather coarse steps are taken in the scale-direction. A *scale-space representation* is a special type of multi-scale representation that comprises a *continuous scale parameter* and preserves the *same spatial sampling* at all scales. The Gaussian scale-space of a signal, as introduced by Witkin (1983), is an *embedding* of the original signal into a one-parameter family of derived signals constructed by convolution with Gaussian kernels of increasing width.

The linear scale-space representation of a continuous signal is constructed as follows. Let $f: \mathbb{R}^N \to \mathbb{R}$ represent any given signal. Then, the scale-space representation $I: \mathbb{R}^N \times \mathbb{R}_+ \to \mathbb{R}$ is defined by letting the scale-space representation at zero scale be equal to the original signal $I(\cdot; 0) = f$ and for $t > 0$

$$I(\cdot; t) = g(\cdot; t) * f, \tag{1.4}$$

where $t \in \mathbb{R}_+$ is the scale parameter, and $g: \mathbb{R}^N \times \mathbb{R}_+\backslash\{0\} \to \mathbb{R}$ is the Gaussian kernel; in arbitrary dimensions ($x \in \mathbb{R}^N, x_i \in \mathbb{R}$) it is written

$$g(x; t) = \frac{1}{(4\pi t)^{N/2}} e^{-x^T x/(4t)} = \frac{1}{(4\pi t)^{N/2}} e^{-\sum_{i=1}^{N} x_i^2/(4t)}. \tag{1.5}$$

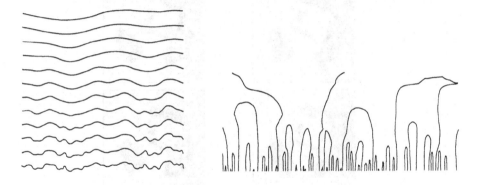

Figure 1.6. (left) The main idea with a scale-space representation of a signal is to generate a family of derived signals in which the fine-scale information is successively suppressed. This figure shows a one-dimensional signal that has been smoothed by convolution with Gaussian kernels of increasing width. (right) Under this transformation, the zero-crossings of the second derivative form paths across scales that are never closed from below. (Adapted from Witkin 1983).

Historically, the main idea behind this construction of the Gaussian scale-space representation is that the fine-scale information should be suppressed with increasing values of the scale parameter. Intuitively, when convolving a signal by a Gaussian kernel with standard deviation $\sigma = \sqrt{2t}$, the effect of this operation is to suppress[4] most of the structures in the signal with a

---

[4]Some care must, however, be taken when expressing such a statement. As we shall in section 3 in next chapter, adjacent structures (e.g. extrema) can be arbitrary close after arbitrary large amounts of smoothing, although the likelihood for the distance between two adjacent structures to be less than some value ϵ decreases with increasing scale.

Scale-Space   *Pyramid*

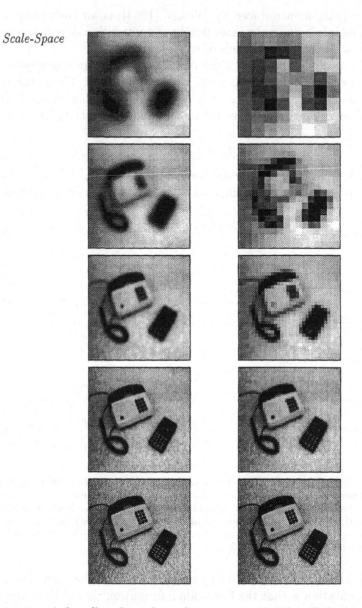

Figure 1.7. A few slices from the scale-space representation of the image used for illustrating the pyramid concept. The scale levels have been selected such that the standard deviation of the Gaussian kernel is approximately equal to the standard deviation of the equivalent convolution kernel corresponding to the combined effect of smoothing and subsampling (from bottom to top; $\sigma^2 = 0.5$, 2.5, 10.5, 42.5 and 270.5). For comparison, the result of applying the Laplacian operator to these images is shown as well. Observe the differences and similarities compared to figure 1.5. (From [223].)

characteristic length less than σ; see figure 1.6(a). Notice how this successive smoothing captures the intuitive notion of the signals becoming gradually smoother. A two-dimensional example is presented in figure 1.7.

5. Towards formalizing the scale-space concept

In this section we shall review some of the most important approaches for formalizing the notion of scale and for deriving the shape of the scale-space kernels in the linear scale-space theory. In a later chapter in this book by Alvarez and Morel (1994) it will be described how these ideas can be extended to non-linear scale-spaces and how the approach relates to mathematical morphology.

5.1. Continuous signals: Original formulation

When Witkin introduced the term scale-space, he was concerned with one-dimensional signals and observed that new local extrema cannot be created in this family. Since differentiation commutes with convolution,

$$\partial_{x^n} I(\cdot; t) = \partial_{x^n}(g(\cdot; t) * f) = g(\cdot; t) * \partial_{x^n} f, \qquad (1.6)$$

this non-creation property applies also to any n^{th}-order spatial derivative computed from the scale-space representation.

Recall that an extremum in a one-dimensional signal I is equivalent to a zero-crossing in the first-order derivative I_x. The non-creation of new local extrema means that the zero-crossings in any derivative of I form closed curves across scales, which will never be closed from below; see figure 1.6(b). Hence, in the one-dimensional case, the zero-crossings form paths across scales, with a set of inclusion relations that gives rise to a tree-like data structure, termed "interval tree". (For higher-dimensional signals, however, new local extrema and zero-crossings *can* be created; see section 5.5.)

An interesting empirical observation made by Witkin was a marked correspondence between the length of the branches in the interval tree and perceptual saliency:

> ... intervals that survive over a broad range of scales tend to leap out to the eye ...

In later work by Lindeberg (1991, 1993, 1994) it has been demonstrated that this observation can be extended to a principle for actually detecting significant image structures from the scale-space representation. That approach is based on the stability and lifetime over scales, the local contrast, and the spatial extent of blob-like image structures.

Gaussian smoothing has been used also before Witkin proposed the scale-space concept, e.g. by Marr and Hildreth (1980) who considered zero-crossings of the Laplacian in images convolved with Gaussian kernels of different standard deviation. One of the most important contributions of Witkins scale-space formulation, however, was the systematic way to *relate* and *interconnect* such representations and image structures at different scales in the sense that a *scale dimension* should be added to the scale-space representation, so that the behaviour of structures across scales can be studied. Some aspects of the resulting "deep structure" of scale-space, i.e. the study of the relations between structures at different scales, will be considered in the next chapter (section 3). See also (Koenderink 1994).

5.2. Inner scale, outer scale, and scale-space

Koenderink (1984) emphasized that the problem of scale must be faced in any imaging situation. Any real-world image has a limited extent determined by two scales, the *outer scale* corresponding to the finite size of the image, and the *inner scale* given by the resolution. For a digital image the inner scale is determined by the pixel size, and for a photographic image by the grain size in the emulsion.

As described in the introduction, similar properties apply to objects in the world, and hence to image features. The outer scale of an object or a feature may be said to correspond to the (minimum) size of a window that completely contains the object or the feature, whereas the inner scale of an object or a feature may loosely be said to correspond to the coarsest scale at which substructures of the object or the feature begin to appear.

Referring to the analogy with cartography given in the introduction, it should be emphasized that an atlas usually contains a set of maps covering some region of interest. Within each map the outer scale typically scales in proportion with the inner scale. A single map is, however, usually not sufficient for us to find our way around the world. We need the ability to zoom in to structures at different scales; i.e., decrease and increase the inner scale of the observation according to the type of situation at hand. (For an excellent illustration of this notion, see the popular book *Powers of Ten* (Morrison and Morrison 1985), which shows pictures of the world over 50 decades of scale from the largest to the smallest structures in the universe known to man.)

Koenderink also stressed that if there is no a priori reason for looking at specific image structures, the visual system should be able to handle image structures at *all* scales. Pyramid representations approach this problem by successive smoothing and subsampling of images. However,

> "The challenge is to understand the image really on all these levels *simultaneously*, and not as unrelated set of derived images at different levels of blurring ..."

Adding a scale dimension to the original data set, as is done in the one-parameter embedding, provides a formal way to express this interrelation.

5.3. Causality

The observation that new local extrema cannot be created when increasing the scale parameter in the one-dimensional case shows that the Gaussian convolution satisfies certain sufficiency requirements for being a smoothing operation. The first proof of the *necessity* of Gaussian smoothing for a scale-space representation was given by Koenderink (1984), who also gave a formal extension of the scale-space theory to higher dimensions. He introduced the concept of *causality*, which means that new level surfaces

$$\{(x, y;\ t) \in \mathbb{R}^2 \times \mathbb{R} : I(x, y;\ t) = I_0\}$$

must not be created in the scale-space representation when the scale parameter is increased (see figure 1.8). By combining causality with the notions of *isotropy* and *homogeneity*, which essentially mean that all spatial positions and all scale levels must be treated in a similar manner, he showed that the scale-space representation must satisfy the diffusion equation

$$\partial_t I = \nabla^2 I. \tag{1.7}$$

This diffusion equation (with initial condition $I(\cdot;\ 0) = f$) is the well-known physical equation that describes how a heat distribution I evolves over time t in a homogeneous medium with uniform conductivity, given an initial heat distribution $I(\cdot;\ 0) = f$ (see e.g. Widder 1975 or Strang 1986). Since the Gaussian kernel is the Green's function of the diffusion equation at an infinite domain, it follows that the Gaussian kernel is the unique kernel for generating the scale-space. A similar result holds, as we shall see later, in any dimension.

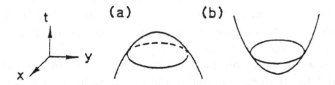

Figure 1.8. The causality requirement means that level surfaces in scale-space must point with their concave side towards finer scale (a); the reverse situation (b) should never occur. (From [223].)

The technique used for proving this necessity result was by studying the level surface through any point in scale-space for which the grey-level function assumes a maximum with respect to the spatial coordinates. If no new level surface is to be created when increasing scale, the level surface should point with its concave side towards decreasing scales. This gives rise to a sign condition on the curvature of the level surface, which assumes the form (1.7) when expressed in terms of derivatives of the scale-space representation with respect to the spatial and scale coordinates. Since points at which extrema are obtained cannot be assumed to be known a priori, this condition must hold in any point, which proves the result.

In the *one-dimensional* case, this level surface condition becomes a level curve condition, and is equivalent to the previously stated non-creation of local extrema. Since any n^{th}-order derivative of I also satisfies the diffusion equation

$$\partial_t I_{x^n} = \nabla^2 I_{x^n}, \tag{1.8}$$

it follows that new zero-crossing curves in I_x cannot be created with increasing scale, and hence, no new maxima.

A similar result was given by Yuille and Poggio (1985, 1986) concerning the zero-crossings of the Laplacian of the Gaussian. Related formulations have also been expressed by Babaud *et al.* (1986) and Hummel (1986, 1987).

5.4. Non-creation of local extrema

Lindeberg (1990, 1991) considered the problem of characterizing those kernels in one dimension that share the property of not introducing new local extrema in a signal under convolution. A kernel $h \in \mathbb{L}_1$ possessing the property that for *any* input signal $f_{in} \in \mathbb{L}_1$ the number of extrema in the convolved signal $f_{out} = h * f_{in}$ is always less than or equal to the number of local extrema in the original signal is termed a *scale-space kernel*:

- scale-space kernel: $\#_{extrema}(h * f_{in}) \leq \#_{extrema}(f_{in}) \quad \forall f_{in} \in \mathbb{L}_1$.

From similar arguments as in section 5.1, this definition implies that the number of extrema (or zero-crossings) in any n^{th}-order derivative is guaranteed never to increase. In this respect, convolution with a scale-space kernel has a strong smoothing property.

Such kernels can easily be shown to be positive and unimodal both in the spatial and the frequency domain. These properties may provide a formal justification for some of the design criteria listed in section 3.2 concerning the choice of filter coefficients for pyramid generation.

If the notion of a local maximum or zero-crossing is defined in a proper manner to cover also non-generic functions, it turns out that scale-space kernels can be completely classified using classical results by Schoenberg (1950). It can be shown that a continuous kernel h is a scale-space kernel if and only if it has a bilateral Laplace-Stieltjes transform of the form

$$\int_{x=-\infty}^{\infty} h(x) e^{-sx} dx = C e^{\gamma s^2 + \delta s} \prod_{i=1}^{\infty} \frac{e^{a_i s}}{1 + a_i s} \qquad (1.9)$$

where $-c < \text{Re}(s) < c$ for some $c > 0$ and where $C \neq 0$, $\gamma \geq 0$, δ and a_i are real, and $\sum_{i=1}^{\infty} a_i^2$ is convergent; see also the excellent books by Hirschmann and Widder (1955) and by Karlin (1968).

Interpreted in the spatial domain, this result means that for continuous signals there are four primitive types of linear and shift-invariant smoothing transformations; convolution with the *Gaussian kernel*,

$$h(x) = e^{-\gamma x^2}, \qquad (1.10)$$

convolution with the *truncated exponential functions*,

$$h(x) = \begin{cases} e^{-|\lambda| x} & x \geq 0, \\ 0 & x < 0, \end{cases} \qquad h(x) = \begin{cases} e^{|\lambda| x} & x \leq 0, \\ 0 & x > 0, \end{cases} \qquad (1.11)$$

as well as trivial *translation* and *rescaling*.

The product form of the expression (1.9) reflects a direct consequence of the definition of a scale-space kernel; the convolution of two scale-space kernels is a scale-space kernel. Interestingly, the reverse holds; a shift-invariant linear transformation is a smoothing operation if and only if it can be decomposed into these primitive operations.

5.5. Semi-group and continuous scale parameter

Another approach to find the appropriate families of scale-space kernels is provided by *group theory*. A natural structure to impose on a scale-space representation is a *semi-group* structure[5], i.e., if every smoothing kernel is associated with a parameter value, and if two such kernels are convolved with each other, then the resulting kernel should be a member of the same family,

$$h(\cdot; t_1) * h(\cdot; t_2) = h(\cdot; t_1 + t_2). \qquad (1.12)$$

In particular, this condition implies that the transformation from a fine scale to any coarse scale should be of the same type as the transformation from the original signal to any scale in the scale-space representation. Algebraically, this property can be written as

$$\begin{aligned} I(\cdot; t_2) \quad &= \{\text{definition}\} = h(\cdot; t_2) * f \\ &= \{\text{semi-group}\} = (h(\cdot; t_2 - t_1) * h(\cdot; t_1)) * f \\ &= \{\text{associativity}\} = h(\cdot; t_2 - t_1) * (h(\cdot; t_1) * f) \\ &= \{\text{definition}\} = h(\cdot; t_2 - t_1) * I(\cdot; t_1). \end{aligned} \qquad (1.13)$$

---

[5]Note that the semi-group describes an essentially one-way process. In general, this convolution operation cannot be reversed.

If a semi-group structure is imposed on a one-parameter family of scale-space kernels that satisfy a mild degree of smoothness (Borel-measurability) with respect to the parameter, and if the kernels are required to be symmetric and normalized, then the family of smoothing kernels is uniquely determined to be a Gaussian (Lindeberg 1990),

$$h(x;\ t) = \frac{1}{\sqrt{4\pi\alpha t}}e^{-x^2/(4\alpha t)} \qquad (t > 0 \quad \delta \in R). \qquad (1.14)$$

In other words, when combined with the semi-group structure, the non-creation of new local extrema means that the smoothing family is uniquely determined.

Despite the completeness of these results, however, they cannot be extended directly to higher dimensions, since in two (and higher) dimensions there are no non-trivial kernels guaranteed to never increase the number of local extrema in a signal. One example of this, originating from an observation by Lifshitz and Pizer (1990), is presented below; see also Yuille (1988) concerning creation of other types of image structures:

> Imagine a two-dimensional image function consisting of two hills, one of them somewhat higher than the other one. Assume that they are smooth, wide, rather bell-shaped surfaces situated some distance apart clearly separated by a deep valley running between them. Connect the two tops by a narrow sloping ridge without any local extrema, so that the top point of the lower hill no longer is a local maximum. Let this configuration be the input image. When the operator corresponding to the diffusion equation is applied to the geometry, the ridge will erode much faster than the hills. After a while it has eroded so much that the lower hill appears as a local maximum again. Thus, a new local extremum has been created.

Notice however, that this decomposition of the scene is intuitively quite reasonable. The narrow ridge is a fine-scale phenomenon, and should therefore disappear before the coarse-scale peaks. The property that new local extrema can be created with increasing scale is inherent in two and higher dimensions.

5.6. Scale invariance and the Pi theorem

A formulation by Florack *et al.* (1992) and continued work by Pauwels *et al.* (1994) show that the class of allowable scale-space kernels can be restricted under weaker conditions, essentially by combining the earlier mentioned conditions about linearity, shift invariance, rotational invariance and semi-group structure with *scale invariance*. The basic argument is taken from physics; physical laws must be independent of the choice of fundamental parameters. In practice, this corresponds to what is known as dimensional analysis[6]; a function that relates physical observables must be independent of the choice of dimensional units.[7] Notably, this condition comprises no direct measure of "structure" in the signal; the non-creation of new structure is only implicit in the sense that physically observable entities that are subject to scale changes should be treated in a self-similar manner.

Since this way of reasoning is valid in arbitrary dimensions and not very technical, we shall reproduce it (although in a modified form and with somewhat different proofs). The main result we shall arrive at is that scale invariance implies that the Fourier transform of the convolution kernel must be of the form

$$\hat{h}(\omega; \sigma) = \hat{H}(\omega\sigma) = e^{-\alpha |\omega\sigma|^p} \qquad (1.15)$$

for some $\alpha > 0$ and $p > 0$. The Gaussian kernel corresponds to the specific case $p = 2$ and arises as a unique choice if certain additional requirements are imposed.

Preliminaries: Semi-group with arbitrary parametrization. When basing a scale-space formulation on scale invariance, some further considerations are needed concerning the assumption about a semi-group structure.

In previous section, the scale parameter t associated with the semi-group (see equation (1.12)) was regarded as an abstract ordering parameter only. *A priori*, *i.e.* in the stage of formulating the axioms, there was no direct connection between this parameter and measurements of scale in terms of units of length. The only requirement was the qualitative (and essential) constraint that increasing values of the scale parameter should somehow correspond to representations at coarser scales. *A posteriori*, *i.e.* after deriving the shape of the convolution kernel, we could conclude that this parameter is related to scale as measured in units of length, *e.g.* *via* the standard deviation of the Gaussian kernel σ. The relationship is $t = \sigma^2/2$ and the semi-group operation corresponds to adding σ-values in the Euclidean norm.

In the derivation in this section, we shall assume that such a relationship exists already in the stage of formulating the axioms. Introduce σ as a scale parameter of dimension [length] associated with each layer in the scale-space representation. To allow for maximum generality in the relation between t and σ, assume that there exists some (unknown) transformation

$$t = \varphi(\sigma) \qquad (1.16)$$

---

[6]The great work by Fourier (1822) *Théorie Analytique de la Chaleur* has become famous for its contribution on Fourier analysis. However, it also contains a second major contribution that has been greatly underestimated for quite some time, *viz.* on the use of dimensions for physical quantities.

[7]For a tutorial on this subject, see e.g. Cooper 1988).

such that the semi-group structure of the convolution kernel corresponds to mere adding of the scale values when measured in terms of t. For kernels parameterized by σ the semi-group operation then assumes the form

$$h(\cdot;\ \sigma_1) * h(\cdot;\ \sigma_2) = h(\cdot;\ \gamma^{-1}(\gamma(\sigma_1) + \gamma(\sigma_2))) \qquad (1.17)$$

where φ^{-1} denotes the inverse of φ. If zero scale should correspond to the original signal it must hold that $\varphi(0) = 0$. To preserve the ordering of scale values $\varphi: \mathbb{R}_+ \to \mathbb{R}_+$ must be monotonically increasing.

Proof: Necessity from scale invariance.   In analogy with the previous scale-space formulations, let us state that the first stages of processing should be linear and be able to function without any *a priori* knowledge of the outside world. Combined with scale invariance, this gives the following basic axioms:

- linearity,
- no preferred location (shift invariance),
- no preferred orientation (isotropy),
- no preferred scale (scale invariance).

Recall that any linear and shift-invariant operator can be expressed as a convolution operator and introduce $\sigma \in \mathbb{R}_+$ to represent an abstract scale parameter of dimension [length]. Then, we can assume that the scale-space representation $I: \mathbb{R}^N \times \mathbb{R}_+ \to \mathbb{R}$ of any signal $f: \mathbb{R}^N \to \mathbb{R}$ is constructed by convolution with some one-parameter family of kernels $h: \mathbb{R}^N \times \mathbb{R}_+ \to \mathbb{R}$

$$I(\cdot;\ \sigma) = h(\cdot;\ \sigma) * f. \qquad (1.18)$$

In the Fourier domain ($\omega \in \mathbb{R}^N$) this convolution becomes a product:

$$\hat{I}(\omega;\ \sigma) = \hat{h}(\omega;\ \sigma)\, \hat{f}(\omega). \qquad (1.19)$$

Part A: Dimensional analysis, rotational symmetry, and scale invariance. From dimensional analysis it follows that if a physical process is scale independent, it should be possible to express the process in terms of dimensionless variables. These variables can be obtained by using a result in physics known as the Pi-theorem (see *e.g.* Olver 1986) which states that if the dimensions of the occurring variables are arranged in a table with as many rows as there are physical units and as many columns as there are derived quantities (see next) then the number of independent dimensionless quantities is equal to the number of derived quantities minus the rank of the system matrix. With respect to the linear scale-space representation, the following dimensions and variables occur:

	$\hat{I}$	$\hat{f}$	ω	σ
[luminance]	+1	+1	0	0
[length]	0	0	-1	+1

Obviously, there are four derived quantities and the rank of the matrix is two. Hence, we can *e.g.* introduce the dimensionless variables $\hat{I}/\hat{f}$ and $\omega\sigma$. Using the Pi-theorem, a necessary requirement for (1.19) to reflect a scale invariant process is that (1.19) can be written on the form

$$\frac{\hat{I}(\omega; \sigma)}{\hat{f}(\omega; \sigma)} = \hat{h}(\omega; \sigma) = \tilde{h}(\omega\sigma) \tag{1.20}$$

for some function $\tilde{h}: \mathbb{R}^N \to \mathbb{R}$. A necessary requirement on $\tilde{h}$ is that $\tilde{h}(0) = 1$. Otherwise $\hat{I}(\omega; 0) = \hat{f}(\omega)$ would be violated. Since we require no preference for orientation, it is sufficient to assume that $\hat{h}$ depends on the magnitude of its argument only and that for some function $\hat{H}: \mathbb{R} \to \mathbb{R}$ with $\hat{H}(0) = 1$ it holds that

$$\frac{\hat{I}(\omega; \sigma)}{\hat{f}(\omega; \sigma)} = \tilde{h}(\omega\sigma) = \hat{H}(|\omega\sigma|). \tag{1.21}$$

In the Fourier domain, the semi-group relation (1.17) with the arbitrary transformation function φ can be written

$$\hat{h}(\omega; \sigma_1) \hat{h}(\omega; \sigma_2) = \hat{h}(\omega; \gamma^{-1}(\gamma(\sigma_1) + \gamma(\sigma_2))) \tag{1.22}$$

and from (1.21) it follows that $\hat{H}$ must obey

$$\hat{H}(|\omega\sigma_1|) \hat{H}(|\omega\sigma_2|) = \hat{H}(|\omega \varphi^{-1}(\varphi(\sigma_1) + \varphi(\sigma_2))|) \tag{1.23}$$

for all $\sigma_1, \sigma_2, \gamma \in \mathbb{R}_+$. It is straightforward to show (see the following paragraph) that scale invariance implies that φ must be of the form

$$\varphi(\sigma) = C \sigma^p \tag{1.24}$$

for some arbitrary constants $C > 0$ and $p > 0$. Without loss of generality we may take $C = 1$, since this parameter corresponds to an unessential linear rescaling of t. Then, with $\varphi(\sigma) = \sigma^p$ and $\tilde{H}(x^p) = \hat{H}(x)$ we obtain

$$\tilde{H}(|\omega\sigma_1|^p) \tilde{H}(|\omega\sigma_2|^p) = \tilde{H}(|\omega\sigma_1|^p + |\omega\sigma_2|^p), \tag{1.25}$$

which can be recognized as the definition of the exponential function $(\psi(\xi_1)\psi(\xi_2) = \psi(\xi_1 + \xi_2) \Rightarrow \psi(\xi) = a^\xi$ for some $a > 0)$. Consequently,

$$\hat{H}(\omega\sigma) = \tilde{H}(|\omega\sigma|^p) = \exp(-\alpha|\omega\sigma|^p) \tag{1.26}$$

for some $\alpha \in \mathbb{R}$, and

$$\hat{h}(\omega; \sigma) = \hat{H}(\omega\sigma) = e^{-\alpha |\omega\sigma|^p}. \tag{1.27}$$

A real solution implies that α must be real. Concerning the sign of α, it is natural to require

$$\lim_{\sigma \to \infty} \hat{h}(\omega; \sigma) = 0 \tag{1.28}$$

rather than $\lim_{\sigma \to \infty} \hat{h}(\omega; \sigma) = \infty$. This means that α must be negative, and we can without loss of generality set $\alpha = -1/2$ to preserve consistency with the definition of the standard deviation of the Gaussian kernel σ in the case when $p = 2$.

Necessity of self-similar parametrization $t = C\sigma^p$. To verify that scale invariance implies that $\varphi(\sigma)$ must be of the form (1.24), we can observe that the left hand side of (1.23) is unaffected if σ_1 and σ_2 are multiplied by an arbitrary constant γ while ω is simultaneously divided by the same constant. Hence, for all $\sigma_1, \sigma_2, \gamma \in \mathbb{R}_+$ the following relation must hold:

$$\gamma\varphi^{-1}(\varphi(\sigma_1/\gamma) + \varphi(\sigma_2/\gamma)) = \varphi^{-1}(\varphi(\sigma_1) + \varphi(\sigma_2)). \tag{1.29}$$

Differentiation with respect to σ_i $(i = 1, 2)$ gives

$$\varphi^{-1'}(\varphi(\sigma_1/\gamma) + \varphi(\sigma_2/\gamma)) \, \varphi'(\sigma_i/\gamma) = \varphi^{-1'}(\varphi(\sigma_1) + \varphi(\sigma_2)) \, \varphi'(\sigma_i) \tag{1.30}$$

where φ' denotes the derivative of φ, etc. Dividing these equations for $i = 1, 2$ and letting $\gamma = \sigma_2/\sigma_3$ gives

$$\varphi'\left(\frac{\sigma_1\sigma_3}{\sigma_2}\right) = \frac{\varphi'(\sigma_1)\,\varphi'(\sigma_3)}{\varphi'(\sigma_2)}. \tag{1.31}$$

Let $C' = \varphi'(1)$ and $\psi(\sigma) = \varphi'(\sigma)/C'$. With $\sigma_2 = 1$ equation (1.31) becomes

$$\psi(\sigma_1\sigma_3) = \psi(\sigma_1)\,\psi(\sigma_3). \tag{1.32}$$

This relation implies that $\psi(\sigma) = \sigma^q$ for some q. (The sceptical reader may be convinced by introducing a new function θ defined by $\psi(\sigma) = \exp(\theta(\sigma))$. This reduces (1.32) to the definition of the logarithm function $\theta(\sigma_1\sigma_3) = \theta(\sigma_1) + \theta(\sigma_3)$ and $\psi'(\sigma) = \exp\log_a \sigma = \sigma^{1/\log a}$ for some a.)

Finally, the functional form of $\varphi(\sigma)$ (equation (1.24)) can be obtained by integrating $\varphi'(\sigma) = C'\sigma^q$. Since $\phi(0) = 0$, the integration constant must be zero. Moreover, the singular case $q = -1$ can be disregarded. The constants C and p must be positive, since φ must be positive and increasing.

Part B: Choice of scale invariant semi-groups. So far, the arguments based on scale invariance have given rise to a one-parameter family of semi-groups. The convolution kernels of these are characterized by having Fourier transforms of the form

$$\hat{h}(\omega; \sigma) = e^{-|\omega\sigma|^p/2} \qquad (1.33)$$

where the parameter $p > 0$ is undetermined. In the work by Florack *et al.* (1992) separability in Cartesian coordinates was used as a basic constraint. Except in the one-dimensional case, this fixates h to be a Gaussian.

Since, however, rotational symmetry combined with separability *per se* are sufficient to fixate the function to be a Gaussian, and the selection of two orthogonal coordinate directions constitutes a very specific choice, it is illuminating to consider the effect of using other choices of p.[8]

In a recent work, Pauwels *et al.* (1994) have analyzed properties of these convolution kernels. Here, we shall review some basic results and describe how different additional constraints on h lead to specific values of p.

Powers of ω that are even integers. Consider first the case when p is an even integer. Using the well-known relation between moments in the spatial domain and derivatives in the frequency domain

$$\int_{x\in\mathbb{R}} x^n h(x)\, dx = (-i)^n \hat{h}^{(n)}(0), \qquad (1.34)$$

it follows that the second moments of h are zero for any $p > 2$. Hence, $p = 2$ is the only even integer that corresponds to a non-negative convolution kernel (recall from section 5.4 that non-creation of local extrema implies that the kernel has to be non-negative).

Locality of infinitesimal generator. An important requirement of a multi-scale representation is that it should be differentiable with respect to the scale parameter. A general framework for expressing differentiability of semi-groups is in terms of *infinitesimal generators* (see section 7.2 for a review and a scale-space formulation based on this notion). In Pauwels *et al.* (1994) it is shown that the corresponding multi-scale representations generated by convolution kernels of the form (1.33) have *local* infinitesimal generators

---

[8]This well-known result can be easily verified as follows: Consider for simplicity the two-dimensional case. Rotational symmetry and separability imply that h must satisfy $h(r\cos\phi, r\sin\phi) = h_1(r) = h_2(r\cos\phi)\, h_2(r\sin\phi)$ for some functions h_1 and h_2 $((r,\phi)$ are polar coordinates). Inserting $\phi = 0$ shows that $h_1(r) = h_2(r)\, h_2(0)$. With $\psi(\xi) = \log(h_2(\xi)/h_2(0))$ this relation reduces to $\psi(r\cos\phi) + \psi(r\sin\phi) = \psi(r)$. Differentiating this relation with respect to r and ϕ and combining these derivatives shows that $\psi'(r\sin\phi) = \psi'(r)\sin\phi$. Differentiation gives $1/r = \psi''(r)/\psi'(r)$ and integration $\log r = \log\psi'(r) - \log b$ for some b. Hence, $\psi'(\xi) = b\xi$ and $h_2(\xi) = a\exp(b\xi^2/2)$ for some a and b.

(basically meaning that the multi-scale representations obey differential equations that can be expressed in terms of differential operators only; see section 7.2) if and only if the exponent p is an even integer.

Specific choice: Gaussian kernel.   In these respects, $p = 2$ constitutes a very special choice, since it is the only choice that corresponds to a local infinitesimal generator and a non-negative convolution kernel.

Similarly, $p = 2$ is the unique choice for which the multi-scale representation satisfies the causality requirement (as will be described in section 7.2, a reformulation of the causality requirement in terms of non-enhancement of local extrema implies that the scale-space family must have an infinitesimal generator corresponding to spatial derivatives up to order two).

5.7. Other special properties of the Gaussian kernel

The Gaussian kernel has some other special properties. Consider for simplicity the one-dimensional case and define normalized second-moments Δx and $\Delta \omega$ in the spatial and the Fourier domain respectively by

$$\Delta x = \frac{\int_{x \in \mathbb{R}} x^T x |h(x)|^2 dx}{\int_{x \in \mathbb{R}} |h(x)|^2 dx}, \quad \Delta \omega = \frac{\int_{\omega \in \mathbb{R}} \omega^T \omega |\hat{h}(\omega)|^2 d\omega}{\int_{\omega \in \mathbb{R}} |\hat{h}(\omega)|^2 d\omega}, \quad (1.35)$$

These entities measure the "spread" of the distributions h and $\hat{h}$, (where the Fourier transform of any function $h \colon \mathbb{R}^N \times \mathbb{R} \to \mathbb{R}$ is given by $\hat{h}(\omega) = \int_{x \in \mathbb{R}^N} h(x) e^{-i\omega^T x} dx$). Then, the *uncertainty relation* states that

$$\Delta x \Delta \omega \geq \frac{1}{2}. \quad (1.36)$$

A remarkable property of the Gaussian kernel is that it is the only real kernel that gives equality in this relation.

The Gaussian kernel is also the frequency function of the normal distribution. The central limit theorem in statistics states that under rather general requirements on the distribution of a stochastic variable, the distribution of a sum of a large number of such variables asymptotically approaches a normal distribution when the number of terms tend to infinity.

6. Gaussian derivative operators

Above, it has been shown that by starting from a number of different sets of axioms it is possible to single out the Gaussian kernel as the unique kernel for generating a (linear) scale-space. From this scale-space representation, *multi-scale spatial derivative operators* can then be defined by

$$I_{i_1 \dots i_n} (\cdot; \; t) = \partial_{i_1 \dots i_n} I(\cdot; \; t) = G_{i_1 \dots i_n} (\cdot; \; t) * I, \quad (1.37)$$

Discrete Gauss   *Sampled Gauss*   *Discrete Gauss*   *Sampled Gauss*

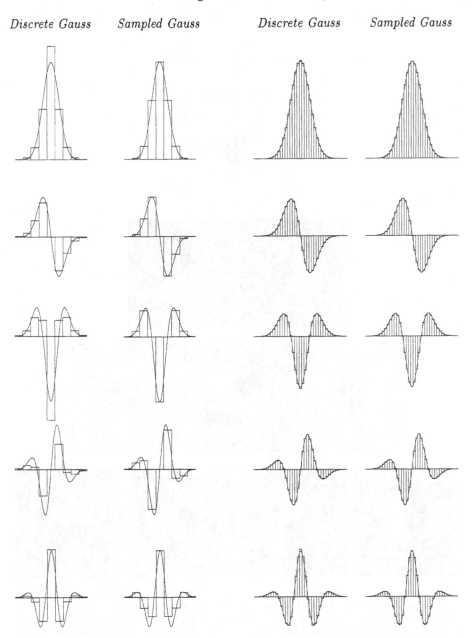

Figure 1.9.   Graphs of the one-dimensional Gaussian derivative kernels $\partial_{x^n} g(x; \; t)$ up to order $n = 4$ at scales $\sigma^2 = 1.0$ (left columns) and $\sigma^2 = 16.0$ (right columns). The derivative/difference order increases from top to bottom. The upper row shows the raw smoothing kernel. Then follow the first-, second-, third- and fourth-order derivative/difference kernels. The continuous curves show the continuous Gaussian derivative kernels and the block diagrams discrete approximations (see section 7.3). (From [223].)

Gaussian derivative kernels

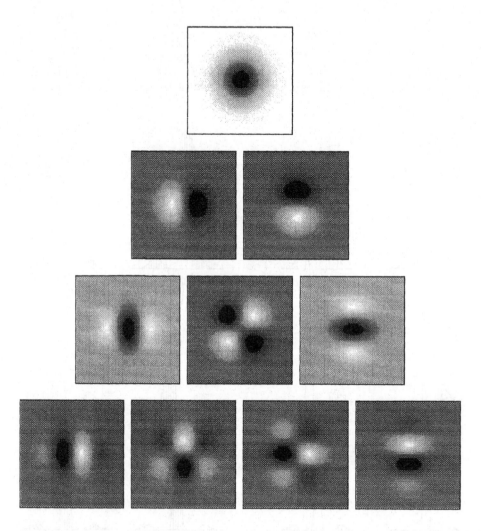

Figure 1.10.   Grey-level illustrations of two-dimensional Gaussian derivative kernels up to order three. (Top row) Zero-order smoothing kernel, T, (inverted). (Second row) First-order derivative kernels, $\delta_x T$ and $\delta_y T$. (Third row) Second-order derivative kernels $\delta_{xx} T$, $\delta_{xy} T$, $\delta_{yy} T$. (Bottom row) Third-order derivative kernels $\delta_{xxx} T$, $\delta_{xxy} T$, $\delta_{xyy} T$, $\delta_{yyy} T$. Qualitatively, these kernels resemble the shape of the continuous Gaussian derivative kernels. In practice though, they are defined as discrete derivative approximations using the canonical discretization framework described in section 7.3. (Scale level $t = 64.0$, image size 127×127 pixels.) (From [223].)

where $G_{i_1...i_n}(\cdot;\ t)$ denotes a (possibly mixed) derivative of some order $n = i_1 + ... i_N$ of the Gaussian kernel. In terms of explicit integrals, the convolution operation (1.37) is written

$$I_{i_1...i_n}(x;\ t) = \int_{x' \in \mathbb{R}^N} g_{i_1...i_n}(x';\ t)\, f(x - x')\, dx'. \qquad (1.38)$$

Graphical illustrations of such *Gaussian derivative kernels* in the one-dimensional and two-dimensional cases are given in figures 1.9 and 1.10.

6.1. Infinite differentiability

This representation where *scale-space derivatives* are defined by *integral operators* has a strong regularizing property. If f is bounded by some polynomial, i.e. if there exist some constants $C_1, C_2 \in \mathbb{R}_+$ such that

$$|f(x)| \le C_1\,(1 + x^T x)^{C_2} \qquad (x \in \mathbb{R}^N), \qquad (1.39)$$

then the integral (1.38) is guaranteed to converge for any $t > 0$. This means that although f may not be differentiable of any order, or not even continuous, the result of the Gaussian derivative operator is always well-defined. According to the theory of generalized functions (or Schwartz distributions) (Schwartz 1951; Hörmander 1963), we can then for any $t > 0$ treat $I(\cdot;\ t) = g(\cdot;\ t) * f$ as infinitely differentiable.

6.2. Multi-scale N-jet representation and necessity

Considering the spatial derivatives up to some order N enables characterization of the local image structure up to that order, e.g., in terms of the Taylor expansion[9] of the intensity function

$$I(x + \delta x) = I(x) + I_i \delta x_i + \frac{1}{2!} I_{ij} \delta x_i \delta x_j + \frac{1}{3!} I_{ijk} \delta x_i \delta x_j \delta x_k + \mathcal{O}(\delta x^4). \quad (1.40)$$

In early work, Koenderink and van Doorn (1987) advocated the use of this so-called *multi-scale N-jet signal representation* as a model for the earliest stages of visual processing[10]. Then, in (Koenderink and van Doorn 1992) they considered the problem of deriving linear operators from the scale-space representation that are to be invariant under scaling transformations.

---

[9]Here paired indices are summed over the spatial dimensions. In two dimensions we have $I_i \delta x_i = I_x dx + I_y dx$.

[10]In section 2 in the next chapter it is shown how this framework can be applied to computational modeling of various types of early visual operations for computing image features and cues to surface shape.

Inspired by the relation between the Gaussian kernel and its derivatives, here in one dimension,

$$\partial_{x^n} g(x; \sigma^2) = (-1)^n \frac{1}{\sigma^n} H_n(\frac{x}{\sigma}) g(x; \sigma^2), \qquad (1.41)$$

which follows from the well-known relation between derivatives of the Gaussian kernel and the Hermite polynomials H_n (see table 1.1)

$$\partial_{x^n}(e^{-x^2}) = (-1)^n H_n(x) e^{-x^2}, \qquad (1.42)$$

they considered the problem of deriving operators with a similar scaling behaviour. Starting from the *Ansatz*

$$\psi^{(\alpha)}(x; \sigma) = \frac{1}{\sigma^\alpha} \varphi^{(\alpha)}(\frac{x}{\sigma}) g(x; \sigma), \qquad (1.43)$$

where the superscript (α) describes the "order" of the function, they considered the problem of determining all functions $\varphi^{(\alpha)}: \mathbb{R}^N \to \mathbb{R}$ such that $\psi^{(\alpha)}: \mathbb{R}^N \to \mathbb{R}$ satisfies the diffusion equation. Interestingly, $\varphi^{(\alpha)}$ must then satisfy the time-independent Schrödinger equation

$$\nabla^T \nabla \varphi(\xi) + ((2\alpha + N) - \xi^T \xi)\varphi(\xi) = 0, \qquad (1.44)$$

where $\xi = x/\sigma$. This is the physical equation that governs the quantum mechanical free harmonic oscillator. It is well-known from mathematical physics that the solutions $\varphi^{(\alpha)}$ to this equation are the Hermite functions, that is Hermite polynomials multiplied by Gaussian functions. Since derivatives of a Gaussian kernel are also Hermite polynomials times Gaussian functions, it follows that the solutions $\psi^{(\alpha)}$ to the original problem are the derivatives of the Gaussian kernel. This result provides a formal statement that Gaussian derivatives are *natural operators* to derive from scale-space. (Figure 1.11 shows a set of Gaussian derivatives computed in this way.)

6.3. Scale-space properties of Gaussian derivatives

As pointed out above, these scale-space derivatives satisfy the diffusion equation and obey scale-space properties, for example, the *cascade smoothing property*

$$g(\cdot; t_1) * g_{x^n}(\cdot; t_2) = g_{x^n}(\cdot; t_2 + t_1). \qquad (1.45)$$

The latter result is a special case of the more general statement

$$g_{x^m}(\cdot; t_1) * g_{x^n}(\cdot; t_2) = g_{x^{m+n}}(\cdot; t_2 + t_1), \qquad (1.46)$$

order	Hermite polynomial
$H_0(x)$	1
$H_1(x)$	x
$H_2(x)$	$x^2 - 1$
$H_3(x)$	$x^3 - 3x$
$H_4(x)$	$x^4 - 6x^2 + 3$
$H_5(x)$	$x^5 - 10x^3 + 15x$
$H_6(x)$	$x^6 - 15x^4 + 45x^2 - 15$
$H_7(x)$	$x^7 - 21x^5 + 105x^3 - 105x$

TABLE 1.1. The first eight Hermite polynomials ($x \in \mathbb{R}$).

whose validity follows directly from the commutative property of convolution and differentiation.

6.4. Directional derivatives

Let $(\cos \beta, \sin \beta)$ represent a unit vector in a certain direction β. From the well-known expression for the nth-order directional derivative ∂_β^n of a function I in any direction β,

$$\partial_\beta^n I = (\cos \beta \, \partial_x + \sin \beta \, \partial_y)^n I. \qquad (1.47)$$

if follows that a directional derivative of order n in any direction can be constructed by linear combination of the partial scale-space derivatives

$$I_{x_1}, I_{x_2}, I_{x_1 x_1}, I_{x_1 x_2}, I_{x_2 x_2}, I_{x_1 x_1 x_1}, I_{x_1 x_1 x_2}, I_{x_1 x_2 x_2}, I_{x_2 x_2 x_2}, \cdots$$

of that order. Figure 1.12 shows *equivalent* derivative approximations kernels of order one and two constructed in this way.
In the terminology of Freeman and Adelson (1990) and Perona (1992), kernels whose outputs are related by linear combinations are said to be "steerable". Note, however, that in this case the "steerable" property is not attributed to the specific choice of the Gaussian kernel. The relation (1.47) holds for any n times continuously differentiable function.

7. Discrete scale-space

The treatment so far has been concerned with continuous signals. Since real-world signals obtained from standard digital cameras are discrete, however, an obvious problem concerns how to discretize the scale-space theory while still maintaining the scale-space properties.

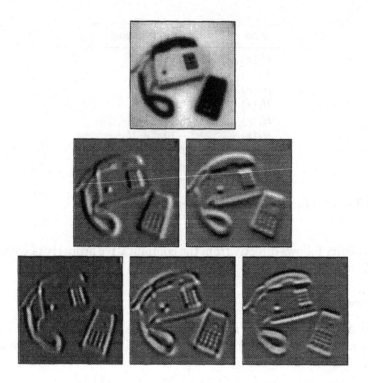

Figure 1.11. Scale-space derivatives up to order two computed from the telephone and calculator image at scale level $\sigma^2 = 4.0$ (image size 128×128 pixels). From top to bottom and from left to right; I, I_x, I_y, I_{xx}, I_{xy}, and I_{yy}.

7.1. Non-creation of local extrema

For one-dimensional signals a complete discrete theory can be based on a discrete analogy to the treatment in section 5.4. Following Lindeberg (1990, 1991), define a discrete kernel $h \in \mathbb{L}_1$ to be a *discrete scale-space kernel* if for any signal f_{in} the number of local extrema in $f_{out} = h * f_{in}$ does not exceed the number of local extrema in f_{in}.

Using classical results (mainly by Schoenberg 1953; see also Karlin 1968 for a comprehensive summary), it is possible to completely classify those kernels that satisfy this definition. A discrete kernel is a scale-space kernel if and only if its generating function $\varphi_h(z) = \sum_{n=-\infty}^{\infty} h(n) \, z^n$ is of the form

$$\varphi_K(z) = c \, z^k \, e^{(q_{-1} z^{-1} + q_1 z)} \prod_{i=1}^{\infty} \frac{(1 + \alpha_i z)(1 + \delta_i z^{-1})}{(1 - \beta_i z)(1 - \gamma_i z^{-1})}, \qquad (1.48)$$

where

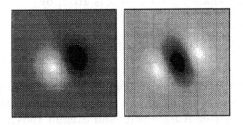

Figure 1.12. First- and second-order directional derivative approximation kernels in the 22.5 degree direction computed by linear combination of partial derivatives according to (1.47). (Scale level $t = 64.0$, image size 127×127 pixels.) (From [223].)

$$c > 0,\ k; \in Z,$$
$$q_{-1}, q_1, \alpha_i, \beta_i, \gamma_i, \delta_i \geq 0,$$
$$\beta_i, \gamma_i < 1, \text{ and}$$
$$\sum_{i=1}^{\infty} (\alpha_i + \beta_i + \gamma_i + \delta_i) < \infty.$$

The interpretation of this result is that there are five primitive types of linear and shift-invariant smoothing transformations, of which the last two are trivial;

— two-point weighted average or *generalized binomial smoothing,*

$$f_{out}(x) = f_{in}(x) + \alpha_i f_{in}(x-1) \qquad (\alpha \geq 0),$$
$$f_{out}(x) = f_{in}(x) + \delta_i f_{in}(x+1) \qquad (\delta_i \geq 0),$$

— moving average or *first-order recursive filtering,*

$$f_{out}(x) = f_{in}(x) + \beta_i f_{out}(x-1) \qquad (0 \leq \beta_i < 1),$$
$$f_{out}(x) = f_{in}(x) + \gamma_i f_{out}(x+1) \qquad (0 \leq \gamma_i < 1),$$

— infinitesimal smoothing or *diffusion smoothing* (explained below),
— rescaling, and
— translation.

It follows that a discrete kernel is a scale-space kernel if and only if it can be decomposed into the above primitive transformations. Moreover, the only non-trivial smoothing kernels of *finite support* arise from generalized binomial smoothing (i.e., non-symmetric extensions of the filter (1.2)). If this definition is combined with a requirement that the family of smoothing transformations must obey a *semi-group* property (1.12) over scales and possesses a *continuous scale parameter*, then there is in principle only one way to construct a scale-space for discrete signals. Given a signal

$f: \mathbb{Z} \to \mathbb{R}$ the scale-space $: \mathbb{Z} \times \mathbb{R}_+ \to \mathbb{R}$ is given by

$$I(x; t) = \sum_{n=-\infty}^{\infty} T(n; 2t) f(x - n), \qquad (1.49)$$

where $T: \mathbb{Z} \times \mathbb{R}_+ \to \mathbb{R}$ is a kernel termed *the discrete analogue of the Gaussian kernel*. It is defined in terms of one type of Bessel functions, the modified Bessel functions $\tilde{I}_n$ (see Abramowitz and Stegun 1964):[11]

$$T(n; 2t) = e^{-2\alpha t} \tilde{I}_n(2\alpha t). \qquad (1.50)$$

This kernel satisfies several properties in the discrete domain that are similar to those of the Gaussian kernel in the continuous domain; for example, it tends to the discrete delta function when $t \to 0$, while for large t it approaches the continuous Gaussian. The scale parameter t can be related to spatial scale from the second-moment of the kernel, which when $\alpha = 1$ is

$$\sum_{n=-\infty}^{\infty} n^2 T(n; 2t) = 2t. \qquad (1.51)$$

The term "diffusion smoothing" can be understood by noting that the scale-space family I satisfies a semi-discretized version of the diffusion equation:

$$\partial_t I(x; t) = I(x + 1; t) - 2I(x; t) + I(x - 1; t) = \nabla_2^2 I x; t) \qquad (1.52)$$

with initial condition $I(x; 0) = f(x)$, i.e., the equation that is obtained if the continuous one-dimensional diffusion equation is discretized in space using the standard second difference operator $\nabla_3^2 I$, but the continuous scale parameter is left untouched.

A simple interpretation of the discrete analogue of the Gaussian kernel is as follows: Consider the time discretization of (1.52) using Eulers explicit method

$$I^{(k+1)(i)} = \Delta t \, I^{(k)}(i + 1) + (1 - 2\Delta t) \, I^{(k)}(i) + \Delta t \, I^{(k)}(i - 1), \qquad (1.53)$$

where the superscript (k) denotes iteration index. Assume that the scale-space representation of I at scale t is to be computed by applying this iteration formula using n steps with step size $\Delta t = t/n$. Then, the

---

[11] The factor 2 in the notation $2t$ arises due to the use of different parameterizations of the scale parameter. In this book, the scale parameter is related to the standard deviation of the Gaussian kernel σ by $\sigma^2 = 2t$ which means that the diffusion equation assumes the form $I_t = \nabla^2 I$. In a large part of the other scale-space literature, the diffusion equation is written $I_t = 1/2 \, \nabla^2 I$ with the advantage that the scale parameter is the square of the standard deviation of the Gaussian kernel $t = \sigma^2$.

discrete analogue of the Gaussian kernel is the limit case of the equivalent convolution kernel

$$(\frac{t}{n}, \quad 1 - \frac{2t}{n}, \quad \frac{t}{n})^n, \tag{1.54}$$

when n tends to infinity, i.e., when the number of steps increases and each individual step becomes smaller. This shows that the discrete analogue of the Gaussian kernel can be interpreted as the limit case of iterative application of generalized binomial kernels.

Despite the completeness of these results, and their analogies to the continuous situation, however, they cannot be extended to higher dimensions. Using similar arguments as in the continuous case it can be shown that there are no non-trivial kernels in two or higher dimensions that are guaranteed to never introduce new local extrema. Hence, a discrete scale-space formulation in higher dimensions must be based on other axioms.

7.2. Non-enhancement and infinitesimal generator

It is clear that the continuous scale-space formulations in terms of causality and scale invariance cannot be transferred directly to discrete signals; there are no direct discrete correspondences to level curves and differential geometry in the discrete case. Neither can the scaling argument be carried out in the discrete situation if a continuous scale parameter is desired, since the discrete grid has a preferred scale given by the distance between adjacent grid points. An alternative way to express the causality requirement in the continuous case, however, is as follows (Lindeberg 1990):

> *Non-enhancement of local extrema:* If for some scale level t_0 a point x_0 is a local maximum for the scale-space representation at that level (regarded as a function of the space coordinates only) then its value must not increase when the scale parameter increases. Analogously, if a point is a local minimum then its value must not decrease when the scale parameter increases.

It is clear that this formulation is equivalent to the formulation in terms of level curves for continuous data, since if the grey-level value at a local maximum (minimum) would increase (decrease) then a new level curve would be created. Conversely, if a new level curve is created then some local maximum (minimum) must have increased (decreased). An intuitive description of this requirement is that it prevents local extrema from being enhanced and from "popping up out of nowhere". In fact, it is closely related to the maximum principle for parabolic differential equations (see, e.g., Widder 1975 and also Hummel 1987).

Preliminaries: Infinitesimal generator. If the semi-group structure is combined with a strong continuity requirements with respect to the scale parameter, then it follows from well-known results in (Hille and Phillips 1957) that the scale-space family must have an *infinitesimal generator* (Lindeberg 1990, 1991). In other words, if a transformation operator $\mathcal{T}_t$ from the input signal to the scale-space representation at any scale t is defined by

$$I(\cdot;\, t) = \mathcal{T}_t f, \tag{1.55}$$

then there exists a limit case of this operator (the infinitesimal generator)

$$\mathcal{A}f = \lim_{h \downarrow 0} \frac{\mathcal{T}_h f - f}{h} \tag{1.56}$$

and

$$\lim_{h \downarrow 0} \frac{I(\cdot, \cdot;\, t+h) - I(\cdot, \cdot;\, t)}{h} = \mathcal{A}(\mathcal{T}_t f) = \mathcal{A}I(\cdot;\, t). \tag{1.57}$$

Non-enhancement of local extrema implies a second-order infinitesimal generator. By combining the existence of an infinitesimal scale-space generator with the non-enhancement requirement, linear shift-invariance, and spatial symmetry it can be shown (Lindeberg 1991, 1992, 1994) that the scale-space family $I: \mathbb{Z}^N \times \mathbb{R}^+ \to \mathbb{R}$ of a discrete signal $f: \mathbb{Z}^N \to \mathbb{R}$ must satisfy the semi-discrete differential equation

$$(\partial_t I)(x;\, t) = (A_{ScSp} I)(x;\, t) = \sum_{\xi \in \mathbb{Z}^N} a_\xi I(x - \xi;\, t), \tag{1.58}$$

for some *infinitesimal scale-space generator* A_{ScSp} characterized by
- the *locality* condition $a_\xi = 0$ if $|\xi|_\infty > 1$,
- the *positivity* constraint $a_\xi \geq 0$ if $\xi \neq 0$,
- the *zero sum* condition $\sum_{\xi \in \mathbb{Z}^N} a_\xi = 0$, as well as
- the *symmetry* requirements $a_{(-\xi_1, \xi_2, \dots, \xi_N)} = a_{(\xi_1, \xi_2, \dots, \xi_N)}$ and $a_{P_k^N(\xi_1, \xi_2, \dots, \xi_N)} = a_{(\xi_1, \xi_2, \dots, \xi_N)}$ for all $\xi = (\xi_1, \xi_2, \dots, \xi_N) \in \mathbb{Z}^N$ and all possible permutations P_k^N of N elements.

Notably, the locality condition means that A_{ScSp} corresponds to the discretization of derivatives of order up to two. In one and two dimensions respectively (1.58) reduces to

$$\partial_t I = \alpha_1 \nabla_3^2 I, \tag{1.59}$$

$$\partial_t I = \alpha_1 \nabla_5^2 I + \alpha_2 \nabla_{\times^2}^2 I, \tag{1.60}$$

for some constants $\alpha_1 \geq 0$ and $\alpha_2 \geq 0$. Here, the symbols, ∇_5^2 and $\nabla_{\times^2}^2$ denote the two common discrete approximations of the Laplace operator; defined by (below the notation $f_{-1,1}$ stands for $f(x - 1, y + 1)$ etc.):

$$(\nabla_5^2 f)_{0,0} = f_{-1,0} + f_{+1,0} + f_{0,-1} + f_{0,+1} - 4f_{0,0},$$

$$(\nabla_{\times^2}^2 f)_{0,0} = 1/2(f_{-1,-1} + f_{-1,+1} + f_{+1,-1} + f_{+1,+1} - 4f_{0,0}).$$

In the particular case when $\alpha_2 = 0$, the two-dimensional representation is given by convolution with the one-dimensional Gaussian kernel along each dimension. On the other hand, using $\alpha_1 = 2\alpha_2$ corresponds to a representation with maximum spatial isotropy in the Fourier domain.

7.3. Discrete derivative approximations

Concerning operators derived from the discrete scale-space, it holds that the scale-space properties transfer to any discrete derivative approximation defined by spatial linear filtering of the scale-space representation. In fact, the converse result is true as well (Lindeberg 1993); if derivative approximation kernels are to satisfy the *cascade smoothing property*,

$$\delta_{x^n} T(\cdot; \; t_1) * T(\cdot; \; t_2) = \delta_{x^n} T(\cdot; \; t_1 + t_2), \tag{1.61}$$

and if similar continuity requirements concerning scale variations are imposed, then by necessity also the derivative approximations must satisfy the semi-discretized diffusion equation (1.58). The specific choice of operators δ_{x^n} is however arbitrary; any linear operator satisfies this relation. Graphs of these kernels at a few levels of scale and for the lowest orders of differentiation are shown in figure 1.9 and figure 1.10.

To summarize, there is a unique and consistent way to define a scale-space representation and discrete analogues to smoothed derivatives for discrete signals, which to a large extent preserves the algebraic structure of the multi-scale N-jet representation in the continuous case.

8. Scale-space operators and front-end vision

As we have seen, the uniqueness of the Gaussian kernel for scale-space representation can be derived in a variety of different ways, non-creation of new level curves in scale-space, non-creation of new local extrema, non-enhancement of local extrema, and by combining scale invariance with certain additional conditions. Similar formulations can be stated both in the continuous and in the discrete domains. The essence of these results is that the scale-space representation is given by a (possibly semi-discretized) parabolic differential equation corresponding to a *second-order* differential operator with respect to the spatial coordinates, and a *first-order* differential operator with respect to the scale parameter.

8.1. Scale-space: A canonical visual front-end model

A natural question now arises: Does this approach constitute the *only* reasonable way to perform the low-level processing in a vision system, and are the Gaussian kernels and their derivatives the only smoothing kernels that can be used? Of course, this question is impossible to answer to without further specification of the purpose of the representation, and what tasks the visual system has to accomplish. In any sufficiently specific application it should be possible to design a smoothing filter that in some sense has a "better performance" than the Gaussian derivative model. For example, it is well-known that scale-space smoothing leads to shape distortions at edges by smoothing across object boundaries, and also in estimation of surface shape using algorithms such as shape-from-texture. Hence, it should be emphasized that the theory developed here is rather aimed at describing the principles of the very first stages of low-level processing in an *uncommitted* visual system aimed at handling a large class of different situations, and in which no or very little a priori information is available.

Then, once initial hypotheses about the structure of the world have been generated within this framework, the intention is that it should be possible to invoke more refined processing, which can compensate for this, and adapt to the current situation and the task at hand (see section 8 in next chapter as well as following chapters). From the viewpoint of such approaches, the linear scale-space model serves as the natural starting point.

8.2. Relations to biological vision

In fact, a certain degree of agreement can be obtained with the result from this solely theoretical analysis and the experimental results of biological vision systems. Neurophysiological studies by Young (1985, 1986, 1987) have shown that there are receptive fields in the mammalian retina and visual cortex, whose measured response profiles can be well modeled by Gaussian derivatives. For example, Young models cells in the mammalian retina by kernels termed 'differences of offset Gaussians' (DOOG), which basically correspond to the Laplacian of the Gaussian with an added Gaussian offset term. He also reports cells in the visual cortex, whose receptive field profiles agree with Gaussian derivatives up to order four.

Of course, far-reaching conclusions should not be drawn from such a qualitative similarity, since there are also other functions, such as Gabor functions (see section 6.2.1 in next chapter) that satisfy the recorded data up to the tolerance of the measurements. Nevertheless, it is interesting to note that operators similar to the Laplacian of the Gaussian (center-surround receptive fields) have been reported to be dominant in the retina. A possible explanation concerning the construction of derivatives of other orders from the output of these operators can be obtained from the observation that the original scale-space representation can always be reconstructed if Laplacian derivatives are available at all other scales. If the scale-space representation tends to zero at infinite scale, then it follows from the diffusion equation that

$$I(x;\ t) = -(I(x;\ \infty) - I(x;\ t)) = -\int_{t'=t}^{\infty} \partial_t I(x;\ t') dt' = -\int_{t'=t}^{\infty} \nabla^2 I(x;\ t') dt'.$$

Observe the similarity with the standard method for reconstructing the original signal from a bandpass pyramid (Burt 1981).

What remains to be understood is whether there are any particular theoretical advantages of computing the Laplacian of the Gaussian in the first step. Of course, such an operation suppresses any linear illumination gradients, and in a physiological system it may lead to robustness to the loss of some fibers because there is substantial integration over all available scales. Furthermore, one can contend that spatial derivatives of the Gaussian can be approximated by differences of Gaussian kernels at different spatial position, and it is therefore, at least in principle, possible to construct any spatial derivative from this representation. Remaining questions concerning the plausibility concerning biological vision are left to the reader's speculation and further research.

8.3. Foveal vision

Another interesting similarity concerns the spatial layout of receptive fields over the visual field[12]. If the scale-space axioms are combined with the assumption of a fixed readout capacity per scale from the visual front-end, it is straightforward to show that there is a natural distribution of receptive fields (of different scales and different spatial position) over the retina such that the minimum receptive field size grows linearly with eccentricity, that is the distance from the center of the visual field (Lindeberg and Florack 1992, 1994). A similar (log-polar) result is obtained when a conformal metric is chosen for a non-linear scale-space (see the chapter by Florack *et al.* (1994) in this book). There are several results in psychophysics, neuroanatomy and electrophysiology in agreement with such an increase (Koenderink and van Doorn 1978; van de Grind *et al.* 1986; Bijl 1991). In fact, physical

---

[12]For an introduction to the neurophysiological findings regarding the front-end visual system, see e.g. the tutorial book by Hubel (1988) and the more recent book by Zeki (1993).

sensors with such characteristics receive increasing interest and are being
constructed in hardware (Tistarelli and Sandini 1992).

LINEAR SCALE-SPACE II: EARLY VISUAL OPERATIONS

Tony Lindeberg

Royal Institute of Technology (KTH)
Computational Vision and Active Perception Laboratory (CVAP)
Department of Numerical Analysis and Computing Science
S-100 44 Stockholm, Sweden

and

Bart M. ter Haar Romeny

Utrecht University, Computer Vision Research Group,
Heidelberglaan 100 E.02.222,
NL-3584 CX Utrecht, The Netherlands

1. Introduction

In the previous chapter a formal justification has been given for using linear filtering as an initial step in early processing of image data (see also section 5 in this chapter). More importantly, a catalogue has been provided of what filter kernels are natural to use, as well as an extensive theoretical explanation of how different kernels of different orders and at different scales can be related. This forms the basis of a theoretically well-founded modeling of visual front-end operators with a smoothing effect.

Of course, linear filtering cannot be used as the only component in a vision system aimed at deriving information from image data; some non-linear steps must be introduced into the analysis. More concretely, some mechanism is required for combining the output from the Gaussian derivative operators of different order and at different scales into some more explicit descriptors of the image geometry.

This chapter continues the treatment of linear scale-space by showing how different types of early visual operations can be expressed within the scale-space framework. Then, we turn to theoretical properties of linear scale-space and demonstrate how the behaviour of image structures over scales can be analyzed. Finally, it is described how access to additional information suggests situations when the requirements about uncommitted processing can be relaxed. This open-ended material serves as a natural starting point for the non-linear approaches considered in following chapters.

2. Multi-scale feature detection in scale-space

An approach that has been advocated by Koenderink and his co-workers is to describe image properties in terms of differential geometric descriptors, i.e., different (possibly non-linear) combinations of derivatives. A basic motivation for this position is that differential equations and differential geometry constitute natural frameworks for expressing both physical processes and geometric properties. More technically, and as we have seen in section 6 in the previous chapter, it can also be shown that spatial derivatives are natural operators to derive from the scale-space representation.

When using such descriptors, it should be observed that a single partial derivative, e.g. I_{x_1}, does not represent any geometrically meaningful information, since its value is crucially dependent on the arbitrary choice of coordinate system. In other words, it is essential to base the analysis on descriptors that do not depend on the actual coordinatization of the spatial and intensity domains. Therefore, it is natural to require the representation to be invariant with respect to primitive transformations such as translations, rotations, scale changes, and certain intensity transformations[1]. In fact, quite a few types of low-level operations can be expressed in terms of such *multi-scale differential invariants* defined from (non-linear) combinations of Gaussian derivatives at multiple scales. Examples of these are feature detectors, feature classification methods, and primitive shape descriptors. In this sense, the scale-space representation can be used as a basis for early visual operations.

2.1. Differential geometry and differential invariants

Florack *et al.* (1992, 1993) and Kanatani (1990) have pursued this approach of deriving differential invariants in an axiomatic manner, and considered image properties defined in terms of directional derivatives along certain preferred coordinate directions. If the direction, along which a directional derivative is computed, can be uniquely defined from the intensity pattern, then rotational invariance is obtained automatically, since the preferred

---

[1]In fact, it would be desirable to directly compute features that are invariant under perspective transformations. Since, however, this problem is known to be much harder, most work has so far been restricted to invariants of two-dimensional Euclidean operations and natural linear extensions thereof, such as uniform rescaling and affine transformations of the spatial coordinates. For an overview of geometric invariance applied to computer vision, see the book edited by Mundy and Zisserman (1992). An excellent discussion of the invariant properties of the diffusion equation is found in Olver (1986). Concerning analysis of differential invariants, see also the chapter by Olver *et al.* (1994) in this book.

direction follows any rotation of the coordinate system. Similarly, any derivative is translationally invariant. These properties hold both concerning transformations of the original signal f and the scale-space representation I of f generated by rotationally symmetric Gaussian smoothing.

Detailed studies of differential geometric properties of two-dimensional and three-dimensional scalar images are presented by Salden *et al.* (1991), who makes use of classical techniques from differential geometry (Spivak 1975; Koenderink 1990), algebraic geometry, and invariant theory (Grace and Young 1965; Weyl 1946) for classifying geometric properties of the N-jet of a signal at a given scale in scale-space.

Here, a short description will be given concerning some elementary results. Although the treatment will be restricted to the two-dimensional case, the ideas behind it are general and can be easily extended to higher dimensions. For more extensive treatments, see also (ter Haar Romeny *et al.* 1993; Florack 1993; Lindeberg 1994).

2.1.1. *Local directional derivatives*

One choice of preferred directions is to introduce a local orthonormal coordinate system (u, v) at any point P_0, where the v-axis is parallel to the gradient direction at P_0, and the u-axis is perpendicular, i.e. $e_v = (\cos \alpha, \sin \alpha)^T$ and $e_u = (\sin \alpha, -\cos \alpha)^T$, where

$$e_v|_{P_0} = \begin{pmatrix} \cos \alpha \\ \sin \alpha \end{pmatrix} = \frac{1}{\sqrt{I_x^2 + IL_y^2}} \begin{pmatrix} I_x \\ I_y \end{pmatrix} \Bigg|_{P_0} . \qquad (2.1)$$

In terms of Cartesian coordinates, which arise frequently in standard digital images, these local directional derivative operators can be written

$$\partial_u = \sin \alpha \, \partial_x - \cos \alpha \, \partial_y. \quad \partial_v = \cos \alpha \, \partial_x + \sin \alpha \, \partial_y, \qquad (2.2)$$

This (u, v)-coordinate system is characterized by the fact that one of the first-order directional derivatives, I_u, is zero.

Another natural choice is a coordinate system in which the mixed second order derivative is zero; such coordinates are named (p, q) by Florack *et al.* (1992). In these coordinates, in which $I_{pq} = 0$, the explicit expressions for the directional derivatives become slightly more complicated (see Lindeberg 1994 for explicit expressions).

A main advantage of expressing differential expressions in terms of such *gauge coordinates* is that the closed-form expressions for many differential invariants become simple since a large number of terms disappear.

2.1.2. *Monotonic intensity transformations*

One approach to derive differential invariants is to require the differential entities to be invariant with respect to arbitrary *monotonic intensity transformations*[2] Then, any property that can be expressed in terms of the *level curves* of the signal is guaranteed to be invariant. A classification by Florack *et al.* (1992) and Kanatani (1990), which goes back to the classical classification of polynomial invariants by Hilbert (1893), shows that concerning derivatives up to order two of two-dimensional images, there are only two irreducible differential expressions that are invariant to these transformations; the isophote curvature κ and the flowline curvature μ (see also figure 2.1 for an illustration).

$$\kappa = \frac{2I_xI_yI_{xy} - I_x^2I_{yy} - I_y^2I_{xx}}{(I_x^2 + I_y^2)^{3/2}} = -\frac{I_{uu}}{I_v}, \tag{2.3}$$

$$\mu = \frac{(I_x^2 - I_y^2)I_{xy} - I_xI_y(I_{yy} - I_{xx})}{(I_x^2 + I_y^2)^{3/2}} = -\frac{I_{uv}}{I_v}. \tag{2.4}$$

A general scheme for extending this technique to higher-order derivatives and arbitrary dimensions has been proposed by Florack *et al.* (1993) and Salden *et al.* (1992).

2.1.3. *Affine intensity transformations*

Another approach is to restrict the invariance to affine intensity transformations. Then, the class of invariants becomes larger. A natural condition to impose is that a differential expression $\mathcal{D}I$ should (at least) be a *relative invariant* with respect to scale changes, i.e., under a rescaling of the spatial coordinates, $I'(x) \hat{=} I(sx)$, the differential entity should transform as $\mathcal{D}I' = s^k\mathcal{D}I$ for some k. Trivially, this relation holds for any product of mixed directional derivatives, and extends to sums (and rational functions) of such expressions provided that the sum of the orders of differentiation is the same for any product of derivatives constituting one term in a sum.

To give a formal description of this property, let $I_{u^m v^n} = I_{w^\alpha}$ denote a mixed directional derivative of order $|\alpha| = m + n$, and let $\mathcal{D}$ be a (possibly non-linear) homogeneous differential expression of the form

$$\mathcal{D}I = \sum_{i=1}^{I} c_i \prod_{j=1}^{J} I_{w^{\alpha_{ij}}}, \tag{2.5}$$

where $|\alpha_{ij}| > 0$ for all $i = [1..I]$ and $j = [1..J]$, and

$$\sum_{j=1}^{J} |\alpha_{ij}| = N \tag{2.6}$$

---

[2]In the chapter by Alvarez *et al.* (1994) this property of invariance under monotonic intensity transformations is referred to as "morphological invariance".

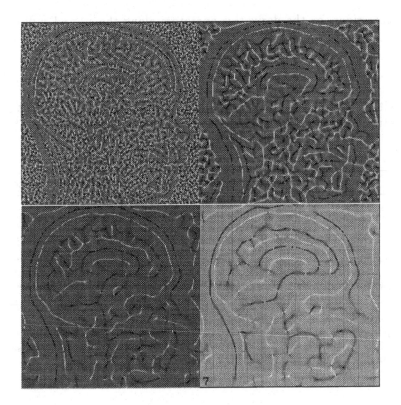

Figure 2.1. The result of computing the isophote curvature κ of a sagittal NMR image a scale levels $\sigma=1$, 3, 5 and 7. (Image size: 256 × 256 pixels. Note the ridge structure at the different scales. (In section 2.2.1 and figure 2.3 it is shown how curves representing such ridges can be extracted using differential operators. These features have been used for matching of medical images of different modalities (van de Elsen *et al.* 1993))

for all $i \in [1..I]$. Then, $\mathcal{D}I$ is invariant with respect to translations, rotations, and affine intensity transformations, and relative invariant under uniform rescalings of the spatial coordinates.

2.1.4. *Tensor notation for invariant expressions*

A useful analytical tool when dealing with differential invariants is to express them in terms of tensor notation (Abraham 1988, Lawden 1962). Adopt the Einstein summation convention that double occurrences of a certain index means that summation is to be performed over that index. Furthermore, let δ_{ij} be the symmetric Kronecker tensor and let $\epsilon_{ij...}$ represent the antisymmetric Levi-Civita connection (see Kay 1988).[3] Then,

---

[3]The Kronecker tensor has unity elements on the diagonal locations (equal indices), whereas the Levi-Civita tensor has zero at these locations, and unity at the others, the

the expressions for the derivative operators along the u- and v-directions (2.1) assume the form

$$\partial_u = \frac{I_i \epsilon_{ij} \partial_j}{\sqrt{I_k I_k}}, \qquad \partial_v = \frac{I_i \delta_{ij} \partial_j}{\sqrt{I_k I_k}}. \tag{2.7}$$

Explicit expressions for a few differential invariants on the different forms respectively are shown in table 2.1 (see also ter Haar Romeny *et al.* (1994)).

Name	Cartesian	Tensor	Gauge
Intensity	I	I	I
Gradient^2	$I_x^2 + I_y^2$	$I_i I_i$	I_v^2
Laplacian	$I_{xx} + I_{yy}$	I_{ii}	$I_{uu} + I_{vv}$
Isophote curvature	$\dfrac{2I_x I_y I_{xy} - I_x^2 I_{yy} - I_y^2 I_{xx}}{(I_x^2 + I_y^2)^{3/2}}$	$\dfrac{I_i I_j I_{ij} - I_i I_i I_{jj}}{(I_k I_k)^{3/2}}$	$-\dfrac{I_{uu}}{I_v}$
Flowline curvature	$\dfrac{I_x I_y (I_{yy} - I_{xx}) + I_{xy}(I_x^2 - I_y^2)}{(I_x^2 + I_y^2)^{3/2}}$	$\dfrac{I_i I_{ij} I_j - I_i I_i I_{jj}}{(I_k I_k)^{3/2}}$	$-\dfrac{I_{uv}}{I_v}$

TABLE 2.1. Some examples of two-dimensional differential invariants under orthogonal transformations, expressed in (i) Cartesian coordinates, (ii) tensor notation, and (iii) gauge coordinates, respectively.

2.2. Feature detection from differential singularities

The *singularities* (zero-crossings) of differential invariants play an important role (Lindeberg 1993). This is a special case of a more general principle of using zero-crossings of differential geometric expressions for describing geometric features; see e.g. Bruce and Giblin (1984) for an excellent tutorial. If a feature detector can be expressed as a zero-crossing of such a differential expression, then the feature will also be absolute invariant to uniform rescalings of the spatial coordinates, i.e. size changes.

Formally, this invariance property can be expressed as follows: Let $\mathcal{S}_{\mathcal{D}} I$ denote the *singularity set* of a differential operator of the form (2.5), i.e.

$$\mathcal{S}_{\mathcal{D}} I = \{(x; t) \in \mathbb{R}^2 \times \mathbb{R}_+ : \mathcal{D} I(x; t) = 0\},$$

and let $\mathcal{G}$ be the Gaussian smoothing operator, i.e., $I = \mathcal{G} f$. Under these transformations of the spatial domain (represented by $x \in \mathbb{R}^2$) and the intensity domain (represented by either the unsmoothed f or the smoothed I) the singularity sets[4] transform as follows:

---

sign determined by sign of the permutation of the indices.

[4]Here, R is a rotation matrix, Δx is a vector ($\in \mathbb{R}^2$), whereas a, b and s are

Transformation	Definition	Invariance
translation	$(\mathcal{T}I)(x;\, t) = I(x + \Delta x;\, t)$	$\mathcal{S}_D \mathcal{G} \mathcal{T} f = \mathcal{S}_D \mathcal{T} \mathcal{G} f = \mathcal{T} \mathcal{S}_D \mathcal{G} f$
rotation	$(\mathcal{R}I)(x;\, t) = I(Rx;\, t)$	$\mathcal{S}_D \mathcal{G} \mathcal{R} f = \mathcal{S}_D \mathcal{R} \mathcal{G} f = \mathcal{R} \mathcal{S}_D \mathcal{G} f$
uniform scaling	$(\mathcal{U}I)(x;\, t) = I(sx;\, t)$	$\mathcal{S}_D \mathcal{G} \mathcal{U} f = \mathcal{S}_D \mathcal{U} \mathcal{G} f = \mathcal{U} \mathcal{S}_D \mathcal{G} f$
affine intensity	$(\mathcal{A}I)(x;\, t) = aI(x;\, t) + b$	$\mathcal{S}_D \mathcal{G} \mathcal{A} f = \mathcal{S}_D \mathcal{A} \mathcal{G} f = \mathcal{S}_D \mathcal{G} f$

In other words, feature detectors formulated in terms of differential singularities by definition commute with a number of elementary transformations, and it does not matter whether the transformation is performed before or after the smoothing step. A few examples of feature detectors that can be expressed in this way are discussed below.

2.2.1. *Examples of feature detectors*

Edge detection. A natural way to define edges from a continuous grey-level image $I : \mathbb{R}^2 \to \mathbb{R}$ is as the union of the points for which the gradient magnitude assumes a maximum in the gradient direction. This method is usually referred to as *non-maximum suppression* (see e.g. Canny (1986) or Korn (1988)). Assuming that the second and third-order directional derivatives of I in the v-direction are not simultaneously zero, a necessary and sufficient condition for P_0 to be a gradient maximum in the gradient direction may be stated as:

$$\begin{cases} I_{vv} = 0, \\ I_{vvv} < 0. \end{cases} \tag{2.8}$$

Since only the sign information is important, this condition can be restated as

$$\begin{cases} I_v^2 I_{vv} = I_x^2 I_{xx} + 2 I_x I_y I_{xy} + I_y^2 I_{yy} = 0, \\ I_v^3 I_{vvv} = I_x^3 I_{xxx} + 3 I_x^2 I_y I_{xxy} + 3 I_x I_y^2 I_{xyy} + I_y^3 I_{yyy} < 0. \end{cases} \tag{2.9}$$

Interpolating for zero-crossings of I_{vv} within the sign-constraints of I_{vvv} gives a straightforward method for sub-pixel edge detection (Lindeberg 1993); see figure 2.2 for an illustration.

---

scalar constants. The definitions of the transformed singularity sets are as follows; $\mathcal{T}\mathcal{S}_D I = \{(x;\, t) : \mathcal{D}I(x + \Delta x;\, t) = 0\}$, $\mathcal{R}\mathcal{S}_D I = \{(x;\, t) : \mathcal{D}I(Rx;\, t) = 0\}$, and $\mathcal{U}\mathcal{S}_D I = \{(x;\, t) : \mathcal{D}I(sx;\, \underline{s^2}t) = 0\}$.

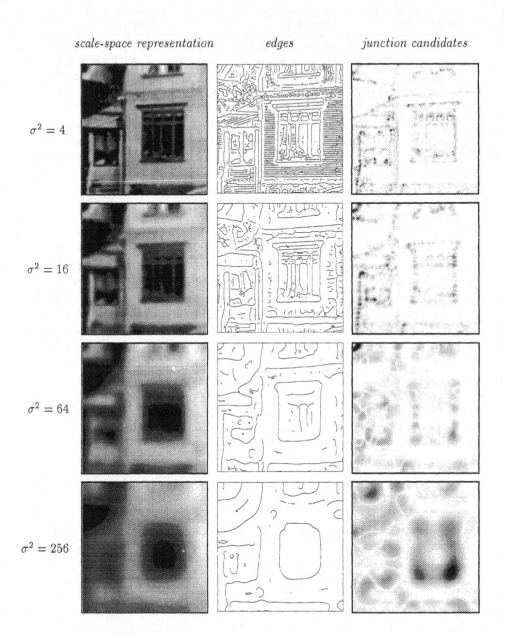

Figure 2.2. Examples of multi-scale feature detection in scale-space using singularities of differential invariants. (left) Smoothed grey-level images. (middle) Edges defined by $I_{vv} = 0$ and $I_{vvv} < 0$. (right) Magnitude of $\tilde{\kappa}$ (see eq. (2.11)). (Scale levels from top to bottom: $\sigma^2 = 4$, 16, 64, and 256. Image size: 256×256 pixels.) (From [222].)

Ridge detection. A ridge detector can be expressed in a conceptually similar way by detecting zero-crossings in I_{uv} that satisfy $I_{uu}^2 - I_{vv}^2 > 0$ (Lindeberg 1994). A natural measure of the strength of the response is given by I_{uv}; points with $I_{uv} > 0$ correspond to dark ridges and points with $I_{uu} < 0$ to bright ridges (see figure 2.3).

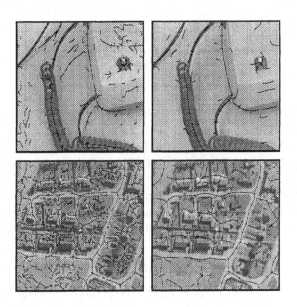

Figure 2.3. Examples of differential geometric ridge detection (without thresholding): (a)–(b) dark ridges from a detail from a telephone image at scale levels $\sigma^2 = 16$ and 64, (c)–(d) bright ridges from an aerial image at scale levels $\sigma^2 = 16$ and 64. (From [222].)

Junction detection. An entity commonly used for junction detection is the curvature of level curves in intensity data, see e.g. Kitchen and Rosenfeld (1982) or Koenderink and Richards (1988). In terms of directional derivatives it can be expressed as

$$\kappa = -\frac{I_{uu}}{I_v}. \tag{2.10}$$

To give a stronger response near edges, the level curve curvature is usually multiplied by the gradient magnitude I_v raised to some power k. A natural choice is $k = 3$. This leads to a polynomial expression (see e.g. Brunnström *et al.* 1992)

$$|\tilde{\kappa}| = |I_v^2 I_{uu}| = |I_y^2 I_{xx} - 2I_x I_y I_{xy} + I_x^2 I_{yy}|. \tag{2.11}$$

Since the sum of the order of differentiation with respect to x and y is the same for all terms in this sum, it follows that junction candidates given by extrema in $\tilde{\kappa}$ also are invariant under skew transformations (Blom 1992) and affine transformations (see the chapter by Olver *et al.* (1994)).

Assuming that the first- and second-order differentials of $\tilde{\kappa}$ are not simultaneously degenerate, a necessary and sufficient condition for a point P_0 to be a maximum in this *rescaled level curve curvature* is that:

$$\begin{cases} \partial_u \kappa = 0, \\ \partial_v \kappa = 0, \\ \mathcal{H}(\kappa) = \kappa_{\mathcal{H}} = \kappa_{uu}\kappa_{vv} - \kappa_{uv}^2 > 0, \\ \text{sign}(\kappa)\kappa_{uu} < 0. \end{cases} \tag{2.12}$$

Interpolating for simultaneous zero-crossings in $\partial_u \kappa$ and $\partial_v \kappa$) gives a sub-pixel corner detector.

Junction detectors of higher order can be derived algebraically (Salden *et al.* (1992)) by expressing the local structure up to some order in terms of its (truncated) local Taylor expansion and by studying the roots (*i.e.*, the discriminant) of the corresponding polynomial. Figure 2.4 shows the result of applying a fourth-order (rotationally symmetric) differential invariant obtained in this way, $D_4 I$, to a noisy image of checkerboard pattern

$$\begin{aligned} D_4 I \ = -&\left(I_{x^4} I_{y^4} - 4 I_{x^3 y} I_{xy^3} + 3 I_{x^2 y^2}^2 \right)^3 \\ +27 &\left(I_{x^4} \left(I_{x^2 y^2} I_{y^4} - I_{xy^3}^2 \right) - I_{x^3 y} \left(I_{x^3 y} I_{y^4} - I_{x^2 y^2} I_{xy^3} \right) \right. \\ &\left. + I_{x^2 y^2} \left(I_{x^3 y} I_{xy^3} - I_{x^2 y^2}^2 \right) \right)^2 . \end{aligned} \tag{2.13}$$

The Mathematica[TM] [387] code by which figure 2.4 is generated is given in the appendix.

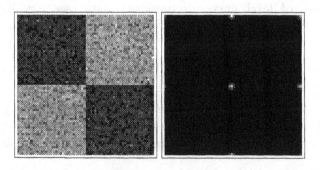

Figure 2.4. Fourth-order junction detector. (left) Input image 64×64 pixels with 20 % added Gaussian noise. (right) Magnitude of the fourth-order differential invariant given by (2.13). (The periodicity is due to an implementation in the Fourier domain.)

Blob detection. Zero-crossings of the Laplacian

$$\nabla^2 I = I_{uu} + I_{vv} = I_{xx} + I_{yy} = 0 \qquad (2.14)$$

have been used for *stereo matching* (see, e.g., Marr 1992) and *blob detection* (see, e.g., Blostein and Ahuja 1987). Blob detection methods can also be formulated in terms of local extrema of the grey-level landscape (Lindeberg 1991, 1993) and extrema of the Laplacian (Lindeberg and Gårding 1993).

Analysis: "Edge detection" using zero-crossings of the Laplacian. Zero-crossings of the Laplacian have been used also for edge detection, although the localization is poor at curved edges. This can be understood from the relation between the Laplace operator and the second derivative in the gradient direction (obtained from (2.10) and (2.14))

$$\nabla^2 I = I_{uu} + I_{vv} = I_{vv} + \kappa I_v. \qquad (2.15)$$

which shows that the deviation between zero-crossings of $\nabla^2 I$ and zero-crossings of I_{vv} increases with the isophote curvature κ. This example constitutes a simple indication of how theoretical analysis of feature detectors becomes tractable when expressed in terms of the suggested differential geometric framework.

2.3. Scale selection

Although the scale-space theory presented so far provides a well-founded framework for dealing with image structures at different scales and can be used for formulating multi-scale feature detectors, it does not directly address the problem of how to *select* appropriate scales for further analysis.

Whereas the problem of selecting "the best scale(s)" for handling a real-world data set may be intractable unless at least some *a priori* information about the scene contents is available, in many situations a mechanism is required for generating hypothesis about interesting scale levels.
One general method for scale selection has been proposed in (Lindeberg 1993, 1994). The approach is based on the evolution over scales of (possibly non-linear) combinations of *normalized derivatives* defined by

$$\partial_{\xi_i} = \sqrt{t}\, \partial_{x_i}, \qquad (2.16)$$

where the *normalized coordinates*

$$\xi = x/\sqrt{t} \qquad (2.17)$$

are the spatial correspondences to the dimensionless frequency coordinates $\omega\sigma$ considered in section 5.6 in the previous chapter. The basic idea of the scale selection method is to *select scale levels from the scales at which differential geometric entities based on normalized derivatives assume local maxima over scales.* The underlying motivation for this approach is to select the scale level(s) where the operator response is as its strongest. A theoretical support can also be obtained from the fact that for a large class of polynomial differential invariants (homogeneous differential expressions of the form (2.5)) such extrema over scales have a nice behaviour under rescalings of the input signal: If a *normalized differential invariant* $\mathcal{D}_{norm}L$ assumes a maximum over scales at a certain point $(x_0; t_0)$ in scale-space, then if a rescaled signal f' is defined by $f'(sx) = f(x)$, a scale-space maximum in the corresponding normalized differential entity $\mathcal{D}_{norm}L'$ is assumed at $(sx_0; s^2 t_0)$.

Example: Junction detection with automatic scale selection. In junction detection, a useful entity for selecting detection scales is the normalized rescaled curvature of level curves,

$$\tilde{\kappa}_{norm} = t^2 \, |\nabla I|^2 \, I_{uu}. \tag{2.18}$$

Figure 2.5 shows the result of detecting *scale-space maxima* (points that are simultaneously maxima with respect to variations of *both* the scale parameter and the spatial coordinates) of this normalized differential invariant. Observe that a set of junction candidates is generated with reasonable interpretation in the scene. Moreover, the circles (with their areas equal to the detection scales) give natural regions of interest around the candidate junctions.

Second stage selection of localization scale. Whereas this junction detector is conceptually very clean, it can certainly lead to poor localization, since shape distortions may be substantial at coarse scales in scale-space. A straightforward way to improve the location estimate is by determining the point x that minimizes the (perpendicular) distance to all lines in a neighbourhood of the junction candidate x_0. By defining these lines with the gradient vectors as normals and by weighting each distance by the pointwise gradient magnitude, this can be expressed as a standard least squares problem (Förstner and Gülch 1987),

$$\min_{x \in \mathbb{R}^2} x^T A x - 2 x^T b + c \quad \Longleftrightarrow \quad A x = b, \tag{2.19}$$

where $x = (x_1, x_2)^T$, w_{x_0} is a window function, and A, b, and c are entities determined by the local statistics of the gradient directions in a

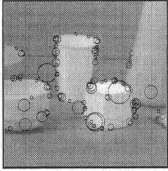

Figure 2.5. Junction candidates obtained by selecting the 150 scale-space maxima having the strongest maximum normalized response. (From [218, 219].)

neighbourhood of x_0,

$$A = \int_{x' \in \mathbb{R}^2} (\nabla I)(x') \, (\nabla I)^T(x') \, w_{x_0}(x') \, dx', \qquad (2.20)$$

$$b = \int_{x' \in \mathbb{R}^2} (\nabla I)(x') \, (\nabla I)^T(x') \, x' \, w_{x_0}(x') \, dx', \qquad (2.21)$$

$$c = \int_{x' \in \mathbb{R}^2} x'^T \, (\nabla I)(x') \, (\nabla I)^T(x') \, x' w_{x_0}(x') \, dx'. \qquad (2.22)$$

Figure 2.6 shows the result of computing an improved localization estimate

Figure 2.6. Improved localization estimates for the junction candidates in figure 2.5. (left) Circle area equal to the *detection scale*. (right) Circle area equal to the *localization scale*. (From [218, 219].)

in this way using a Gaussian window function with scale value equal to the detection scale and by selecting the localization scale that minimizes the *normalized residual*

$$d_{min} = (c - b^T A^{-1} b)/\text{trace} A \qquad (2.23)$$

over scales (Lindeberg 1993, 1994). This procedure has been applied iteratively five times and those points for which the procedure did not converge after five iterations have been suppressed. Notice that a sparser set of junction candidates is obtained and how the localization is improved.

2.4. Cues to surface shape (texture and disparity)

So far we have been concerned with the theory of scale-space representation and its application to feature detection in image data. A basic functionality of a computer vision system, however, is the ability to derive information about the three-dimensional shape of objects in the world.

Whereas a common approach, historically, has been to compute two-dimensional image features (such as edges) in a first processing step, and then combining these into a three-dimensional shape description (e.g., by stereo or model matching), we shall here consider the problem of deriving shape cues *directly from image data*, and by using only the types of front-end operations that can be expressed within the scale-space framework.

Examples of work in this direction have been presented by (Jones and Malik 1992; Lindeberg and Gårding 1993; Malik and Rosenholtz 1993; Gårding and Lindeberg 1994). A common characteristic of these methods is that they are based on measurements of the distortions that surface patterns undergo under perspective projection; a problem which is simplified by considering the locally linearized component, leading to computation of cues to surface shape from measurements of local affine distortions.

Measuring local affine distortions. The method by Lindeberg and Gårding (1993) is based on an image texture descriptor called the *windowed second-moment matrix*. With I denoting the image brightness it is defined by

$$\mu_I(q) = \int_{x' \in \mathbb{R}^2} (\nabla I)(x') \, (\nabla I)^T(x') \, g(q - x') \, dx'. \qquad (2.24)$$

With respect to measurements of local affine distortions, this image descriptor transforms as follows: Define R by $I(\xi) = R(B\xi)$ where B is an invertible 2×2 matrix representing a linear transformation. Then, we have

$$\mu_I(q) = B^T \mu_R(p) \, B, \qquad (2.25)$$

where $\mu_R(p)$ is the second-moment matrix of R expressed at $p = Bq$ computed using the "backprojected" normalized window function $w'(\eta - p) = (\det B)^{-1} w(\xi - q)$.

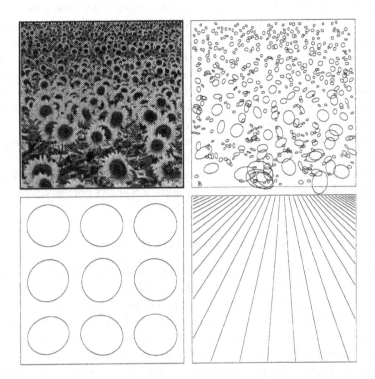

Figure 2.7. Local surface orientation estimates computed by a shape-from-texture method based on measurements of local affine distortions using the windowed second-moment matrix (2.24). (top left) An image of a sunflower field. (top right) Blobs detected by scale-space extrema of the normalized Laplacian. (bottom left) Surface orientation computed under the assumption of *weak isotropy*. (bottom right) Surface orientation computed under the assumption of *constant area*. (From [229].)

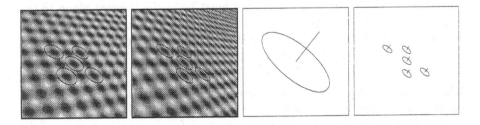

Figure 2.8. Local surface orientation estimated from the gradient of horizontal disparity in a synthetic stereo pair with 5% noise. (left and middle left) Right and left images with ellipse representation of five second-moment matrices. (middle right) Reference surface orientation. (right) Estimated surface orientation at five manually matched points. (From [127].)

Shape-from-texture and disparity gradients. Given two measurements of μ_I and μ_R, the relation (2.25) can be used for recovering B (up to an arbitrary rotation). This gives a direct method for deriving surface orientation from monocular cues, by imposing specific assumptions on μ_R, e.g., that μ_R should be a constant times the unit matrix, $\mu_R = cI$ (weak isotropy) or that $\det \mu_R$ should be locally constant (constant area). Similarly, if two cameras fixate the same surface structure, a direct estimate of surface orientation can be obtained provided that the vergence angle is known.

Figure 2.7 shows surface orientation estimates computed in this way. Note that for this image the weak isotropy assumption gives the orientation of the individual flowers, whereas the constant area assumption reflects the orientation of the underlying surface. Figure 2.8 shows corresponding results for stereo data.

3. Behaviour across scales: Deep structure

The treatment so far has been concerned with the formal definition of the scale-space representation and the definition of image descriptors at any single scale. An important problem, however, concerns how to relate structures at different scales. This subject has been termed *deep structure* by Koenderink (1984). When a pattern is subject to scale-space smoothing, its shape changes. This gives rise to the notion of *dynamic shape*, which as argued by Koenderink and van Doorn (1986) is an essential component of any shape description of natural objects.

3.1. Iso-intensity linking

An early suggestion by Koenderink (1984) to relate structures at different scales was to identify points across scales that have the same grey-level and correspond to paths of steepest ascent along level surfaces in scale-space.

Since the tangent vectors of such paths must be in the tangent plane to the level surface and the spatial component must be parallel to (I_x, I_y), these *iso-intensity paths* are the integral paths of the vector field

$$(I_x I_t, \quad I_y I_t, \quad -(I_x^2 + I_y^2)). \tag{2.26}$$

Lifshitz and Pizer (1990) considered such paths in scale-space, and constructed a multi-scale "stack" representation, in which the grey-level at which an extremum disappeared was used for defining a region in the original image by local thresholding on that grey-level.

Although the representation was demonstrated to be applicable for certain segmentation problems in medical image analysis, Lifshitz and Pizer observed the serious problem of *non-containment*, which basically means that a point, which at one scale has been classified as belonging to a certain region (associated with a local maximum), can escape from that region when the scale parameter increases. Moreover, such paths can be intertwined in a rather complicated way.

3.2. Feature based linking (differential singularities)

The main cause to problem in the iso-intensity linking is that grey-levels corresponding to a feature tracked over scales change under scale-space smoothing. For example, concerning a local extremum it is a necessary consequence of the diffusion equation that the grey-level value at the maximum point must decrease with scale. For this reason, it is more natural to identify *features* across scales rather than grey-levels. A type of representation defined in this way is the *scale-space primal sketch* of blob-like image structures (extrema with extent) defined at all scales in scale-space and linked into a tree-like data structure (Lindeberg 1991, 1993). More generally, consider a feature that at any level of scale is defined by

$$h(x;\, t) = 0 \qquad (x \in \mathbb{R}^N, t \in \mathbb{R}_+) \tag{2.27}$$

for some function $h: \mathbb{R}^N \times \mathbb{R}_+ \to \mathbb{R}^M$. (For example, the differential singularities considered in section 2.2 are of this form.) Using the implicit function theorem it is then formally easy to analyze the dependence of x on t in the solution to (2.27). Here, some simple examples will be presented of how such analysis can be performed; see Lindeberg (1992, 1994) for a more extensive treatment. Consider, for simplicity, data given as two-dimensional images. Then, it is sufficient to study the cases when M is either 1 or 2.

3.2.1. *Pointwise entities*

If $M = 2$ the features will in general be isolated points. The implicit function theorem states that these points form smooth paths across scales (one-dimensional curves in three-dimensional scale-space) provided that the Jacobian $\partial_x h$ is non-degenerate. The drift velocity along such a path can be written

$$\partial_t x = -(\partial_x^T h)^{-1} \partial_t h.$$

Critical points. Concerning critical points in the grey-level landscape, we have $h = (I_x, I_y)^T$ and the drift velocity can be written

$$\partial_t x = -\frac{1}{2}(\mathcal{H}I)^{-1}\nabla^T\nabla(\nabla I),$$

where $\mathcal{H}I$ denotes the Hessian matrix of I and the fact that the spatial derivatives satisfy the diffusion equation has been used for replacing derivatives of I with respect to t by derivatives with respect to x.

Other structures. A similar analysis can be performed for other types of point structures, e.g. junctions given as maxima in $\tilde{\kappa}$, although the expressions then contain derivatives of higher-order. Concerning $\tilde{\kappa}$, the drift velocity contains derivatives up to order five (Lindeberg 1994).

This result gives an estimate of the drift velocity of features due to scale-space smoothing, and provides a theoretical basis for relating and, hence, linking corresponding features across scales in a well-defined manner.

3.2.2. *Curve entities*

If $M = 1$, then the set of feature points will in general be curves when treated at a single scale and surfaces when treated at all scales. Hence, there is no longer any unique correspondence between points at adjacent scales. This ambiguity is similar to the so-called "aperture problem" in motion analysis. Nevertheless, the normal component of the drift can be determined. If s represents a coordinate along the normal direction, then the drift velocity can be expressed as

$$\partial_t s = -\tilde{h}_s^{-1}\,\tilde{h}_t = -\frac{h_t}{|\nabla h|} .$$

Curved edges. For example, concerning an edge given by non-maximum suppression ($\alpha = I_{vv} = 0$), the drift velocity in the normal direction assumes the form

$$(\partial_t u, \partial_t v) = -\frac{I_v(I_{uuvv} + I_{vvvv}) + 2I_{uv}(I_{uuu} + I_{uvv})}{2((I_v I_{uvv} + 2I_{uv}I_{uu})^2 + (I_v I_{vvv} + 2I_{uv}^2)^2)}\Big(\frac{\alpha_u}{I_v}, \frac{\alpha_v}{I_v}\Big), \quad (2.28)$$

where

$$\begin{aligned}
\alpha_u &= I_v^2 I_{uvv} + 2I_v I_{uv}I_{uu}, \\
\alpha_v &= I_v^2 I_{vvv} + 2I_v I_{uv}^2,
\end{aligned} \qquad (2.29)$$

represent the components of the normal vector (α_u, α_v) to the edge expressed in the (u, v) coordinate system. Unfortunately, this expression cannot be further simplified unless additional constraints are posed on I.

Straight edges. For a straight edge, however, where all partial derivatives with respect to v are zero, it reduces to

$$(\partial_t u, \partial_t v) = -\frac{1}{2}\frac{I_{vvvv}}{I_{vvv}}(0, 1). \qquad (2.30)$$

This analysis can be used for stating a formal description of the *edge focusing* method developed by Bergholm (1987), in which edges are detected at a coarse scale and then tracked to finer scales; see also Clark (1989) and Lu and Jain (1989) concerning the behaviour of edges in scale-space. (An extension of the edge focusing idea is also presented in the chapter by Richardson and Mitter (1994) in this book.)

Linking in pyramids. Note the qualitative difference between linking across scales in the scale-space representation of a signal and the corresponding problem in a pyramid. In the first case, the linking process can be expressed in terms of differential equations, while in the second case it corresponds to a combinatorial matching problem. It is well-known that it is a hard algorithmic problem to obtain stable links between features in different layers in a pyramid.

3.3. Bifurcations in scale-space

The previous section states that the scale linking is well-defined whenever the appropriate submatrix of the Jacobian of h is non-degenerate, When the Jacobian degenerates, *bifurcations* may occur.

Concerning critical points in the grey-level landscape, the situation is simple. In the one-dimensional case, the generic bifurcation event is the annihilation of a pair consisting of a local maximum and a minimum point, while in the two-dimensional case a pair consisting of a saddle point and an extremum can be both annihilated and created[5] with increasing scale. A natural model of this so-called *fold singularity* is the polynomial

$$I(x;\ t) = x_1^3 + 3x_1(t - t_0) + \sum_{i=2}^{N} \pm (x_i^2 + t - t_0), \qquad (2.31)$$

which also satisfies the diffusion equation; see also Poston and Stewart (1978), Koenderink and van Doorn (1986), Lifshitz and Pizer (1990), and Lindeberg (1992, 1994). The positions of the critical points are given by

$$x_1(t) = \pm\sqrt{t_0 - t} \qquad (x_i = 0,\ i > 1) \qquad (2.32)$$

i.e. the critical points merge along a parabola, and the drift velocity tends to infinity at the bifurcation point.

Johansen (1994) gives a more detailed differential geometric study of such bifurcations, covering also a few cases that are generically unstable when treated in a single image. Under more general parameter variations, however, such as in image sequences, these singularities can be expected to be stable in the sense that a small disturbance of the original signal causes the singular point to appear at a slightly different time moment.

4. Scale sampling

Although the scale-space concept comprises a continuous scale parameter, it is necessary to actually compute the smoothed representations at some

---

[5] An example of a creation event is given at the end of section 5.5 in previous chapter.

discrete set of sampled scale levels. The fact that drift velocities may (momentarily) tend to infinity indicates that in general some mechanism for adaptive scale sampling must be used. Distributing scale levels over scales is closely related to the problem of measuring scale differences. From the dimensional analysis in section 5.6 in previous chapter it follows that the scale parameter σ provides a unit of length unit at scale $t = \sigma^2$. How should we then best parametrize the scale parameter for scale measurements? As we shall see in this section, several different ways of reasoning, in fact, lead to the same result.

4.1. Natural scale parameter: Effective scale

Continuous signals. For continuous signals, a natural choice of transformed scale parameter τ is given by

$$\tau = \log \frac{\sigma}{\sigma_0} \tag{2.33}$$

for some σ_0 and τ_0. This can be obtained directly from scale invariance: If the scale parameter σ (which is measured by dimension [length]) is to be parametrized by a dimensionless scale parameter τ, then scale-invariance or self-similarity implies that $d\sigma/\sigma$ must be the differential of a dimensionless variable (see section 5.6 in the previous chapter and Florack *et al.* 1992). Without loss of generality one can let $d\sigma/\sigma = d\tau$ and select a specific reference scale σ_0 to correspond to $\tau = 0$. Hence, we obtain (2.33).

In a later chapter in this book by Eberly (1994), this idea is pursued further and he considers the problem of combining measurements of scale differences in terms of $d\tau = d\sigma/\sigma$ with measurements of normalized spatial differences in terms of $d\xi = dx/\sigma$.

Discrete signals. Some more care must be taken if the lifetime of a structure in scale-space is to be used for measuring significance in discrete signals, since otherwise a structure existing in the original signal (assigned scale value zero) would be assigned an infinite lifetime. An analysis in (Lindeberg 1991, 1993) shows that a natural way to introduce such a scale parameter for discrete signals is by

$$\tau(t) = A + B \log p(t), \tag{2.34}$$

where $p(t)$ constitutes a measure of the "amount of structure" in a signal at scale t. For practical purposes, this measure is taken as the density of local extrema in a set of reference data.

Continuous vs. discrete models.   Under rather general conditions on a one-dimensional continuous signal it holds that the number of local extrema in a signal decrease with scale as $p(t) \sim \frac{1}{t^\alpha}$ for some $\alpha > 0$ (see section 7). This means that $\tau(t)$ given by (2.34) reduces to (2.33). For discrete signals, on the other hand, $\tau(t)$ is approximately linear at fine scales and approaches the logarithmic behaviour asymptotically when t increases.

In this respect, the latter approach provides a well-defined way to express the notion of *effective scale* to comprise both continuous and discrete signals as well as for modeling the transition from the genuine discrete behaviour at fine scales to the increasing validity of the continuous approximation at coarser scales.

5. Regularization properties of scale-space kernels

According to Hadamard, a problem is said to be well-posed if: (i) a solution exists, (ii) the solution is unique, and (iii) the solution depends continuously on the input data. It is well-known that several problem in computer vision are ill-posed; one example is differentiation. A small disturbance in a signal, $f(x) \mapsto f(x) + \varepsilon \sin \omega x$, where ε is small and ω is large, can lead to an arbitrarily large disturbance in the derivative

$$f_x(x) \mapsto f_x(x) + \omega \varepsilon \cos \omega x, \tag{2.35}$$

provided that ω is sufficiently large relative to $1/\epsilon$.

Regularization is a technique that has been developed for transforming ill-posed problems into well-posed ones; see Tikhonov and Arsenin (1977) for an extensive treatment of the subject. Torre and Poggio (1986) describe this issue with application to one of the most intensely studied subproblems in computer vision, edge detection, and develop how regularization can be used in this context. One example of regularization concerning the problem "given an operator $\mathcal{A}$ and data y find z such that $\mathcal{A}z = y$" is the transformed problem "find z that minimizes the following functional"

$$\min_z \ (1 - \lambda) \, ||\mathcal{A}z - y||^2 + \lambda \, ||\mathcal{P}z||^2, \tag{2.36}$$

where $\mathcal{P}$ is a stabilizing operator and $\lambda \in [0, 1]$ is a regularization parameter controlling the compromise between the degree of regularization of the solution and closeness to the given data. Variation of the regularization parameter gives solutions with different degree of smoothness; a large value of λ may give rise a smooth solution, whereas a small value increases the accuracy at the cost of larger variations in the estimate. Hence, this parameter has a certain interpretation in terms of spatial scale in the result. (It should be observed, however, that the solution to the regularized problem

is in general not a solution to the original problem, not even in the case of
ideal noise-free data.)

In the special case when $\mathcal{P} = \partial_{xx}$, and the measured data points are discrete,
the solution of the problem of finding $S: \mathbb{R} \to \mathbb{R}$ that minimizes

$$\min_S \ (1 - \lambda) \sum (f_i - S(x_i))^2 + \lambda \int |S_{xx}(x_i)|^2 dx \qquad (2.37)$$

given a set of measurements f_i is given by approximating cubic splines;
see de Boor (1978) for an extensive treatment of the subject. Interestingly,
this result was first proved by Schoenberg (1946), who also proved the
classification of Pólya frequency functions and sequences, which are the
natural concepts in mathematics that underlie the scale-space kernels
considered in sections 5.4–5.5 and section 7.1 in previous chapter. Torre
and Poggio made the observation that the corresponding smoothing filters
are very close to Gaussian kernels.

The strong regularization property of scale-space representation can be
appreciated in the introductory example. Under a small high-frequency
disturbance in the original signal $f(x) \mapsto f(x) + \varepsilon \cos \omega x$, the propagation of
the disturbance to the first-order derivative of the scale-space representation
is given by

$$I_x(x; \ t) \mapsto I_x(x; \ t) + \varepsilon \omega e^{\omega^2 t/2} \cos \omega x. \qquad (2.38)$$

Clearly, this disturbance can be made arbitrarily small provided that the
derivative of the signal is computed at a sufficiently coarse scale t in scale-
space. (The subject of regularization is also treated in the chapters by
Mumford, Nordström, Leaci and Solimini in this book.)

6. Related multi-scale representations

6.1. Wavelets

A type of multi-scale representation that has attracted a great interest in
both signal processing, numerical analysis, and mathematics during recent
years is *wavelet representation*, which dates back to Strömberg (1983) and
Meyer (1989, 1992). A (two-parameter) family of translated and dilated
(scaled) functions

$$h_{a,b}(x) = |a|^{-1/2} h(\frac{x-b}{a}) \qquad a, b \in \mathbb{R}, a \neq 0 \qquad (2.39)$$

defined from a single function $h: \mathbb{R} \to \mathbb{R}$ is called a *wavelet*. Provided that h
satisfies certain admissibility conditions

$$\int_{\omega=-\infty}^{\infty} \frac{|\hat{h}(\omega)|^2}{|\omega|} \, d\omega < \infty, \qquad (2.40)$$

the representation $\mathcal{W}f: \mathbb{R}\backslash\{0\} \times \mathbb{R} \to \mathbb{R}$ given by

$$(\mathcal{W}f)(a,b) = \langle f, h_{a,b} \rangle = |a|^{-1/2} \int_{x \in \mathbb{R}} f(x)\, h(\frac{x-b}{a})\, dx \qquad (2.41)$$

is called the *continuous wavelet transform* of $f: \mathbb{R} \to \mathbb{R}$. From this background, scale-space representation can be considered as a *special case of continuous wavelet representation, where the scale-space axioms imply that the function h must be selected as a derivative of the Gaussian kernel.* In traditional wavelet theory, the zero-order derivative is not permitted; it does not satisfy the admissibility condition, which in practice implies that

$$\int_{x=-\infty}^{\infty} h(x)\, dx = 0. \qquad (2.42)$$

There are several developments of this theory concerning different special cases. A particularly well studied problem is the construction of *orthogonal wavelets* for discrete signals, which permit a compact non-redundant multi-scale representation of the image data. This representation was suggested for image analysis by Mallat (1989, 1992). We shall not attempt to review any of that theory here. Instead, the reader is referred to the rapidly growing literature on the subject; see e.g. the books by Daubechies (1992) and Ruskai et al. (1992).

6.2. Tuned scale-space kernels

Interestingly, it is possible to obtain other scale-space kernels by expanding the scale-space axioms in section 5.6 in the previous chapter by providing additional information. This operation is called *tuning* (Florack et al. 1992). Although the Gaussian family is complete in the sense that Gaussian derivatives up to some order n completely characterize the local image structure in terms of its nth-order Taylor expansion at any scale, this family may not always be the most convenient one. When dealing with time-varying imagery, for example, local optic flow may be obtained directly from the output of a Gaussian family of space-time filters, but it may be more convenient to first tune these filters to the parameter of interest (in this case the velocity vector field). Another example is to tune the low-level processing to a particular spatial frequency, which leads to the family of Gabor functions.

6.2.1. *Gabor functions*
If a family of kernels is required to be selective for a particular spatial frequency, the dimensional analysis must be expanded to include the wavenumber k of that spatial frequency:

	$\hat{I}$	$\hat{f}$	ω	σ	k
Luminance	1	1	0	0	0
Length	0	0	-1	-1	-1.

According to the Pi-theorem, there are now three dimensionless independent entities; $\hat{I}/\hat{f}$, $\omega\sigma$, and $k\omega$. Reasoning along similar lines lines as in section 5.6 in the previous chapter results in the family of Gabor filters,

$$G_b(x; t; k) = \frac{1}{\sqrt{2\pi t}} e^{-x^T x/2t} e^{-ik^T x}, \qquad (2.43)$$

which are essentially sine and cosine functions modulated by a Gaussian weighting function. Historically, these functions have been extensively used in e.g. texture analysis (see figure 2.9 for graphical illustrations).

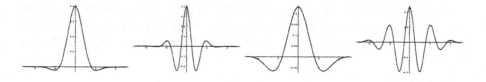

Figure 2.9.   Examples of Gabor functions; (left) ($t = 1, k = 1$), (middle left) ($t = 1, k = 3$), (middle right) ($t = 3, k = 1$), (right) ($t = 3, k = 3$).

6.2.2. *Velocity tuned kernels*
It the family of kernels is tuned to a certain spatial velocity c, the Pi-theorem diagram is expanded with both time and velocity:

	$\hat{I}$	$\hat{f}$	ω	σ_x	t	σ_t	c
Luminance	1	1	0	0	0	0	0
Length	0	0	-1	1	0	0	1
Time	0	0	0	0	1	1	-1

Here, the spatial width of the kernel is denoted by σ_x and t represents time.[6] We obtain a family of velocity-tuned spatio-temporal kernels comprising a

---

[6]Here, the temporal Gaussian kernel has infinite tails, which in principle extend from minus infinity (the past) to plus infinity (the future). To cope with this (time) causality problem, Koenderink (1988) proposed to reparameterize the time scale in a logarithmic way so as to map the present (unreachable) moment to infinity.

temporal scale σ_τ indicating the temporal width of the kernel.

$$G_b(x,t;\ \sigma_x,\sigma_\tau;\ c) = \frac{1}{\sqrt{2\pi\sigma_x^2}}\frac{1}{\sqrt{2\pi\sigma_\tau^2}}e^{-(x-ct)^2/2\sigma_x^2}\,e^{-t^2/2\sigma_\tau^2}. \qquad (2.44)$$

Example. Consider a point stimulus $I_0(\vec{x},t) = \delta(x - c_0 t)$ moving with a constant velocity c_0. Convolving this input image sequence with the velocity-tuned kernels yields

$$I_{c,\sigma_x,\sigma_\tau}(x,t) = \frac{A}{\sqrt{2\pi\sigma_\Delta^2}^D}e^{(x-c_0 t)^2/2\sigma_\Delta^2} \qquad (2.45)$$

where $\Delta = |c - c_0|/c_0$ and $\sigma_\Delta = \sigma\sqrt{1+\Delta^2}$. This is an ensemble of Gaussian blobs centered at the location of the stimulus and with the most pronounced member being the kernel with the tuning velocity equal to the stimulus velocity ($\Delta = 0$). This framework is well suited for the detection of simultaneous velocities (motion transparency) (Florack [107] 1992).

7. Behaviour across scales: Statistical analysis

In section 3 we analyzed the qualitative behaviour over scales of image features using differential geometric techniques. When to describe global properties, such as the evolution properties over scales of the number of local extrema, or irregular properties, such as the behaviour of noise in scale-space, other tools are needed. Here, we shall exemplify how such analysis can be performed statistically or based on statistical approximations.

7.1. Decreasing number of local extrema

Stationary processes. According to a result by Rice (1945), the density of local maxima μ for a stationary normal process can be estimated by the second- and fourth-order moments of the spectral density S

$$\mu = \frac{1}{2\pi}\sqrt{\frac{\int_{-\infty}^{\infty}\omega^4 S(\omega)\,d\omega}{\int_{-\infty}^{\infty}\omega^2 S(\omega)\,d\omega}}. \qquad (2.46)$$

Using this result it is straightforward to analyze how the number of local extrema can be expected to decrease with scale in the scale-space representation of various types of noise (Lindeberg 1991, 1993). For noise with a self-similar spectral density of the form

$$S(\omega) = \omega^{-\beta} \qquad (0 \le \beta < 3), \qquad (2.47)$$

the density of local extrema decreases with scale as

$$p_\beta(t) = \frac{1}{2\pi}\sqrt{\frac{3-\beta}{2}}\frac{1}{\sqrt{t}},\tag{2.48}$$

showing that the density is basically inversely proportional to the value of the scale parameter provided that the scale parameter is measured in terms of $\sigma = \sqrt{t}$.

Dimensional analysis. A corresponding result can be obtained from dimensional analysis. Assuming that an input image is sufficiently "generic" such that, roughly speaking, structures at all scales within the scale range of the image are equally represented, let us assume that this data set contains an equal amount of structure per natural volume element independent of scale. This implies that the density of "generic local features" $N_\tau(\tau)$ will be proportional to the number of samples $N(\tau)$, and we can expect the decrease of local extrema during a short scale interval (measured in terms of effective scale τ) to be proportional to the number of features at that scale, i.e.,

$$\partial_\tau N_\tau = -\alpha D N_\tau,\tag{2.49}$$

where D denotes the dimensionality of the signal. It follows immediately that the density of local extrema as function of scale is given by

$$N_\tau(\tau) = -e^{-D\tau},\tag{2.50}$$

and in terms of the regular scale parameter, t, we have

$$N_t(t) = N_0\,t^{-D/2}.\tag{2.51}$$

Experiments. Figure 2.10 shows experimental results from real image data concerning the evolution over scales of the number of local extrema and the number of elliptic regions (connected regions satisfying $I_{xx}I_{yy} - I_{xy}^2 > 0$) respectively. Note the qualitative similarities between these graphs. Observe also that a straight-line approximation is only valid in the interior part of the interval; at fine scales there is interference with the inner scale of the image given by its sampling density and at coarse scales there is interference with the outer scale of the image given by its finite size.

7.2. Noise propagation in scale-space derivatives

Computing higher-order spatial derivatives in the presence of noise is known to lead to computational problems. In the Fourier domain this is effect is usually explained in terms of amplification of higher frequencies

$$\mathcal{F}\left\{\frac{\partial^{n_1+\ldots+n_D} I}{\partial x_1^{n_1}\ldots\partial x_D^{n_D}}\right\} = (i\omega_1)^{n_1}\ldots(i\omega_D)^{n_D}\,\mathcal{F}\{I\}. \qquad (2.52)$$

Given that the noise level is known *a priori*, several studies have been presented in the literature concerning design of "optimal filters" for detecting or enhancing certain types of image structures while amplifying the noise as little as possible. From that viewpoint, the underlying idea of scale-space representation is different. Assuming that no prior knowledge is available, it follows that noise must be treated as part of the incoming signal; there is *no way* for an uncommitted front-end vision system to discriminate between noise and "underlying structures" unless specific models are available.

In this context it is of interest to study analytically how sensitive the Gaussian derivative kernels are to noise in the input. Here, we shall summarize the main results from a treatment by Blom (1993) concerning additive pixel-uncorrelated Gaussian noise with zero mean. This noise is completely described by its mean (assumed to be zero) and its variance. The ratio between the variance $< M^2_{m_x m_y} >$ of the output noise, and the variance $< N^2 >$ of the input is a natural measure of the noise attenuation. In summary, this ratio as a function of the scale parameter σ and the orders of differentiation, m_x and m_y, is given by

$$\frac{< M^2_{m_x m_y} >}{< N^2 >} = \frac{\epsilon^2}{4\pi\sigma^2}\cdot\frac{Q_{2m_x}Q_{2m_y}}{(4\sigma^2)^{m_x+m_y}} \qquad (2.53)$$

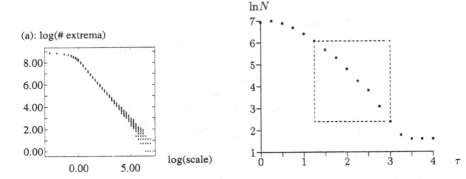

Figure 2.10. Experimental results showing the number of image features as function of scale in log-log scale. (left) The number of *local extrema* in a noisy pattern. (right) The number of *elliptic regions* (connected regions satisfying $I_{xx}I_{yy} - I_{xy}^2 > 0$).

where Q_n are defined by

$$Q_n = \begin{cases} 1 & n = 0 \\ 0 & n \text{ odd} \\ \prod_{i=1}^{n/2}(2i-1) & n \text{ even} \end{cases} \tag{2.54}$$

and the factor $\epsilon^2/4\pi\sigma^2$ describes the influence of the kernel width and ϵ is a measure of the extent of spatial correlation.

It can be seen that Q_n increases rapidly with n; we have $Q_0 = 1$, $Q_2 = 1$, $Q_4 = 3$, $Q_6 = 15$, $Q_8 = 105$, and $Q_{10} = 945$. In table 2.2 explicit values are given up to order four. Figure 2.11 shows graphical illustrations for G, G_{xx}, G_{xxxx} and G_{xxyy}. Note the marked influence of increasing the scale.

derivative	$\frac{8\pi}{\epsilon^2}\frac{<M^2>}{<N^2>}$	derivative	$\frac{8\pi}{\epsilon^2}\frac{<M^2>}{<N^2>}$
I	$\frac{1}{\sigma^2}$		
I_x, I_y	$\frac{1}{2\sigma^4}$	I_{xxy}, I_{xyy}	$\frac{3}{8\sigma^8}$
I_{xx}, I_{yy}	$\frac{3}{4\sigma^6}$	I_{xxxx}, I_{yyyy}	$\frac{105}{16\sigma^{10}}$
I_{xy}	$\frac{1}{4\sigma^6}$	I_{xxxy}, I_{xyyy}	$\frac{15}{16\sigma^{10}}$
I_{xxx}, I_{yyy}	$\frac{15}{8\sigma^8}$	I_{xxyy}	$\frac{9}{16\sigma^{10}}$

TABLE 2.2. Ratio between the variance of the output noise and the variance of the input noise for Gaussian derivative kernels up to order four. (Adapted from Blom 1993.)

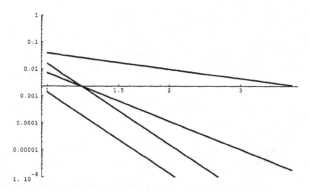

Figure 2.11. Decrease of output of Gaussian derivative filter as function of scale (log-log scale with $\sigma \in [1,4]$). At $\sigma = 1$ from top to bottom: G, G_{xxxx}, G_{xx} and G_{xxyy}.

8. Non-uniform smoothing

Whereas the linear scale-space concept and the associated theory for feature detection provides a conceptually clean model for early visual computations which can also be demonstrated to give highly useful results, there are certain limitations in basing a vision system on rotationally symmetric Gaussian kernels only. For example, smoothing across "object boundaries" may affect both the shape and the localization of edges in edge detection. Similarly, surface orientation estimates computed by shape from texture algorithms are affected, since the anisotropy of a surface pattern may decrease when smoothed using a rotationally symmetric Gaussian. These are some basic motivations for considering non-linear extensions of the linear scale-space theory, as will be the subject of the following chapters in this book.

8.1. Shape distortions in computation of surface shape

It is illuminating to consider in more detail the problem of deriving cues to surface shape from noisy image data. For simplicity, let us restrict the analysis to the monocular case, the shape-from-texture problem. The underlying ideas are, however, of much wider validity and apply to problems such as shape-from-stereo-cues and shape-from-motion.

Model. Following (Lindeberg 1994; Lindeberg and Gårding 1994) consider a non-uniform Gaussian blob

$$f(x_1, x_2) = g(x_1; \, l_1^2) \, g(x_2; \, l_2^2) \quad (l_1 \geq l_2 > 0), \tag{2.55}$$

as a simple linearized model of the projection of a rotationally symmetric Gaussian blob, where l_1 and l_2 are characteristic lengths in the x_1- and x_2-coordinate directions and g (here) is the one-dimensional Gaussian,

$$g(x_1; \, t) = \frac{1}{\sqrt{4\pi t}} e^{-x_1^2/4t}. \tag{2.56}$$

The slant angle (the angle between the visual ray and the surface normal) σ and the foreshortening $\epsilon = \cos \sigma$ are given by

$$\epsilon = \cos \sigma = l_2/l_1, \tag{2.57}$$

and the tilt direction (the direction of the projection of the surface normal onto the image plane) is

$$\theta = \pi/2. \tag{2.58}$$

Effects due to scale-space smoothing. From the semi-group property of the Gaussian kernel $g(\cdot;\, t_1) * g(\cdot;\, t_2) = g(\cdot;\, t_1 + t_2)$, it follows that the scale-space representation of f at scale t is

$$I(x_1, x_2;\, t) = g(x_1;\, l_1^2 + t)\, g(x_2;\, l_2^2 + t). \tag{2.59}$$

Thus, under scale-space smoothing the estimate of foreshortening varies as

$$\hat{\epsilon}(t) = \sqrt{\frac{l_2^2 + t}{l_1^2 + t}}, \tag{2.60}$$

i.e., it increases and tends to one, which means that after a sufficiently large amount of smoothing the image will eventually be interpreted as flat.

Hence, in cases when non-infinitesimal amounts of smoothing are necessary (e.g., due to the presence of noise), the surface orientation estimates will by necessity be *biased*. Observe in this context that no assumptions have been made here about what actual method should be used for computing surface orientation from image data. The example describes essential effects of the smoothing operation that arise in *any* shape-from-X method that contains a smoothing module and interprets a non-uniform Gaussian blob as the projection of a rotationally symmetric one.

Shape adaptation of the smoothing kernels. If, on the other hand, we have initial estimates of the slant angle and the tilt direction $(\hat{\sigma}, \hat{\theta})$ computed, say by using rotationally symmetric Gaussian smoothing, a straightforward compensation technique is to let the scale parameter in the (estimated) tilt direction, denoted $t_{\hat{t}}$, and the scale parameter in the perpendicular direction, denoted $t_{\hat{b}}$, be related by

$$t_{\hat{t}} = t_{\hat{b}} \cos^2 \hat{\sigma}. \tag{2.61}$$

If this estimate is correct, then the slant estimate will be *unaffected* by the non-uniform smoothing operation. To illustrate this property, assume that the tilt estimate is correct $(\hat{\theta} = \theta = \pi/2)$ and convolve the signal with a non-uniform Gaussian kernel

$$g(x_1, x_2;\, t_{\hat{t}}, t_{\hat{b}}) = g(x_1;\, t_{\hat{t}})\, g(x_2;\, t_{\hat{b}}), \tag{2.62}$$

which gives

$$I(x_1, x_2;\, t) = g(x_1;\, l_1^2 + t_{\hat{b}})\, g(x_2;\, l_2^2 + t_{\hat{t}}). \tag{2.63}$$

Then, the new foreshortening estimate is

$$\hat{\epsilon} = \epsilon(\hat{\sigma};\, t_{\hat{t}}, t_{\hat{b}}) = \sqrt{\frac{l_2^2 + t_{\hat{t}}}{l_1^2 + t_{\hat{b}}}} = |\cos\sigma| \sqrt{1 + \frac{t_{\hat{b}}}{l_1^2 + t_{\hat{b}}}\left(\frac{\cos^2 \hat{\sigma}}{\cos^2 \sigma} - 1\right)}. \tag{2.64}$$

Clearly, $\hat{\epsilon} = \epsilon$ if $\hat{\sigma} = \sigma$. In practice, we cannot of course assume that true values of (σ, θ) are known, since this requires knowledge about the solution to the problem we are to solve. A more realistic formulation is therefore to first compute initial surface orientation estimates using rotationally symmetric smoothing (based on the principle that in situations when no *a priori* information is available, the first stages of visual processes should be as uncommitted as possible and have no particular bias). Then, when a hypothesis about a certain surface orientation $(\hat{\sigma}_0, \hat{\theta}_0)$ has been established, the estimates can be improved iteratively.

More generally, it can be shown that by extending the linear scale-space concept based on the rotationally symmetric Gaussian kernel towards an *affine scale-space* representation based on the *non-uniform Gaussian kernel* with its shape specified by a (symmetric positive semi-definite) covariance matrix, Σ_t,

$$g(x_1; \Sigma_t) = \frac{1}{2\pi\sqrt{\det \Sigma_t}} e^{-x^T \Sigma_t^{-1} x/2} \quad \text{where} \quad x \in \mathbb{R}^2, \qquad (2.65)$$

a shape-from-texture method can be formulated such that up to first order of approximation, *the surface orientation estimates are unaffected by the smoothing operation*. In practice, this means that the accuracy is improved substantially, typically by an order of magnitude in actual computations.

Affine scale-space.   A formal analysis in (Lindeberg 1994; Lindeberg and Gårding 1994) shows that with respect to the above treated sample problem, the true value of $\hat{\sigma}$ corresponds to a *convergent fixed point* for (2.64). Hence, for the pattern (2.55) the method is guaranteed to converge to the true solution, provided that the initial estimate is sufficiently close to the true value.

The essential step in the resulting shape-from-texture method with *affine shape adaptation* is to adapt the kernel shape according to the local image structure. In the case when the surface pattern is weakly isotropic (see section 2.4) a useful method is to measure the second-moment matrix μ_I according to (2.24) and then letting $\Sigma_t = \mu_I^{-1}$.

8.2. Outlook

General.   An important point with the approach in the previous section is that *the linear scale-space model is used as an uncommitted first stage of processing*. Then, when *additional* information has become available (here, the initial surface orientation estimates) this information is used for tuning the front-end processing to the more specific tasks at hand.

Adapting the shape of the smoothing kernel in a linear way constitutes the presumably most straightforward type of geometry-driven processing that can be performed. In the case when the surface pattern is weakly isotropic, this shape adaptation has a very simple geometric interpretation; it corresponds to rotationally symmetric smoothing in the tangent plane to the surface.

Edge detection. Whereas the affine shape adaptation is conceptually simple, it has an interesting relationship to the non-linear diffusion schemes that will be considered in later chapters. If the shape adaptation scheme is applied at edge points, then it leads to a larger amount of smoothing along the edge than in the perpendicular direction. In this respect, the method constitutes a first step towards linking processing modules based on sparse edge data to processing modules based on dense filter outputs.

Biological vision. Referring to the previously mentioned relations between linear scale-space theory and biological vision one may ask: Are there corresponding geometry-driven processes in human vision? Besides the well-known fact that top-down expectations can influence visual processes at rather low levels, a striking fact is the abundance of fibers projecting backwards (*feedback*) between the different layers in the visual front-end, being more the rule then exception (Zeki 1993; Chapter 31).

At the current point, however, it is too early to give a definite answer to this question. We leave the subject of non-linear scale-space to the following chapters where a number of different approaches will be presented.

9. Appendix: Mathematica example 4th-order junction detector

```
(* Gaussian Derivatives on discrete images                           *)
(* Detection 4-junction in noisy checkerboard images                 *)
(* Author: Bart M. ter Haar Romeny, May, 1994                        *)

Needs["Statistics'ContinuousDistributions'"]; Off[General::spell1];
(* Resolution image = 64x64, scale operator = 3 pixels               *)
xres=yres=64; sc=3;

(* Chessboard testimage with additive Gaussian noise, sigma=2.1      *)
im:=Table[If[(x<xres/2+1 && y<yres/2+1) || (x>xres/2 && y>yres/2),10,0]+
Random[NormalDistribution[0,2.1]],{y,yres},{x,xres}];

(* Gaussian derivative definition GD[f,n,m,s]                        *)
(* n,m: order of differentiation to x resp. y; s: scale              *)
gd[n_,m_,s_]:=D[1/(2 Pi s^2) Exp[-(x^2+y^2)/(2 s^2)],{x,n},{y,m}];
gdk[n_,m_,s_]:=N[Table[Evaluate[gd[n,m,s]],{y,-(yres-1)/2,(yres-1)/2},
{x,-(xres-1)/2,(xres-1)/2}]];
GD[im_,n_,m_,s_]:=Chop[N[Pi s^2 InverseFourier[
Fourier[im] Fourier[RotateLeft[gdk[n,m,s],{yres/2,xres/2}]]]]];

(* Partial derivatives needed for invariant expression               *)
Iyyyy=GD[im,0,4,sc]; Ixyyy=GD[im,1,3,sc]; Ixxyy=GD[im,2,2,sc];
Ixxxx=GD[im,4,0,sc]; Ixxxy=GD[im,3,1,sc];

fourjunction = -(Ixxxx Iyyyy - 4 Ixxxy Ixyyy + 3 Ixxyy^2)^3 +
    27 (Ixxxx (Ixxyy Iyyyy - Ixyyy^2) - Ixxxy (Ixxxy Iyyyy -
    Ixxyy Ixyyy) + Ixxyy (Ixxxy Ixyyy - Ixxyy^2))^2

ListDensityPlot[im, PlotRange->All]
ListDensityPlot[fourjunction,PlotRange->All]
```

This code generates figure 2.4.

Figure 2.12. Anisotropic diffusion of a basket of fruits (implementation: Perona). Detail of a painting by Zurbaran to be found in the Norton-Simon Museum of Pasadena. The diffusion nonlinearity is the exponential, 20 iterations. The value of 'K' was about 10% of the dynamic range. Details of anisotropic diffusion are discused in next chapter.

ANISOTROPIC DIFFUSION

Pietro Perona

California Institute of Technology
Dept. of Electrical Engineering
Pasadena, CA 91125, USA

and

Takahiro Shiota

Kyoto University
Math Department, School of Science
Sakyo-ku, Kyoto 606-01, Japan

and

Jitendra Malik

University of California at Berkeley
Department of Electrical Engineering and Computer Science
Berkeley, CA 94720, USA

1. Introduction

The importance of multi–scale descriptions of images has been recognized from the early days of computer vision e.g. Rosenfeld and Thurston [309]. A clean formalism for this problem is the idea of scale-space filtering introduced by Witkin [386] and further developed in Koenderink[187], Babaud, Duda and Witkin[24], Yuille and Poggio[396], and Hummel[163, 164] and reviewed in the earlier sections of this book.

The essential idea of this approach is quite simple: embed the original image in a family of derived images $I(x, y, t)$ obtained by convolving the original image $I_0(x, y)$ with a Gaussian kernel $G(x, y; t)$ of variance t:

$$I(x, y, t) = I_0(x, y) * G(x, y; t) \tag{3.1}$$

Larger values of t, the scale-space parameter, correspond to images at coarser resolutions (see Fig. 3.1 and 3.2).

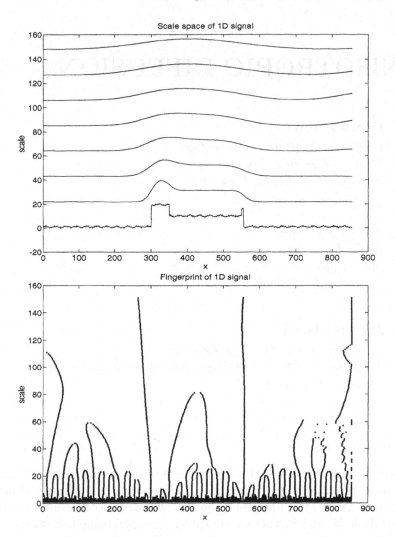

Figure 3.1. (Left) A family of 1-D signals, $I(x, t)$, obtained by convolving the original one (lowest one) with Gaussian kernels whose variance increases from bottom to top. (Right) 'Fingerprint' of $I(x, t)$; it is the locus of the zero crossings of $\nabla^2 I(x, t)$. The zero crossings that are associated to strong gradients may be used to represent the boundaries of the main regions at each scale. Notice, however, that their position changes with scale.

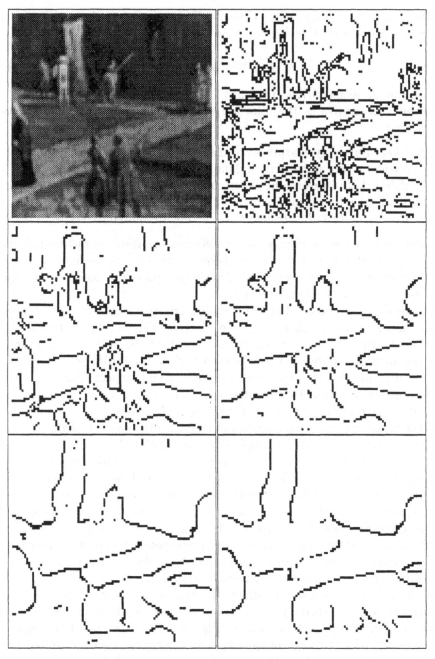

Figure 3.2. Sequence of boundary images produced by the Canny detector run at scales 1, 3, 5, 7, 9 pixels. The images are 100 × 100 pixels.

As pointed out by Koenderink [187], and Hummel [163] this one parameter family of derived images may equivalently be viewed as the solution of the heat conduction or diffusion equation

$$I_t = \nabla^2 I = I_{xx} + I_{yy} \tag{3.2}$$

with the initial condition $I(x, y, 0) = I_0(x, y)$, the original image. Koenderink motivates the diffusion equation formulation by stating two criteria :

1. *Causality* : Any feature at a coarse level of resolution is required to possess a (not necessarily unique) "cause" at a finer level of resolution although the reverse need not be true. In other words, no spurious detailshould be generated when the resolution is diminished.
2. *Homogeneity and Isotropy* : The blurring is required to be space invariant.

These criteria lead naturally to the diffusion equation formulation. It may be noted that the second criterion is only stated for the sake of simplicity; if blurring was allowed to be data-driven and space-varying one would have useful degrees of freedom to play with.

It should also be noted that the causality criterion does not force uniquely the choice of a Gaussian to do the blurring, though this is perhaps the simplest. Hummel [163] has made the important observation that a version of the maximum principle from the theory of parabolic differential equations is equivalent to causality, therefore one would expect that a number of (possibly nonlinear) differential equations would satisfy causality and possibly have useful behaviours for vision applications.

Is the standard scale-space paradigm adequate for vision tasks which need 'semantically meaningful' multiple scale descriptions?

Surfaces in nature usually have a hierarchical organization composed of a small discrete number of levels [247]. At the finest level, a tree is composed of leaves with an intricate structure of veins. At the next level, each leaf is replaced by a single region, and at the highest level there is a single blob corresponding to the treetop. There is a natural range of resolutions (intervals of the scale-space parameter) corresponding to each of these levels of description. Furthermore at each level of description, the regions (leaves, treetops or forests) have well-defined boundaries.

In the linear scale-space paradigm the true location of a boundary at a coarse scale is not directly available at the coarse scale image. This can be seen clearly in the 1-D example in Figure 3.1 and in the 2D example in Figure 3.2. The locations of the edges at the coarse level are shifted from their true locations. In 2-D images there is the additional problem that edge junctions , which contain much of the spatial information of the edge drawing, are destroyed. The only way to obtain the true location of the edges that have been detected at a coarse scale is by tracking them across the scale-space to their position in the original image. This technique proves to be somewhat complicated and expensive [31, 62].

The reason for this spatial distortion is quite obvious – Gaussian blurring does not 'respect' the natural boundaries of objects. Suppose we have the picture of a treetop with the sky as background. The Gaussian blurring process would result in the green of the leaves getting 'mixed' with the blue of the sky, long before the treetop emerges as a feature (after the leaves have been blurred together). In Figure 3.2 the boundaries in a sequence of coarsening images obtained by Gaussian blurring illustrate this phenomenon. With this as motivation, we enunciate the criteria which we believe any candidate paradigm for generating multi-scale 'semantically meaningful' descriptions of images must satisfy:

1. **Causality** : As pointed out by Witkin and Koenderink, a scale-space representation should have the property that no 'spurious detail' should be generated passing from finer to coarser scales.

2. **Immediate Localization**: At each resolution, the region boundaries each resolution, the region boundaries should be sharp and coincide with the semantically meaningful boundaries at that resolution.

3. **Piecewise Smoothing**: At all scales, intra-region smoothing should occur preferentially over inter-region smoothing. In the tree example mentioned earlier, the leaf regions should be collapsed to a treetop *before* being merged with the sky background.

2. Anisotropic diffusion

There is a simple way of modifying the scale-space paradigm to achieve the objectives that we have put forth in the previous section. In the diffusion equation framework of looking at linear scale-space, the diffusion coefficient c is assumed to be a constant independent of the spatial location. There is no fundamental reason why this must be so. To quote Koenderink [187], (pg. 364, left column, l. 19 from the bottom) "... I do not permit space variant blurring. Clearly this is not essential to the issue, but it simplifies the analysis greatly". We argue that a suitable choice of space- and scale-varying $c(x, y, t)$ (t indicates the scale) will enable us to satisfy the second and third criteria listed in the previous section. Furthermore this can be done without sacrificing the causality criterion.

Consider the "anisotropic diffusion" equation [291, 293]

$$I_t = div(c(x, y, t)\nabla I) = c(x, y, t)\nabla^2 I + \nabla c \cdot \nabla I \qquad (3.3)$$

where we indicate with *div* the divergence operator, and with ∇, and ∇^2 the gradient, and Laplacian operators with respect to the space variables.

It reduces to the isotropic heat diffusion equation $I_t = c\nabla^2 I$ if $c(x,y,t)$ is a constant. Suppose at the 'time' t (the scale is in this chapter sometimes called 'time' as time is the physical evolution parameter of the diffusion), we knew the locations of the region boundaries appropriate for that scale. We would want to encourage smoothing *within* a region in preference to smoothing *across* the boundaries. This could be achieved by setting the conduction coefficientto be 1 in the interior of each region and 0 at the boundaries. The blurring would then take place separately in each region with no interaction between regions. The region boundaries would remain sharp. Of course, we do *not* know in advance the region boundaries at each scale (if we did the problem would already have been solved!). What can be computed is a current best estimate of the location of the boundaries (edges) appropriate to that scale.

Let $\mathbf{E}(x,y,t)$ be such an estimate: a vector-valued function defined on the image which ideally should have the following properties:

1. $\mathbf{E}(x,y,t) = \mathbf{0}$ in the interior of each region.
2. $\mathbf{E}(x,y,t) = K\mathbf{e}(x,y,t)$ at each edge point, where $\mathbf{e}$ is a unit vector normal to the edge at the point, and K is the local contrast (difference in the image intensities on the left and right) of the edge.

Note that the word *edge* as used above has not been formally defined – we mean here the perceptual subjective notion of an edge as a region boundary. A completely satisfactory formal definition is likely to be part of the solution, rather than the problem definition!

If an estimate $\mathbf{E}(x,y,t)$ is available, the conduction coefficient $c(x,y,t)$ can be chosen to be a function $c = g(\|\mathbf{E}\|)$ of the magnitude of $\mathbf{E}$. According to the previously stated strategy $g(\cdot)$ has to be a nonnegative monotonically decreasing function with $g(0) = 1$ (see Fig. 3.3). This way the diffusion process will mainly take place in the interior of regions, and it will not affect the region boundaries where the magnitude of $\mathbf{E}$ is large.

It is intuitive that the success of the diffusion process in satisfying the three scale-space goals of section 1 will greatly depend on how accurate the estimate $\mathbf{E}$ is as a "guess" of the edges. Accuracy though is computationally expensive and requires complicated algorithms. Fortunately it turns out that the simplest estimate of the edge positions, the gradient of the brightness function i.e. $\mathbf{E}(x,y,t) = \nabla I(x,y,t)$, gives excellent results in many useful cases

$$c(x,y,t) = g(\|\nabla I(x,y,t)\|) \tag{3.4}$$

There are many possible choices for $g(\cdot)$, the most obvious being a binary valued function. In the next sections we argue that in case we use the edge estimate $\mathbf{E}(x,y,t) = \nabla I(x,y,t)$ the choice of $g(\cdot)$ is restricted to a subclass of the monotonically decreasing functions and that g should be smooth.

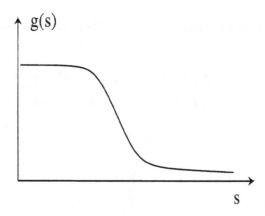

Figure 3.3. The qualitative shape of the nonlinearity $g(\cdot)$.

3. Implementation and discrete maximum principle

Equation (3.3) may be discretized on a square lattice, with brightness values associated to the vertices, and conduction coefficients to the arcs (see Fig. 3.4). A 4-nearest-neighbours discretization of the Laplacian operator may be used:

$$I_{i,j}^{t+1} = I_{i,j}^{t} + \lambda[c_N \cdot \Delta_N I + c_S \cdot \Delta_S I + c_E \cdot \Delta_E I + c_W \cdot \Delta_W I]_{i,j}^{t} \qquad (3.5)$$

where $0 \leq \lambda \leq \frac{1}{4}$ for the numerical scheme to be stable, N,S,E,W are the mnemonic subscripts for North, South, East, West, the superscript and subscripts on the square bracket are applied to all the terms it encloses, and the symbol Δ indicates nearest-neighbour differences:

$$\Delta_N I_{i,j} \equiv I_{i-1,j} - I_{i,j} \qquad \Delta_S I_{i,j} \equiv I_{i+1,j} - I_{i,j}$$
$$\Delta_E I_{i,j} \equiv I_{i,j+1} - I_{i,j} \qquad \Delta_W I_{i,j} \equiv I_{i,j-1} - I_{i,j} \qquad (3.6)$$

The conduction coefficients are updated at every iteration as a function of the brightness gradient:

$$c_{N_{i,j}}^{t} = g(\|(\nabla I)_{i+\frac{1}{2},j}^{t}\|) \qquad c_{S_{i,j}}^{t} = g(\|(\nabla I)_{i-\frac{1}{2},j}^{t}\|)$$
$$c_{E_{i,j}}^{t} = g(\|(\nabla I)_{i,j+\frac{1}{2}}^{t}\|) \qquad c_{W_{i,j}}^{t} = g(\|(\nabla I)_{i,j-\frac{1}{2}}^{t}\|) \qquad (3.7)$$

The value of the gradient can be computed on different neighbourhood structures achieving different compromises between accuracy and locality. The simplest choice consists in approximating the norm of the gradient at

each arc location with the absolute value of its projection along the direction
of the arc:

$$c^t_{N_{i,j}} = g(|\Delta_N I^t_{i,j}|) \qquad c^t_{S_{i,j}} = g(|\Delta_S I^t_{i,j}|)$$
$$c^t_{E_{i,j}} = g(|\Delta_E I^t_{i,j}|) \qquad c^t_{W_{i,j}} = g(|\Delta_W I^t_{i,j}|) \qquad (3.8)$$

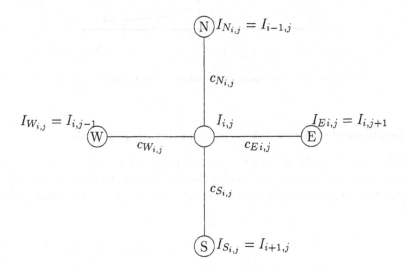

Figure 3.4.   The structure of the discrete computational scheme for simulating the
diffusion equation. The brightness values $I_{i,j}$ are associated to the nodes of a lattice, the
conduction coefficients c to the arcs. One node of the lattice and its 4 North, East, West,
and South neighbours are shown.

This scheme is not the exact discretization of equation (3.3), but of a similar
diffusion equation in which the conduction tensor is diagonal with entries
$g(|I_x|)$ and $g(|I_y|)$ instead of $g(\|\nabla I\|)$ and $g(\|\nabla I\|)$. This discretization
scheme preserves the property of the continuous equation (3.3) that the total
amount of brightness in the image is preserved. Additionally the "flux" of
brightness through each arc of the lattice only depends on the values of the
brightness at the two nodes defining it, which makes the scheme a natural
choice for analog VLSI implementations [292, 149].
Less crude approximations of the gradient yielded perceptually similar results
at the price of increased computational complexity.
It is possible to show that, whatever the choice of the approximation of the
gradient, the discretized scheme still satisfies the maximum (and minimum)
principle provided that the function g is bounded between 0 and 1.

```
function [outimage] = anisodiff(inimage,iterations,K)

lambda = 0.25;
outimage = inimage;        [m,n] = size(inimage);

rowC = [1:m];        rowN = [1 1:m-1];        rowS = [2:m m];
colC = [1:n];        colE = [1 1:n-1];        colW = [2:n m];

for i=1:iterations,
deltaN = outimage(rowN,colC) - outimage(rowC,colC);
deltaE = outimage(rowC,colE) - outimage(rowC,colC);

fluxN = deltaN .* exp( - (1/K) * abs(deltaN) );
fluxE = deltaE .* exp( - (1/K) * abs(deltaE) );

outimage = outimage + lambda *
(fluxN - fluxN(rowS,colC) + fluxE - fluxE(rowC,colW));
end;
```

Figure 3.5. A Matlab implementation of anisotropic diffusion with adiabatic boundary conditions and exponential nonlinearity.

Theorem 1 (Discrete maximum-minimum principle) *If $c \in [0,1]$, the maxima/minima of solutions of (3.5) on a bounded cylinder $(i,j,t) \in (i_0,i_1) \times (j_0,j_1) \times (t_0,t_1)$ belong either to the 'bottom' face $(i,j,t) \in (i_0,i_1) \times (j_0,j_1) \times t_0$ or to the sides $(j = j_0 \times (t_0,t_1),\ j = j_1 \times (t_0,t_1)$ etc).*

Proof 1 *We may show this directly from equation (3.5), using the facts $\lambda \in [0,\frac{1}{4}]$, and $c \in [0,1]$, and defining $I_{M_{i,j}^t} \doteq max\{(I, I_N, I_S, I_E, I_W)_{i,j}^t\}$, and $I_{m_{i,j}^t} \doteq min\{(I, I_N, I_S, I_E, I_W)_{i,j}^t\}$, the maximum and minimum of the neighbours of $I_{i,j}$ at iteration t. We can see that*

$$(I_m)_{i,j}^t \le I_{i,j}^{t+1} \le (I_M)_{i,j}^t \tag{3.9}$$

i.e. no (local) maxima and minima are possible in the interior of the discretized scale-space:

$$
\begin{aligned}
I_{i,j}^{t+1} &= I_{i,j}^t + \lambda[c_N \cdot \Delta_N I + c_S \cdot \Delta_S I + c_E \cdot \Delta_E I + c_W \cdot \Delta_W I]_{i,j}^t \\
&= I_{i,j}^t(1 - \lambda(c_N + c_S + c_E + c_W)_{i,j}^t) + \\
&\quad + \lambda(c_N \cdot I_N + c_S \cdot I_S + c_E \cdot I_E + c_W \cdot I_W)_{i,j}^t \\
&\le I_{M_{i,j}^t}(1 - \lambda(c_N + c_S + c_E + c_W)_{i,j}^t) + \\
&\quad + \lambda I_{M_{i,j}^t}(c_N + c_S + c_E + c_W)_{i,j}^t
\end{aligned}
$$

$$= I_{M_{i,j}^t} \tag{3.10}$$

and, similarly:

$$I_{i,j}^{t+1} \ge I_{m_{i,j}^t}(1 - \lambda(c_N + c_S + c_E + c_W)_{i,j}^t) +$$

$$+\lambda I_{m\,i,j}^{t}(c_N + c_S + c_E + c_W)_{i,j}^{t}$$
$$= I_{m\,i,j}^{t} \qquad\qquad\qquad\qquad\qquad (3.11)$$

The numerical scheme used to obtain the pictures in this chapter is the one given by equations (3.5), (3.6), (3.8), using the original image as the initial condition, and adiabatic boundary conditions, i.e. setting the conduction coefficient to zero at the boundaries of the image (see in Fig. 3.5 the `Matlab` implementation that was actually used). A constant value for the conduction coefficient c (i.e. $g(\cdot) \equiv 1$) leads to Gaussian blurring.

Different functions may be used for $g(\cdot)$ (see Eq. (3.4)) giving anisotropic diffusion behaviour. The images in this chapter were obtained using

$$g(\nabla I) = e^{-(\frac{\|\nabla I\|}{K})^2} \qquad\qquad\qquad (3.12)$$

however there are other good possibilities, e.g.

$$g(\nabla I) = \frac{1}{1 + (\frac{\|\nabla I\|}{K})^2}$$

The scale-spaces generated by these two functions are different: the first privileges high-contrast edges over low-contrast ones, the second privileges wide regions over smaller ones. The role of the nonlinearity g in determining the edge-enhancing behaviour of the diffusion is studied in a later section. It may be worth noticing that the discrete and the continuous model may have different limit behaviours unless the nonlinearity g is modified in one of the two cases. Since the initial condition is bounded (images are bounded below by 0 – negative brightness does not exist – and above by the saturation of the sensors) and this, in turn, bounds the scale-space, the maximum value for the discrete gradient is bounded above by the dynamic range of the image divided by the spacing of the sampling lattice. Therefore the conduction coefficient will never be zero if the nonlinearities g defined above are used. This will not be the case for the continuous model, where the magnitude of the brightness gradient may diverge and conduction may therefore stop completely at some location.

The constant K may either be set by hand at some fixed value, or set using using the "noise estimator" described by Canny [51]: a histogram of the absolute values of the gradient throughout the image is computed, and K is set equal to the, say, 90% value of its integral at every iteration.

A model of the primary visual cortex due to Cohen and Grossberg [67] aims at achieving the same no-diffusion-across boundaries behaviour. The model contains an explicit representation of the boundaries and is much more ambitious and sophisticated than the simple PDE scheme that we discuss in this chapter.

The computational scheme described in this section has been chosen for its simplicity. Other numerical solutions of the diffusion equation, and multiscale algorithms may be considered for efficient software implementations.

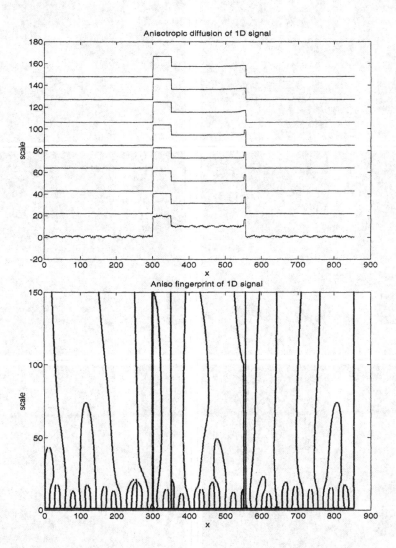

Figure 3.6. (Top) A family of 1-D signals, $I(x, t)$, obtained by running anisotropic diffusion (code in figure 3.5) on the original one (lowest one). (Bottom) Fingerprint of the anisotropic diffusion family. It is the locus of the zero crossings of $\nabla^2 I(x, t)$. Notice that the zero-crossings corresponding to high brightness gradients (i.e. edges) have no drift in scale (they are located at positions 300, 350, 550, 555). Around scale 100 the zero-crossing at position 550 undergoes a transition: it ceases having edge value and its position starts drifting in scale-space. Compare with figure 3.1

Figure 3.7. Sequence of images produced by anisotropic diffusion. The code presented in figure 3.5 was run on the image at the top-left corner for 10, 20, 30, 60, 100 iterations. The original image has pixel values between 0 (black) and 255 (white) and had a size of 100×100 pixels. The coefficient K was set equal to $K = 10$.

Figure 3.8. Sequence of boundary images obtaining by thresholding the gradient of the images in figure 3.7. The threshold was equal to K. Compare with figure 3.2.

4. Edge Enhancement

With conventional low-pass filtering and diffusion the price paid for eliminating the noise, and for performing scale-space, is the blurring of edges. This causes their detection and localization to be difficult. An analysis of this problem is presented in [51].

Edge enhancement and reconstruction of blurry images can be achieved by high-pass filtering or running the diffusion equation backwards in time. This is an ill-posed problem, and gives rise to numerically unstable computational methods, unless the problem is appropriately constrained or reformulated [165].

If the conduction coefficient is chosen to be an appropriate function of the image gradient we can make the anisotropic diffusion enhance edges while running *forward* in time, thus enjoying the stability of diffusions which is guaranteed by the seen in the previous section.

We may study the problem in 1D and model an edge as a step function convolved with a Gaussian. This corresponds to a straight 2D edge that is aligned with the y axis (for curved edges the notation becomes more complicated).

The expression for the divergence operator simplifies to:

$$div(c(x,t)\nabla I) = \frac{\partial}{\partial x}(c(x,t)I_x)$$

We choose c to be a function of the gradient of I: $c(x,t) = g(I_x(x,t))$ as in the previous sections. Let $\phi(I_x) \doteq g(I_x) \cdot I_x$ denote the flux $c \cdot I_x$.

Then the 1-D version of the diffusion equation (3.3) becomes

$$I_t = \frac{\partial}{\partial x}\phi(I_x) = \dot{\phi}(I_x) \cdot I_{xx} \qquad (3.13)$$

We are interested in looking at the variation in time of the slope of the edge: $\frac{\partial}{\partial t}(I_x)$. If the function $I(\cdot)$ is regular enough the order of differentiation may be inverted:

$$\frac{\partial}{\partial t}(I_x) = \frac{\partial}{\partial x}(I_t) = \frac{\partial}{\partial x}(\frac{\partial}{\partial x}\phi(I_x)) = \ddot{\phi} \cdot I_{xx}^2 + \dot{\phi} \cdot I_{xxx} \qquad (3.14)$$

Suppose the edge is oriented in such a way that $I_x > 0$ (see Fig.3.9). At the point of inflection $I_{xx} = 0$, and $I_{xxx} \ll 0$ since the point of inflection corresponds, to the point with maximum slope (see figure 3.9).

Then in a neighbourhood of the point of inflection $\frac{\partial}{\partial t}(I_x)$ has sign opposite to $\dot{\phi}(I_x)$. If $\dot{\phi}(I_x) > 0$ the slope of the edge will decrease with time; if, on the contrary $\dot{\phi}(I_x) < 0$ the slope will increase with time.

Figure 3.9. (Left to right) A mollified step edge and its 1st, 2nd, and 3rd derivatives.

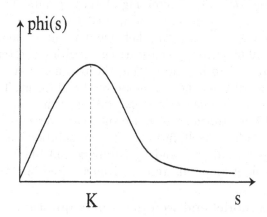

Figure 3.10. A choice of the function $\phi(\cdot)$ that leads to edge enhancement. See sec. 4.

Notice that this increase in slope cannot be caused by a scaling of the edge, because this would violate the maximum principle. The edge becomes sharper.

There are several possible choices for $\phi(\cdot)$, for example, $g(I_x) = C(1 + (I_x/K))^{-(1+\alpha)}$ with $\alpha > 0$ (see fig. 3.10). Then there exists a certain threshold value related to K, and α, below which $\phi(\cdot)$ is monotonically increasing, and beyond which $\phi(\cdot)$ is monotonically decreasing, giving the desirable result of blurring small discontinuities *and* sharpening edges. Notice also that in a neighbourhood of the steepest region of an edge the diffusion appears to run 'backwards' since $\dot{\phi}(I_x)$ in equation (3.13) is negative. This may be a source of concern since it is known that constant-coefficient diffusions running backwards are unstable and amplify noise generating ripples. In our case this concern is unwarranted: the maximum principle guarantees that ripples are not produced. Experimentally one observes that the areas where $\dot{\phi}(I_x) < 0$ quickly shrink, and the process keeps stable.

Notice that while no ripples are created in I, this is not true for I_x. As discussed above the derivative of the image *does* diverge at some points. This is necessary for the edge sharpening behaviour to happen (contrary to the intuition that informs the initial discussion in [57]). If the regions where $\dot{\phi}(I_x) < 0$ are broad these will tend to segment into portions where I_x diverges, separated by portions where $I_x \to 0$. We call this behaviour *staircasing*; it is rather rare in common images, however one may notice it on very blurry images where the brightness gradient varies slowly. Staircasing arises when at the steepest point of the edge the curvature does not have maximum rate of variation; in this case sharpening of the edge may stop after a while because I_{xxx} might cease to be negative at the inflection point. At that point I_x starts decreasing. Of course in this case I_{xx} will have other zero-crossings, and $|I_x|$ starts increasing at those points. The edge will not disappear, but it can be split into multiple edges.

One may reproduce the staircasing behaviour using a Gaussian-smoothed step edge ($\sigma = 100$ pixels, say) as a starting condition (better if some white noise is superimposed to it) and setting the parameter K of the diffusion to a fraction (e.g. 10%) of the maximum of the gradient. The blurry edge slowly converges to a sequence of steps, rather than a single step. A way to prevent this from happening is to use smoothed versions of the gradient, rather than the gradient itself, in Eq. (3.4), this solution was independently proposed by Catté et al [57], and by Nitzberg and Shiota [271] (see also chapter by Niessen et al. on discrete implementations in this book).

5. Continuous model and well-posedness question

In this section we consider issues of stability, existence and uniqueness of solutions of the anisotropic diffusion scheme that uses the magnitude of the gradient as an estimate of the local 'edginess', i.e. equation (3.3) with conduction (3.4).

We will study a continuous model, although a discrete one gives a better approximation of the algorithm. The analysis of discretized schemes poses additional problems since discretization introduces another scale parameter (the lattice spacing) and directionality (e.g., vertical and horizontal edges may be preferred to slanted edges). Such effects may be reduced by making the grid size fine enough, *if* the continuous model is well-posed and stable. Ill-posed problems often appear in applied mathematics, e.g., as inverse problems in control theory. For example, the initial value problem, instead of the boundary value problem, for the laplacian:

$$\left(\frac{\partial^2}{\partial x^2} + \frac{\partial^2}{\partial y^2} \right) u = 0, \quad 0 < x < a, -\infty < y < \infty,$$

$$u\big|_{x=0} = u_0$$

$$\frac{\partial u}{\partial x}\bigg|_{x=0} = u_1$$

is an ill-posed problem. The Euler scheme and other usual methods for initial value problem quickly accumulate errors, and do not give sensible results. But more careful method of constructing a solution using Poisson's kernel does work, at least for a small enough 'time' a (which may be the thickness of the wall of a pipe, etc.). This example shows that the choice of algorithm is very important when solving an ill-posed problem, even if the problem does allow a solution. See [147] and literature therein.

Some ill-posed problems are more ill-posed than others, and may not have solutions at all. In such a case the behavior of the solution of the algorithm is highly dependent on the algorithm, and the model is not very helpful to conceptually understand the algorithm. A typical example of this may be the problem of solving the heat equation backward. One must limit the frequency range to do the task, and the behavior of the solution is very sensitive to the frequency limit. In the fixed grid approximation, this means the result is highly dependent on the grid size.

The nature of the equation (3.3) with conduction (3.4) (continuous case) is not well-understood (we will call it Perona-Malik equation in the following). At the time of writing we don't know whether the equation is well-posed, or, if not, how bad the ill-posedness is. One possible way to study this is to have a regularization of the equation to make it well-posed, and then reduce the amount of regularization to observe the behavior of the solution. If the solution converges, the limit would solve the original equation. If the solution is very sensitive to the amount of regularization and fails to converge, there may be some sort of ill-posedness.

Catté and collaborators [57] have showed existence of solutions to (3.3) with conduction coefficient given by $c(x, y, t) = g(|\nabla(G_\sigma * u)|)$, i.e. a variation of (3.4) where the gradient of the image is substituted with the gradient of the image smoothed by a Gaussian of variance σ (recall the 'physical derivatives' from chapter 1). J.-M. Morel (personal communication) has suggested that one may take the limit of the solution as a function of sigma as a definition in the weak sense of the solution to the Perona-Malik equation. Since the a norm of the solution (defined as the integral of $|\nabla u|$) is bounded above (e.g. by the norm of the initial condition) the sequence must have an accumulation point when $\sigma \to 0$. This accumulation point is a good candidate for a weak solution of the equation.

In a private communication to Mumford, P.-L. Lions showed the well-posedness of the modified Perona-Malik equation in one space dimension with time delay regularization

$$\frac{\partial I}{\partial t} = \left(\frac{I_x}{1 + v/\kappa^2}\right)_x, \tag{3.15}$$

$$\frac{\partial v}{\partial t} = \omega(|I_x|^2 - v) \tag{3.16}$$

(larger ω gives a smaller time delay). This suggests, as also noted by Perona elsewhere in this chapter, that it may be physically more natural for the equation to have some time delay in the heat conduction coefficient. Note also that the equations are still local (i.e., differential equations rather than integral or difference equations) with (3.16) giving time delay, like every law of physics. It is not local if, e.g., the time delay is instead given by

$$v(x,t) = |I_x(x, t - 1/\omega)|^2. \tag{3.17}$$

Actually, this regularization has an advantage that the solution of (3.15) with (3.17) is smooth, as long as the initial data is smooth, while the solution may be much less regular with the regularization (3.16).

Shiota, in collaboration with Mumford and Nitzberg, has attempted some numerical experiments to see if the solution of (3.15), with (3.16) and/or the regularization given by the spatial smoothing of v, converges as the amount of regularization is decreased. These experiments seem to suggest a negative result: with the amount of regularization decreased, the graph of the solution became more and more jagged instead of developing a small number of sharp edges, and the numerical scheme failed more and more quickly as excessive grid subdivision led to floating point exception. These experiments are not conclusive; the jaggedness of solutions and other hints of instability that were observed on the numerical results might be artefacts of the numerical scheme, since as the amount of regularization decreased and the solution became more and more jagged the numerical errors may have become significant.

Useful properties of the Perona-Malik equation may be proved once the existence of a smooth solution is guaranteed (in a sense, our discussions in section 3 is a special case of it, where the regularization is given by discretization). For example, Nirenberg's theorem on maximum principle [269, Sect. 3, theorem 4] may be applied to prove maximum principle on a solution of the regularized equation, and any C^2 solution of the original equation. We conjecture that a maximum principle applies to all solutions of the Perona-Malik equation as well, however we have been unable to prove this.

A physical analogy may help in understanding this conjecture. The Perona-Malik equation models a physical diffusion process where the conduction coefficient is *never* negative. Although the conduction coefficient is coupled to the brightness/temperature distribution u during its evolution in time, one could think of the problem in the following way. Consider the set of all possible (not necessarily continuous) choices for time-varying conduction coefficients with values between zero and one. To each one of these, given the same initial condition u_0, would correspond a solution to the diffusion equation (3.3). None of these would be ill-behaved, e.g. no divergence of u would be possible, all of these, we conjecture, satisfy a maximum principle.

One or more of these solutions would correspond to the solution to the Perona-Malik equation. Therefore the solution to the Perona-Malik equation is bounded and probably satisfies a maximum principle.

Some aspects of the Perona-Malik equation which make the analytic study of the equation difficult:

- *Lack of comparison theorem*: Because of nonlinearity, the maximum principle discussed above does *not* imply the comparison theorem like

$$I_1(x,0) > I_2(x,0)\ (\forall x) \Rightarrow I_1(x,t) > I_2(x,t)\ (\forall x).$$

  It is easy to make a counterexample, using the way the heat conductivity depends on the gradient. This means estimation by an inequality, a basic technique in analysis, may fail to apply.

- *Difficulty of applying variational formalism*: For the Perona-Malik equation, the corresponding energy functional (i.e., the one for which the equation gives the gradient descent) has no dependence on the input image, and thus the set of minima of the functional does not reflect the information of the input image. For the Nordstrom equation [274, 273], the infimum of the energy functional is never achieved (at the limit of any 'minimizing sequence' of the energy functional, the energy value jumps up) unless the input image itself is a step function.

- *Possible instability*: Given a solution I of the Perona-Malik equation, the linearization (infinitesimal perturbation) of the equation at I gives a backward linear heat flow where $|\nabla I| > K$ (or some other function of K, this depends on the nonlinearity g used). This means there is no nontrivial smooth family of solutions through I. Because of nonlinearity, this infinitesimal instability cannot logically imply finite instability (a small difference of initial data may become a huge difference of the solution), but it is likely to be the case. It thus suggests that errors of numerical scheme may be very difficult to control, and that, with the lack of approximability, many useful tools of analysis would fail to apply.

Although there is no general theory of nonlinear differential equations, a certain kind of nonlinear evolution equations are relatively well-understood. This theory is not applicable to the Perona-Malik equation, but it may shed some light on it.

A nonlinear operator $A : H \to H$, where H is a Hilbert space, is said to be *monotone* if $\langle Au - Av, u - v \rangle \geq 0$ for any $u, v \in H$. This notion is a nonlinear analogue of positive operator. If A is monotone and cannot be extended further as a monotone operator, it is called *maximal monotone*. If A is maximal monotone, the equation $Au = f$ is the nonlinear analogue of an elliptic equation, and $du/dt = -Au + f$ is the nonlinear analogue of a parabolic equation. Those equations are very well understood; e.g., the unique solvability is known. See, e.g., [43]. One can check by integration by part that the operator of the form $A := -\nabla \cdot (C(\|\nabla I\|)\nabla I)$ (resp. $-A$) on a bounded domain with Dirichlet boundary condition is monotone if $C(y)y$ is a monotone increasing function (resp. monotone decreasing function) of y. Although this is a functional analytic notion and thus cannot be 'localized', the above observation suggests that the Perona-Malik equation might be 'locally well-posed' in a region where $\|\nabla I\| < \kappa$; while the *backward* Perona-Malik equation might be well-posed where $\|\nabla I\| > \kappa$. If this can be made rigorous, it may prove the nonexistence of C^1 solution to the forward equation with initial data which satisfies $\|\nabla I\| > \kappa$ in a nonempty region. (This would not exclude the existence of less regular solution: appearance of jagged solution in numerical experiments suggests that the situation is far more complicated as the "unstable" region may instantly break up into many thin pieces separated by "stable" regions.)

To conclude, here is a summary of our current understanding of the behaviour of the solutions to the equation:

Existence – Yes, for many regularized versions, but unknown for the original equation. It appears to be possible to prove the existence of weak solutions (see discussion above).

Uniqueness – There is none. Is is easy to construct examples of bifurcations using either blurry edges with arbitrarily small sinusoids superimposed, or by studying a blurry edge which is at the limit between the enhancing and the diffusing behavior of the equation. In all these cases arbitrarily close initial conditions will generate different solutions.

Maximum principle – Easy to prove for C^2 solutions. Conjectured but not proven for discontinuous solutions.

VECTOR-VALUED DIFFUSION

Ross Whitaker

University of North Carolina
Department of Computer Science
Chapel Hill, NC 27514, USA

and

Guido Gerig

Communication Technology Laboratory
Image Science Division, ETH-Zentrum
Gloriastr.35, CH-8092 Zürich, Switzerland

1. Introduction

In this chapter we generalize the ideas of previous chapters in order to construct variable-conductance diffusion processes that are sensitive to higher-order image structure. To do this we consider *anisotropic diffusion* as one instance of a whole class of processes that blur images while preserving certain kinds of interesting structure.

In particular the anisotropic diffusion process uses the *homogeneity* of neighborhoods (measured at some scale) in order to to determine the amount of blurring that should be applied to those neighborhoods. Previous chapters have dealt with equations that define local structure in terms of the properties of a single input image. In these cases structural inhomogeneity is quantified by the gradient magnitude. Such processes seem to be well suited to finding "edges" (discontinuities in luminance) in digital images. In this chapter we extend the definition of an "image" to include vector-valued or multi-valued images and we describe a general framework for quantifying the homogeneity of local image structure for such functions.

We will begin by developing a general framework for variable-conductance diffusion which incorporates vector-valued functions. We will then show how this framework can apply to the specific case of vector-valued images which result from imaging devices that measure multiple properties at each point in space. We will apply the same framework to vector-valued images consisting of sets of *geometric features* that derive from a single scalar image. We will examine two strategies that operate within this framework. The first uses

93

sets of partial derivatives as the input to a vector-valued diffusion equation. We call this approach *geometry-limited diffusion*, and it appears to be useful for detecting certain kinds of higher-order geometric features. The second uses a local frequency decomposition as the input to a vector-valued diffusion process. We call this approach *spectra-limited diffusion*, and it appears to be useful for characterizing the boundaries of patches which contain different textures.

2. Vector-valued diffusion

2.1. Vector-valued images

A vector-valued image is a smooth mapping from the image domain $I \subset \mathbb{R}^n$ to an m-dimensional range, $\vec{F} : I \mapsto D$, where $D \subset \mathbb{R}^m$ is also called the *feature space*. Each point in the domain is characterized by its position in the feature space, and this position is recorded at each point in I as a finite array of measurements, $\vec{F}(\vec{x}) = F_1(\vec{x}), \ldots, F_m(\vec{x})$, where $\vec{x} \in I$. A vector-valued image is thus a *set* or a *stack* of single-valued images that share the same image space. Note that the set of numbers used to record the position of a pixel in the range is not indicative of the shape or nature of D. More specifically Euclidean distance is not necessarily the appropriate measure for comparing positions in the feature space.

Perhaps the most apparent examples of vector-valued images are the data that result from imaging devices that measure several physical quantities at the same location in space. Multi-echo magnetic resonance imaging data (Sect. 2.4), multispectral LANDSAT data, and the red-green-blue (RGB) channels from a color camera are examples. One could treat each physical measurement in these data as a separate image and perform anisotropic diffusion process separately for each measurement. However, if the ultimate goal of a diffusion process is to detect structures that arise from a number of measurements, then a "structure sensitive" diffusion process should reflect the collection of measurements. The following section gives a precise definition of such a vector-valued diffusion process.

2.2. Diffusion with multiple features

The diffusion process introduces a time or evolution parameter $t \in T \subseteq \mathbb{R}^+$ into the function $\vec{F}$, such that $\vec{F} : I \times T \mapsto D$, and there is a vector-valued function at each point in time or each level of processing.

The vector-valued diffusion equation is

$$\nabla \cdot g(\mathcal{D}_s \vec{F}) \nabla \vec{F} = \frac{\partial \vec{F}}{\partial t} , \qquad (4.1)$$

where $\mathcal{D}$ is the *dissimilarity* operator, a generalized form of the gradient magnitude which is defined in the following section. The composition, $\mathcal{D}_s\vec{F} : I \mapsto \mathbb{R}^+$, assigns a degree of dissimilarity to every point in the image space. The scale s indicates the isotropic (Gaussian) scale used to make the dissimilarity measurements so that $\mathcal{D}_s\vec{F} \stackrel{\text{def}}{=} \mathcal{D}(G(\vec{x}, s) \otimes \vec{F}(\vec{x}, t))$. The scale s may (and probably should [381]) change over time. The derivative $\nabla \vec{F}$ takes the form of a matrix,

$$\nabla \vec{F} = \begin{pmatrix} \frac{\partial F_1}{\partial x_1} & \cdots & \frac{\partial F_1}{\partial x_n} \\ \vdots & \vdots & \vdots \\ \frac{\partial F_m}{\partial x_1} & \cdots & \frac{\partial F_m}{\partial x_n} \end{pmatrix},$$

which may be called the generalized Jacobian matrix of $\vec{F}$. The conductance, $g : \mathbb{R} \mapsto \mathbb{R}^+$, is a scalar function, and the operator "$\nabla \cdot$" is a vector that is applied to the matrix $g(\mathcal{D}_s\vec{F})\nabla \vec{F}$ using the usual convention of matrix multiplication.

$$\nabla \cdot g\nabla \vec{F} = \begin{pmatrix} \frac{\partial}{\partial x_1} & \cdots & \frac{\partial}{\partial x_n} \end{pmatrix} \cdot g \begin{pmatrix} \frac{\partial F_1}{\partial x_1} & \cdots & \frac{\partial F_m}{\partial x_1} \\ \vdots & & \vdots \\ \frac{\partial F_1}{\partial x_n} & \cdots & \frac{\partial F_m}{\partial x_n} \end{pmatrix}$$

Equation (4.1) is a system of separate single-valued diffusion processes, evolving simultaneously and sharing a common conductance modulating term.

$$\nabla \cdot g(\mathcal{D}_s\vec{F})\nabla F_1 = \frac{\partial F_1}{\partial t}$$

$$\vdots$$

$$\nabla \cdot g(\mathcal{D}_s\vec{F})\nabla F_m = \frac{\partial F_m}{\partial t}$$

The boundaries, which modulate the flow within a single image, are not defined on any one image but are shared among (and possibly dependent on) all of the images in the system as in fig. 4.1.

2.3. Diffusion in a feature space

The behavior of the system (4.1) is clearly dependent on the choice of the dissimilarity operator $\mathcal{D}$. In the single feature case, $m = 1$, the gradient magnitude proves to be a useful measure, that is

$$\mathcal{D}F \stackrel{\text{def}}{=} (\nabla F \cdot \nabla F)^{\frac{1}{2}},$$

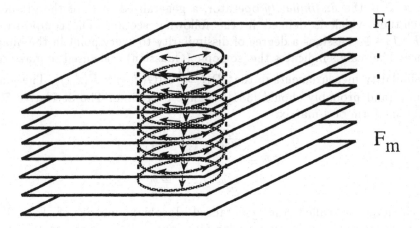

Figure 4.1. Multi-valued diffusion is a system of diffusion processes that are coupled through the conductance function.

where $F(\vec{x}, 0) = I$, where I is the input image. For higher dimensions the dissimilarity is evaluated on the basis distances in D, the feature space.

Applications of statistical pattern recognition to image segmentation are available in the literature [89, 166, 65]. Such approaches apply multivariate statistics in order to find clusters of pixels in these feature spaces. Boundaries between clusters can be defined in a way that optimizes certain metrics, such as the Hotelling trace. Pixels are classified on the basis of their distance to groups of pixels that are nearby in the feature space. Effective use of such techniques requires an appropriate measure of distance within the feature space. Typically, linear transformations are performed on the feature space in order to produce meaningful clusters. These transformations can be determined by statistical measures made on the available data.

The dissimilarity operator is constructed to capture the manner in which neighborhoods in the image space (n-dimensional) map into the feature space. If $\vec{F}$ is a well behaved function (Lipschitz continuous for example), then for a point $x_0 \in I$ and a neighborhood $\mathcal{N}_{x_0}^{\epsilon}$, of size ϵ, there is a point $y_0 = \vec{F}(x_0) \in D$ and a corresponding neighborhood which contains the mapping of $\mathcal{N}_{x_0}^{\epsilon} : \mathcal{N}_{y_0} \supseteq \vec{F}(\mathcal{N}_{x_0}^{\epsilon})$. The dissimilarity at x_0 is a measure of the relative *size* of these neighborhoods. If $\mathcal{N}_{y_0}$ is relatively small for a given ϵ, this indicates that the neighborhood of x_0 has low dissimilarity as shown in figure 4.2.

In the limit as $\epsilon \to 0$ and for $\vec{F} \in C^1$, the relative size of these neighborhoods is described by the first-order derivatives of $\vec{F}$. The matrix of first-order derivatives is the Jacobian matrix, $J = \nabla \vec{F}$.

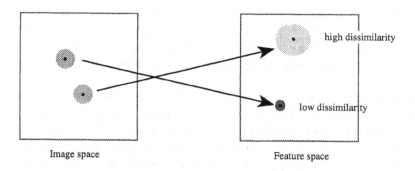

Figure 4.2.   Every neighborhood in the image space maps onto a neighborhood in the feature space. The "size" of the neighborhood in the feature space indicates dissimilarity.

We propose a dissimilarity that is the Euclidean (root sum of squares) norm of the Jacobian matrix. If J is the Jacobian of $\vec{F}$, with elements J_{ij}, then the dissimilarity is

$$\mathcal{D}\vec{F} \stackrel{\text{def}}{=} \|J\| = \left(\sum_{j=1}^{n} \sum_{i=1}^{m} J_{ij}{}^2 \right)^{\frac{1}{2}} . \tag{4.2}$$

We choose this norm for several reasons. First, this matrix norm is induced from the Euclidean vector norm so that in the case of $m = 1$, this expression is the gradient magnitude, as in anisotropic diffusion. Second, the square of this norm (as it often appears in the conductance functions [273]) is differentiable. Finally, when this norm is applied to geometric objects (tensors), it produces a scalar which is invariant to orthogonal coordinate transformations (see Sect. 7.1).

This approach has several advantages above those methods that incorporate techniques of statistical pattern recognition but which treat individual pixels as separate patterns. First, because it uses feature space position in conjunction with the diffusion equation, the vector-valued diffusion process combines information about the image space with position in the feature space. As the process evolves, pixels that are nearby in image space are drawn together in the feature space, except in cases where the dissimilarity is large enough to reduce the conductance. The result is a clustering that is sensitive to the spatial cohesiveness of the feature space. Several authors [133, 388] have shown experimentally that a pair of coupled nonuniform diffusion equations applied to dual-echo MRI can improve the distinction between clusters of pixels in the two-dimensional feature space (see figure 4.3). Vector-valued diffusion as described here constructs a generalized framework for such systems of coupled diffusion equations.

The second advantage is that the dissimilarity depends only on the differential structure of $\vec{F}$ and therefore does not rely on the global structure of the feature space. The dependence on the local structure of the feature space allows for a great deal of flexibility for transformations within that space.

The dissimilarity measure can be generalized to allow for coordinate transformations that are *local* in the feature space. One could include an arbitrary (differentiable) transformation on $\vec{F}$ that is used in the process of computing dissimilarity. This is consistent with the idea that the relative importance of the features $F_1, \ldots, F_m$ might not be reflected in their relative magnitudes. Allow a transformation to a new set of features $\vec{G}(\vec{y}) = \vec{w}$, where $G : \mathbb{R}^m \mapsto \mathbb{R}^k$. Then the dissimilarity of $\vec{F}$ which accounts for this mapping $\vec{G}$ is

$$\mathcal{D}\vec{F} \stackrel{\text{def}}{=} \|\nabla\vec{G}(\vec{F})\| = \|\frac{\partial}{\partial\vec{w}}\vec{G}(\vec{F})\nabla\vec{F}\| = \|\phi(\vec{F})\nabla\vec{F}\|, \qquad (4.3)$$

where ϕ is an $m \times k$ matrix that indicates a local linear transformation in the feature space. The derivative with respect to feature-space coordinates takes the form of a vector:

$$\frac{\partial}{\partial\vec{w}} \stackrel{\text{def}}{=} \frac{\partial}{\partial w_1} \cdots \frac{\partial}{\partial w_m},$$

and we reserve the use of ∇ for derivatives in the image space. If one considers coordinate transformations (rotations and rescaling of axis) in feature space to represent changes in the relative importance of features, then the local coordinate transformation $\phi(\vec{y})$, for $\vec{y} \in \mathbb{R}^m$, allows the relative importance of various features to vary depending on the position in the feature space.

The practical implications of this local transformation can be imagined by considering a vector-valued data set gathered by some imaging device. Suppose that the relative importance of each feature depends not only on its own value but the values of the other features. Such behavior could be accounted for by choosing the proper $\phi(\vec{y})$. At this point it is still not evident how one would go about choosing ϕ given a set of image data. It is conceivable that $\phi(\vec{y})$ should depend on the statistical properties (local and global) of the feature space, the nature of the imaging device, and the task at hand.

For feature spaces that consist of geometric features, this metric is essential for comparing incommensurate quantities associated with axes of these spaces. It is also necessary in order to construct dissimilarity measures that are invariant to certain transformations of one's initial choice of spatial coordinates.

The remainder of this chapter will deal with applications of these principles to various kinds of vector-valued data. In this section we present an application of vector-valued diffusion to images that result from a medical imaging device that produces multiple values at every gridpoint in a 3D image space. Subsequent sections will address the issue of vector-valued images that consist of sets of features that are derived from grey-level input data.

2.4. Application: Noise reduction in multi-echo MRI

Despite significant improvements in image quality over the past several years, the full exploitation of magnetic resonance image (MRI) data is often limited by low signal to noise ratio (SNR) or contrast to noise ratio (CNR). In implementing new MR techniques, the criteria of acquisition speed and image quality are usually paramount. To decrease noise during the acquisition either time averaging over repeated measurements or enlarging voxel volume are employed. However these methods either substantially increase the overall acquisition time or scan a spatial volume in only coarse intervals. In contrast to acquisition-based noise reduction methods a post processing based on variable-conductance diffusion was implemented and tested [133]. Extensions of this diffusion process support multi-echo and 3-D MRI, incorporating higher spatial and spectral dimensions.

To obtain more information from a single clinical MR exam, multiple echoes are frequently measured. The different measurements at one voxel location represent vector-valued information $\vec{F}$, in which the two different components describe different physical properties. They can lead to a better discrimination of tissue characteristics if analyzed together, provided that the anatomical structure can be distinguished in all of the channels. Assuming perfect spatial coincidence, the multi-echo image data suggest the use of a vector-valued diffusion process running simultaneously on multiple channels.

The variable-conductance diffusion performs a piecewise smoothing of the original signal. The propagation of information between discontinuities results in regions of constant intensity or linear variations of low slope. The assumption of piecewise constant or slowly varying intensities is a good estimate to model MR image data, which comprise smooth regions separated by discontinuities, representing various tissue categories characterized by different proton densities and relaxation properties.

As described previously, the single-valued diffusion processes associated with each of two channels share a common conductance term $g(\mathcal{D}\vec{F})$. The dissimilarity $\mathcal{D}\vec{F}$ combined from the multiple image channels is chosen as the Frobenius norm of the Jacobian J (equation 4.2). The vector-valued diffusion has been applied to 2-D and 3-D MR image data with one or multiple channels (echoes).

2.4.1. Three-dimensional 2-channel filtering of double-echo MRI data

Spin-echo image data with thin slices, acquired with a half-Fourier acquisition technique are especially apt to postprocessing by 3-D 2-channel filtering because both the structural correlation in the slicing direction and the correlation among the two echoes support an efficient smoothing and enhancement of region boundaries (see figure 4.3). To obtain optimal information about 3-D anatomical structures, the filtering was applied to 120 double echo contiguous slices of a spin echo acquisition (TR=4000ms, TE=30/80ms) with voxels of dimension $1.8mm * 1mm * 1mm$. To speed up the measurement, a half-Fourier sampling was used. The decrease in the overall acquisition time by a factor of two (which is for the benefit of the patient) is counterbalanced by a worsening of signal to noise ratio by $\sqrt{2}$.

Using three filter iterations with setting the parameter K of the conduction function to 7.3, noise variance in homogeneous tissue regions was reduced by a factor of $\approx$ 13 in the first and second echo channels. The noise elimination is illustrated by generating 2-D histograms (scatterplot) of dual echo pairs. Figure 4.3 bottom left and right demonstrate the significant noise reduction indicated by sharpened clusters. This result is important for a later segmentation of double-echo images by statistical classification methods. Sharp clusters with only minimal overlap allow a clear discrimination of tissue categories with segmentation based on multi-variate pixel properties [134, 388].

Figure 4.4 illustrates enlarged parts of the original and filtered double-echo images of another case. One can clearly recognize the sharpening of tissue boundaries, especially the gray-white matter boundary, and the significant noise reduction within homogeneous regions. Small structures with high contrast like blood vessels (black dots) and the central fissure remain unchanged.

The vector-valued nonlinear diffusion has been an important part of a multi-stage processing system for the segmentation of soft tissue structures from 3-D double-echo MR data [134]. It was furthermore part of the analysis protocol in a large NIH-study (NIH-NINDS-90-03) for investigation of MRI in multiple sclerosis (MS). The study has included over 800 MR examinations acquired at the Brigham and Women's Hospital in Boston and was comprising a careful validation and reliability analysis [176]. After the study period, the large data-base will allow a conclusion as to the clinical utility and sufficiency of the proposed filtering technique, both for visual reporting and for computerized quantitative analysis.

Recent improvements in MR acquisition techniques suggest new application areas, but these advances do not always result in images of better quality. High spatial resolution and high speed imaging can increase noise significantly. However, the results above clearly demonstrate that noisy

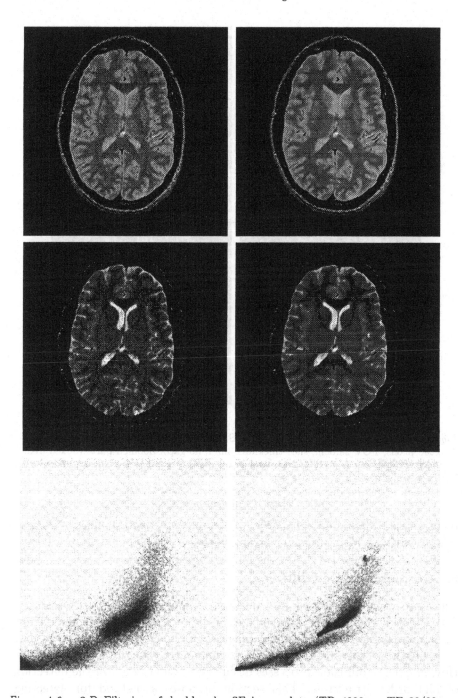

Figure 4.3.   3-D Filtering of double-echo SE image data (TR 4000ms, TE 30/80ms, 256x256 pixels, FOV 24.0cm, slice thickness 1.8mm, no gap). Left column: Original first echo (top), second echo (middle), and scatterplot (bottom); Right column: Filtered first echo (top), filtered second echo (middle) and scatterplot of filtered data (bottom). MR data provided by Dr. R. Kikinis, Brigham and Women's Hospital, Harvard Medical School.

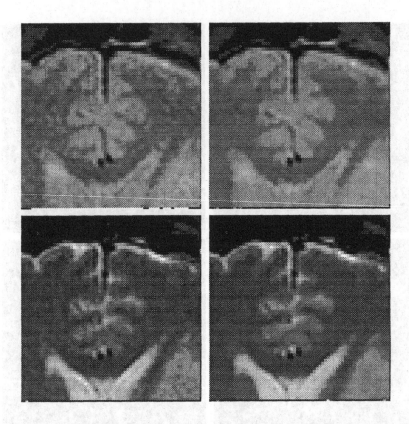

Figure 4.4. 3-D Filtering of double-echo SE image data (TR 3000ms, TE 30/80ms, 256x256 pixels, FOV 24.0cm, slice thickness 3mm, no gap). Zoomed parts (80x80 pixels, 4x pixel replication) of double-echo axial brain slice, original (left column) and filtered (right column) first and second echo.

images can be restored efficiently by a nonlinear, vector-valued diffusion, thus potentially favoring faster acquisitions.

3. Geometry-limited diffusion

The anisotropic diffusion process has been proposed as a means of enhancing and detecting edges in greyscale images, while vector-valued diffusion offers a similar capability in vector-valued images. Yet higher-order image structures including lines, corners, and ridges also allude detection in noisy situations. How might the diffusion processes discussed previously aid in the detection of high-order structures? This section proposes a method for detecting structures. The strategy is to capture the local higher-order geometry of a grey-scale image through a set of measurements, which are implemented as filters. The outputs of these filters are processed by a vector-valued diffusion equation in order to obtain reliable characterizations of image geometry.

4. The local geometry of grey-scale images

A scalar image I is a smooth mapping from the image domain $I \subset \mathbb{R}^n$ to some subset of the set of real numbers, $L \subset \mathbb{R}$, so that $I : I \mapsto L$, where the range L is called the *intensity* or sometimes *luminance*. The graph of this function is the *intensity surface*, as shown in figure 4.5. By this definition scalar images are assumed to be "greylevel" like those obtained from a black and white camera or from a single channel of a color camera.

Each point in the image space has a corresponding point on the intensity surface, and each point on that surface has a local structure which is defined over some neighborhood. This is the *local shape* of an image. We are purposely vague here because we want to allow for a wide range of possibilities for characterizing this structure.

For example, Koenderink [187] describes the "deep structure"of images and argues that local image structure is captured by the differential structure of I. Thus, one option is to capture the local shape at each point in the image by measuring the derivatives of the intensity surface up to order N. These derivatives can e.g. be used in the local Taylor series approximation to each point on the intensity surface. The strategy of encoding local surface structure as sets of derivatives is adopted in this section, but there are other possibilities. In Sect. 8, for example, we explore an alternative representation of local shape that resembles a local frequency or Fourier analysis.

The local shape of the intensity surface at every point in the image is represented or "encoded" through a finite set of scalar values as shown if figure 4.5. At each point in the image there is a set of values $F_1, \ldots, F_m$ that results from measurements of the local image structure. The inserts in figure 4.5 depict the differences in local image structure. This collection of measurements is a *geometric description* of the image. These values constitute a set of geometric descriptors and they create an associated feature space. The goal is to create a diffusion process which operates on geometric descriptions of scalar images.

5. Diffusion of Image Derivatives

In this section we choose as our geometric description sets of derivatives of the intensity surface measured in Cartesian coordinates. That is, a geometric description is a set of derivatives of the original image. For example, one could choose a first-order description:

$$\vec{F} = F_1, F_2 = I_x(\vec{x}), I_y(\vec{x}), \tag{4.4}$$

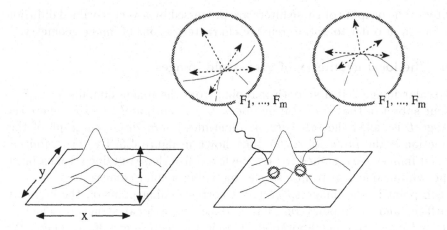

Figure 4.5. The intensity surface is the graph of an image. A finite set of measurements at every point in the image space represent the local shape of the intensity surface. The cut-aways illustrate that each place on the surface has a local surface patch and a set of descriptors which describe that patch (to within some approximation).

where $I_x(\vec{x})$ is the derivative of $I(\vec{x})$ with respect to x at a point $\vec{x} \in I$, and $I_y(\vec{x})$ is the y derivative at that same point. We assume that x and y are orthogonal. In this sense, anisotropic diffusion described in chapter 2 uses a *zero-order* description of shape, i.e.,

$$\vec{F} = F_1 = I(\vec{x}). \tag{4.5}$$

Using sets of derivatives to describe local shape has several useful properties. First, derivatives of the intensity surface can be measured directly from a digital image using linear filters as described in [358]. Second, in analyzing the behavior of this system it is useful to be able to rely on the linearity of these operators. Finally, the use of complete sets of derivatives (all of the derivatives of a particular order treated as a tensor) provides means of constructing processes which are invariant to orthogonal group transformations.

This section combines the concepts of anisotropic diffusion and local shape in order to define a process that forms patches which are locally homogeneous in higher-order information. The differential measurements that comprise the geometric description of local image structure form feature vectors that become the initial conditions of a vector-valued diffusion equation. In the feature space of geometric features, each point corresponds to a different shape, and the metric ϕ quantifies differences in local shape. The strategy is to choose ϕ so that it creates boundaries that are of interest and choose

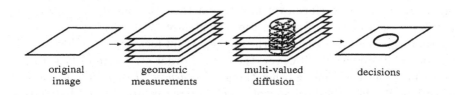

original image   geometric measurements   multi-valued diffusion   decisions

Figure 4.6.   Geometry-limited diffusion in which features are geometric measurements made on a single valued image.

the scale, $s(t)$, in the diffusion equation so that unwanted fluctuations in the feature measurements (noise) are eliminated.

As the anisotropic scale increases, measurements are taken from the resulting set of features with progressively smaller Gaussian kernels [381], and decisions about presence of specific geometric properties are made on the basis of those measurements. This strategy is depicted in figure 4.6. Information about $(N+1)$th-order derivatives can be obtained in three steps. First, make differential measurements of the image up to Nth-order using derivative-of-Gaussian filters of a scale that does not compromise interesting image structures. Second, smooth these derivatives in a way that breaks these "images" into patches that form regions of local homogeneity and become nearly piecewise constant as the process evolves. Third, study the boundaries of these patches by looking at the derivatives of the features and make decisions about the presence of higher-order features which include $(N+1)$th-order derivatives.

6. First-order geometry and creases

6.1. Creases

This approach has already been studied in the zero-order case (anisotropic diffusion as in chapter 2) and a logical next step is to include first-order derivatives as features. For a two-dimensional image space the resulting feature space is three-dimensional. This section investigates the use of first-order geometric features for the purpose of finding "creases" or "ridges".

Pizer and others [295, 211] have shown that creases and the corresponding flank regions are very powerful for forming hierarchical segmentations of medical images. Although the notion of creases has no one precise definition [94, 310, 200], there is an intuition that one can test via a simple experiment. Imagine an image as an intensity surface, and suppose that the intensity axis is aligned with a gravitation field so that the direction of increasing intensity is "up". Place a drop of water on the surface. Creases are places where a drop of water tends to split and run in two or more different directions. Likewise, one could turn the surface over and repeat the experiment. The results in the first case are "ridges" and in the second case "valleys". If the surface is continuous, then every place on the image, except local extrema, has exactly one downhill direction. However, the second order structure (a larger drop of water) describes, among other things, the *degree of variation* in gradient directions in the neighborhood of a point. Creases are places that have a large disparity of gradient directions in an immediate neighborhood. If one characterizes flanks as contiguous regions that have the same (or nearly the same) uphill direction, then creases are the places where flank regions meet. This definition of a crease is independent of image intensity; the height of the intensity surface does not enter into this definition. Therefore, it is appropriate and convenient for this discussion to exclude the zero-order term as one of the features. For a two-dimensional image space encode the local image geometry as two features F^x and F^y, the pair of first partial derivatives of intensity. The *raised indices* denote the fact that these features are undergoing nonlinear diffusion and are no longer the derivatives of the initial image but have the image derivatives as their initial conditions;

$$F_1(x, y, t) = F^x(x, y, t), \text{ and } F^x(x, y, 0) = \frac{\partial I(x, y)}{\partial x},$$

while

$$F_2(x, y, t) = F^y(x, y, t), \text{ and } F^y(x, y, 0) = \frac{\partial I(x, y)}{\partial y}.$$

Thus, the raised indices are merely labels, and these labels indicate the initial conditions of a particular feature in terms of derivatives of the input image. Subscript indices, however, are used to indicate differentiation in the direction of the axis associated with that index. For instance, $I_x(x, y) \stackrel{\text{def}}{=} \frac{\partial I(x,y)}{\partial x}$ while $F^x(x, y, 0) \stackrel{\text{def}}{=} I_x(x, y)$. The pair of first-order measurements comprise a two-dimensional feature space. Effectively, this choice of features models each local surface patch as a plane of a particular orientation.

The features, F^x and F^y, are derivatives of $I(x, y)$ only at $t = 0$. For $t > 0$ it might be that there exists no $I(x, y, t)$ such that

$$\partial I(x, y, t) / \partial x = F^x(x, y, t)$$

and

$$\partial I(x, y, t) / \partial y = F^y(x, y, t).$$

Derivatives do not commute with the nonlinear diffusion operator, so, in general, the relationship $F^x_y = F^y_x$ does not hold. However, the reconstruction of a diffused luminance function, $I(x, y, t)$ is not the goal of this type of processing—the goal is to characterize changes in first-order information—so the lack of $I(x, y, t)$ is of little consequence. This reasoning brings up a very important question about the *geometry* that results from a nonlinear diffusion of first derivatives. Does such geometry make sense? Are differential invariants constructed from $F^x(x, y, 0)$ and $F^y(x, y, 0)$ invariant for $t > 0$? These questions will be addressed in Sect. 7.1.

6.2. A shape metric for creases

Choose the coordinate transformation ϕ, which determines the way in which distances are measured (locally) in the feature space, specifically to capture the notion of creases. The appropriate choice of ϕ is best understood by considering the feature space F^x, F^y as it is described by polar coordinates (see figure 4.7). Each position (x, y) in the image has a position, $(v, w) = \vec{F}(x, y)$ in this feature space. The polar coordinates of each point in this feature space have a geometric interpretation. The magnitude of the gradient vector ($|\nabla f|$) at the point (x, y) is $\rho = (v^2 + w^2)^{\frac{1}{2}}$, and $\theta = \arctan(w/v)$ is the direction of image gradient.

Now define a dissimilarity operator, ϕ, which captures changes only in the direction of the vector, $\vec{F}$. We express ϕ as a rotation R followed by a scaling S. The matrix R is a rotation of the local feature space coordinates into a coordinate system which is aligned with $d\rho$ and $d\theta$ in the polar coordinates. The matrix S controls the scaling and thereby the relative effects of each of these properties. This allows for control of the amount of influence that the gradient magnitude and gradient direction have on the dissimilarity operator. The notion of creases does not depend on the strength of the gradient magnitude, only on direction. Therefore, differences between pixels in the radial direction do not contribute to the dissimilarity. This has the effect of locally collapsing the feature space onto a circle with its center at the origin. The perpendicular component, $d\theta$, is scaled by the inverse of the radius so that small changes in the $d\theta$ direction capture changes in the angle. Use the local metric

$$
\begin{aligned}
\phi &= R^{-1} S R \qquad\qquad\qquad\qquad\qquad\qquad\qquad (4.6) \\
&= \begin{bmatrix} \cos(\theta) & -\sin(\theta) \\ \sin(\theta) & \cos(\theta) \end{bmatrix} \begin{bmatrix} \frac{1}{\rho} & 0 \\ 0 & 0 \end{bmatrix} \begin{bmatrix} \cos(\theta) & \sin(\theta) \\ -\sin(\theta) & \cos(\theta) \end{bmatrix} \\
&= \begin{bmatrix} \frac{\cos^2(\theta)}{\rho} & \frac{-\sin(\theta)\cos(\theta)}{\rho} \\ \frac{-\sin(\theta)\cos(\theta)}{\rho} & \frac{\sin^2(\theta)}{\rho} \end{bmatrix} = \frac{1}{((F^x)^2 + (F^y)^2)^{\frac{3}{2}}} \begin{bmatrix} (F^x)^2 & F^x F^y \\ -F^x F^y & (F^y)^2 \end{bmatrix}
\end{aligned}
$$

When this transformation is used to compute dissimilarity $\mathcal{D}\vec{F} = \|\phi(\vec{F})\nabla\vec{F}\|$, it has the same effect (in continuous images, but necessarily in discrete approximations [379]) as computing the Jacobian on a pair of intermediate features $\vec{G} = G_1 \; G_2$ which are normalized with respect to the

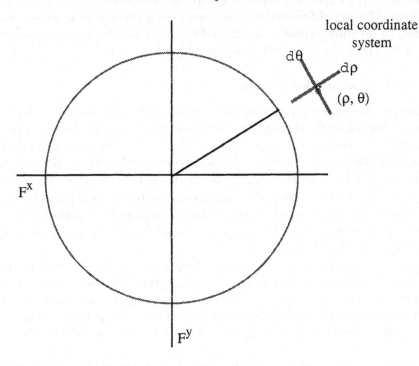

Figure 4.7. First-order feature space as represented in polar coordinates. Every point in this feature space has a local coordinate system that is aligned with the $d\theta$ and $d\rho$ directions.

Euclidean length of the vector $\vec{F}$:

$$G_1 = \frac{F_1}{|\vec{F}|} = \frac{F^x}{(F^{x2} + F^{y2})^{\frac{1}{2}}}, \tag{4.7}$$

$$G_2 = \frac{F_2}{|\vec{F}|} = \frac{F^y}{(F^{x2} + F^{y2})^{\frac{1}{2}}} \text{ and} \tag{4.8}$$

$$\mathcal{D}\vec{F} \stackrel{\text{def}}{=} \|\nabla \vec{G}\|$$

This normalization makes explicit the fact that this distance measure is invariant to any monotonic intensity transformation although the pair of features, F^x and F^y, are clearly not.

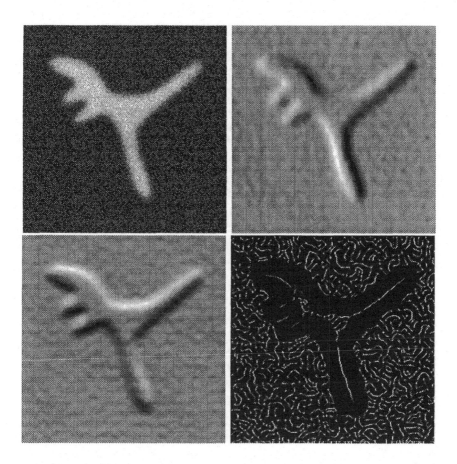

Figure 4.8. From the top left to right: (a) A noisy blob is created by blurring a white on black figure and then adding uniformly distributed random noise. Shown here are the initial values of the first-order features (b) F^x and (c) F^y and (d) the dissimilarity measure on these features.

6.3. Results

Experiments [380] show this first-order geometry-limited diffusion process forms sets of homogeneous regions that have similar gradient directions. These regions correspond to "flanks" or "hillsides" in the intensity surface of the original image.

The test image in figure 4.8a (hereafter the *blob* image) is created by drawing a white figure on a black background and then blurring the result. Uniformly distributed noise is added to the image so that the range of the noise at each pixel is equal to the overall intensity of the foreground, and the size of the image is 256 × 256 pixels. Figures 4.8b–d show the

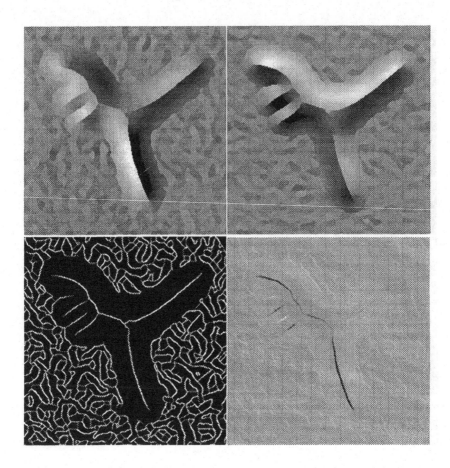

Figure 4.9. From the top left to right: The result values of the first-order features (a) F^x and (b) F^y after the geometry-limited diffusion process, as well as (c) the dissimilarity measure $\mathcal{D}\vec{F}$, and (d) the isophote curvature computed from these features.

initial values of the features, F^x and F^y, and the dissimilarity measure at start of the diffusion process. Because the dissimilarity measure normalizes features with respect to the gradient magnitude, areas of the image that were initially flat are susceptible to noise. Figure 4.9 shows the features and the dissimilarity measure after processing. The areas of high dissimilarity indicate boundaries between flank regions. The pixels within flank regions have been "regularized" so that they have virtually the same gradient values. The boundaries of these regions appear to correspond well with the intensity ridges in the original image. Because of the tendency of this process to produce piecewise constant solutions, the boundaries are distinct and allow for easy detection using either a threshold of the dissimilarity or a zero crossing of a higher-order derivative, i.e. a singularity. The results of this process appear to be robust with respect to noise. The use of scale in measuring dissimilarity controls the size of these regions or patches without distorting the shapes of the boundaries between patches. Figure 4.9d shows another second-order measure that reflects the sign of the "curvature" associated with each boundary. This is analogous to the isophote curvature of a scalar function $I(\vec{x}, t)$ except that it is constructed from *diffused* derivatives. For G_1 and G_2 defined above, this measure is

$$\frac{\partial G_1}{\partial x} + \frac{\partial G_2}{\partial y}, \tag{4.9}$$

and the sign of this measure distinguishes "ridges" (dark) from "valleys" (light). The first-order diffusion is essentially a piecewise linear approximation to the original image which ignores the height differences between patches (the "dc" component). The sign of the curvature at boundaries indicates the manner in which these linear patches meet. If the boundary of two patches "sticks up", the curvature is negative, and the boundary is a ridge. If the two patches slope down toward the place where they meet, the boundary "sticks down", the curvature is positive, and the boundary is a valley. The fact that $\vec{G}$ is normalized means that the steepness of the patches does not enter into the value of (4.9); this measure captures only the isophote curvature.

6.4. The diffused Hessian

The second-order information derived from first-order patches provides a number of interesting measurements. Consider the diffused Hessian,

$$H^{(1)} = \begin{bmatrix} F^x{}_x & F^x{}_y \\ F^y{}_x & F^y{}_y \end{bmatrix}, \tag{4.10}$$

where the superscript (1) indicates that this measurement results from a first-order diffusion process. Unlike the conventional Hessian, this matrix is not necessarily symmetric. The trace of this matrix, $\text{Tr}[H^{(1)}] = F^x{}_x + F^y{}_y$, is a kind of Laplacian, and $\text{Tr}[H^{(1)}] = 2K^{(1)}$, where $K^{(1)}$ might be called the *mean curvature*. In the context of this first-order diffusion process it describes the curvature at patch boundaries. Unlike the isophote curvature it includes information about the steepness of the patches. This measure gives sign information, which indicates whether creases are ridges or valleys, and its magnitude reflects the angle (in two dimensions) at which two linear patches meet. The Euclidean norm of this matrix $\|H^{(1)}\|$ is sometimes called deviation from flatness [199], and it indicates the overall change in first-order information; magnitude and direction. These two measures, the

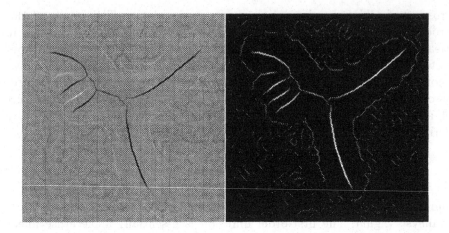

Figure 4.10. The trace (left) and norm (right) of the diffused Hessian for the *blob* image.

mean curvature and the Euclidean norm of H, can be used to sort creases
on the basis of their "sharpness" and thereby exclude, if desired, creases
that result from noise in otherwise very flat regions. Figure 4.10 shows
these two measures computed with the features of figure 4.9. Some creases
respond more strongly than others based on the magnitude of the gradients
within patches. Unlike the dissimilarity these measurements are sensitive to
monotonic intensity transformations.

6.5. Application: Characterization of blood vessel centerlines

Digital subtraction angiography (DSA) is an X-ray imaging technique
that enhances blood vessels by subtracting a static image from images
measured after injection of contrast agent. The segmentation of blood
vessels from 2-D DSA image data is important in a variety of applications
including the location of landmarks in multi-modality matching and the 3-D
reconstruction of vessel trees from a few projections. The central problem
is the segmentation of vessel structures of different width.
A common approach to line detection is matched filtering. A typical analysis
by Canny [51] presented an "optimal" operator for finding ridge profiles. The
symmetric operator can be represented as the second derivative of a finite
kernel that resembles a Gaussian. A matched filter gives an optimal result if
it matches perfectly the structure to be measured, but it fails if the structure
is much smaller or larger than the filter width. DSA images would have to
be filtered with line filters of different width, but the integration of filters
across multiple scales is still an unsolved problem.
Diffusion of first-order geometry as proposed in this chapter overcomes

the scale problem for detecting center lines. The definition of creases is independent of image intensity and also independent of the width of a ridge structure. A crease is located on top of a line structure where the gradient direction undergoes large changes. The vertical and horizontal components of the gradient are smoothed within slope regions but diffusion is stopped at creases. Smooth line structures become progressively sharper and obtain a "roof-like" profile. The geometry-driven first-order diffusion is iteratively applied to the DSA image (figure 4.11a). The K parameter of the conduction function was set to $\pi/8$ (recall that conduction is controlled by the dissimilarity of gradient directions), the initial Gaussian regularization within the conduction function was set to $\sigma = 2.0$, linearly decreasing to 0.0 with increasing iteration number (10 iterations). The diffused Hessian was calculated from the two diffused first-order images (equation 4.10) characterizing creases at locations where the gradient change has become very sharp. The sign of the mean curvature measure $\text{Tr}[H^{(1)}]$ allows us to separate ridge structures from valley structures. Finally, a hysteresis thresholding was applied to the Euclidean norm of the Hessian, only considering ridge-type structures. The result demonstrates that the crease structures are found independent of the image intensity, and that first-order geometry diffusion could locate the center lines of bloodvessels over a large range of widths, illustrating the "multi-scale" behavior. Although there is no explicit correlation between the medial axis and the ridge in intensity, the result seems to approximate the center lines or *medial axes* of the object. Therefore, one can also compare the multi-valued diffusion on first-order geometric features as a kind of *gray-valued medial axis transform (MAT)*. In contrast to the binary MAT, the input are not perfectly segmented binary figures but noisy, blurred gray-valued structures. Note that gaps occur at junctions and crossings in the processed images. These types of complex features are not well characterized by changes in first-order geometry, and they suggest the need to incorporate higher-order geometry.

6.6. Application: Characterization of sulcogyral pattern of the human brain

Another application extracts the gyral and sulcal pattern from a 3-D surface rendering of the human brain segmented from Magnetic Resonance volume data (figure 4.12 top). Such a procedure could by useful for brain researchers for representing morphological abnormalities of the sulcogyral pattern of human brains. It is the goal of cortical pattern analysis to find systematic differences between a group of normals and a group affected by a specific disorder (schizophrenics [344], Alzheimer, e.g.), possibly linking structural alterations with the disorder and providing a new quantitative measure for diagnosis. A structural representation of the sulcogyral pattern could also be

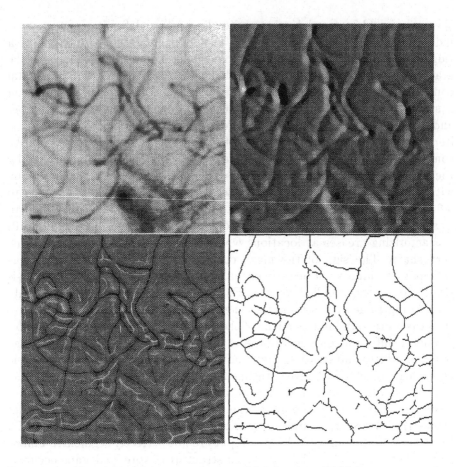

Figure 4.11. From left to right, top to bottom: (a) Subregion of DSA image representing bloodvessels (256x256 pixels), (b) diffused horizontal component of gradient, (c) Trace of diffused Hessian and (d) Hysteresis thresholding of ridge structures.

useful for neurosurgery, where surgical planning requires the comparison of the patient's individual brain structures with anatomical models, stored in a digital anatomical atlas. The individual variability does not allow a point-by-point matching of two images, but favors a structural analysis based on the topological relationships between flexible features.

The vector-valued diffusion on first-order features was applied to 3-D renderings of the brain surface. The parameters were the same as described in the previous application (section 6.5). Besides its ability to characterize the narrow dark sulci, the procedure demonstrates its capacity to sharpen and extract the axes of symmetry of highly smooth structures like the gyral pattern (figure 4.12 middle). Experiments proved that this performance was

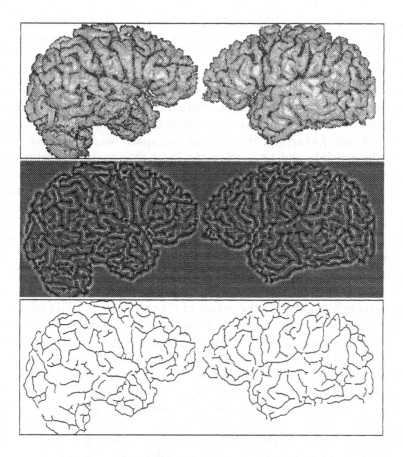

Figure 4.12. Application of geometry-limited diffusion using first-order geometry. From top to bottom: Procedure applied to a 3-D rendering of the human brain segmented from MR data (top). The trace of the diffused Hessian (middle) illustrates the sharp valleys corresponding to sulcal structures (dark) and the sharp ridges representing the centerlines of the gyral structures (bright). Hysteresis thresholding of the valleys results in a binary map of sulcal structures (bottom), which is input to subsequent structural analysis.

by far superior than compared to other line detection methods based on matched filtering or simple ridge following.

6.7. Combining information of zero and first-order - corners

In the examples of Sect. 6.3, intensity is not used as one of the features. There may exist a three-dimensional feature space, a corresponding vector-valued function $\vec{F} = (F^x,\ F^y,\ F)$, and an associated shape metric $\phi(F^x, F^y, F)$ which together would allow a segmentation to produce visually interesting regions and boundaries. This section explores an alternative method for combining zero- and first-order information for the specific purpose of detecting corners in images.

Consider the visual feature called a "corner". The difficulty of finding corners using geometry-limited diffusion is that corners in two dimensions are not contours or lines, but points. Therefore, they cannot form boundaries between geometric patches. If one were to define a dissimilarity measure to capture corners, features in the surrounding space would most likely flow around these points and undermine the local changes in geometry which indicate a corner. An alternate strategy is to compute and process zero and first-order feature spaces separately and then combine the information that results from each of these processes. A reasonable description of a corner is a place on the edge of an object where the boundary turns very sharply. In order for a boundary to turn suddenly, the gradients along that boundary must undergo a large turn in a small neighborhood. This is precisely the definition used previously for creases. *A corner is a place in the image that is both a crease and an edge.* This suggests that one could obtain corners by combining the results of the crease calculations with edges discerned from zero-order diffusion. Because a corner must have both a high gradient (derivative of zero-order quantity) and a large variation in gradient direction (derivative of a first-order quantity), it is natural to multiply the measurements which result from the zero and first-order diffusion processes. To this end, Blom [36] proposes a corner detector of the form $f_{vv} f_w{}^2$ which has some very useful properties, including affine invariance. By analogy a strategy for locating corners is to produce zero and first-order segmentations and then combine the results of the separate processes using the product of zero- and first-order dissimilarity measures (depicted in figure 4.13).

The sample image in figure 4.14a is a white hexagon on a black background. Uncorrelated uniformly distributed random noise is added to the image. The value of $\|H^{(1)}\|$ which results from the first-order diffusion processing and the value of $|\nabla F|$ which results from the zero-order processing are shown in figures 4.14b–c respectively. The image of figure 4.14d is the result of multiplying the pixel values of the images in figures 4.14b and c and simple thresholding. The corners that result from a simple threshold of figure 4.14d are accurate to within 3 pixels.

7. Invariance

7.1. Geometric invariance

Any image processing algorithm should produce results that are independent of the coordinate system that is used to describe the image space. In the particular case of orthogonal transformations, the results should not be

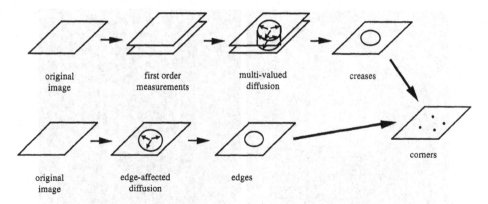

Figure 4.13. The strategy for finding corners is to isolate the zero-order and first order patches separately and then combine the results.

influenced by translations and rotations of the original image. The encoding of local geometry (section 5) in terms of differentials of the luminance function forces the choice of a coordinate system. The absolute positions of points in the feature space depend on this choice. This is because the axes of the feature space are associated with sets of directional derivatives and the directions of these derivatives are aligned with the coordinate axis in the image space as described in Sect. 4 (x and y in two dimensions). This presents a serious question. How can one construct a system which produces invariant results but which relies on a feature space that is intimately tied to our choice of coordinate systems?

The answer lies in the dissimilarity measure, which defines distance in the feature space. Positions in the feature space need not be invariant to orthogonal transformations in order to obtain invariant results. It *is* essential, however, that the *relative positions* of points and the distances between those points be invariant. To ensure this invariance, express the dissimilarity as a geometric invariant of the intensity surface of the original image. Such a dissimilarity is invariant at the start of the process because it is a geometric invariant of the intensity surface. As the process progresses, however, the terms of the invariant change so that they are no longer the derivatives of the intensity surface. Does dissimilarity remain invariant (under the orthogonal group) as the nonlinear, nonuniform diffusion progresses?

The following theorem states that the behavior of the geometry limited diffusion process is invariant to transformations of the original coordinate system. We are primarily concerned with *polynomial invariants* as described by ter Haar Romeny and Florack [358], where polynomials of image

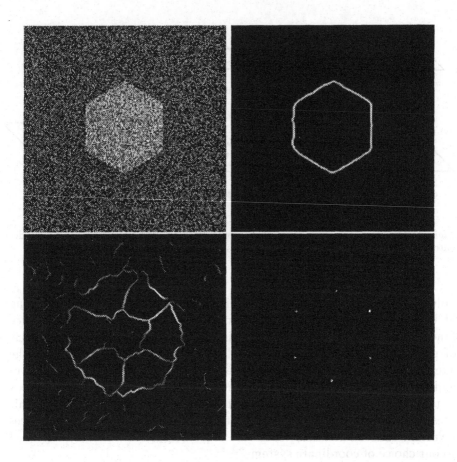

Figure 4.14. Clockwise from the upper left: (a) A test image composed of white hexagon on a black background with additive random noise (pixelwise, uniformly distributed, 50%). (b) The gradient squared after zero-order diffusion. (c) The creases that result from first-order diffusion. (d) The result of multiplying the edge and crease measures in order to detect corners.

derivatives are expressed using the Einstein summation convention of summing expressions like indices over the basis. In geometry-limited diffusion we are concerned not only with invariant expressions of derivatives of scalar functions, but also with *diffused* derivatives (indicated by the raised indices). Diffused derivatives are functions that evolve over time according to the vector-valued diffusion equation and have image derivatives as their initial conditions.

Theorem 1 *Given a geometry-limited diffusion system as described in Sect. 5 with*

1. a set of features, $\vec{F} = F_1, ..., F_m$, *with initial values that are complete sets of derivatives of a smooth image and*

2. *a dissimilarity measure, $\mathcal{D}\vec{F}$, that has the form*

$$\mathcal{D}\vec{F} = \|\nabla \vec{G}(\vec{F})\|,$$

where $\vec{G}$ is a set of intermediate features that form a tensor (i.e. $\vec{G}$ transforms as a tensor under rotations of the initial coordinates).

then solutions are invariant to orthogonal group transformations for all $t \geq 0$.

A proof of this theorem given in [379].

Theorem 1 suggests that geometry-limited diffusion is a diffusion process that operates on *tensors*. The features associated with a geometry-limited diffusion process can be treated in groups according to the order of derivative that they represent, and these groups, or tensors, transform in the same manner as the initial conditions. Because of this property, a large number of invariants can be created by treating the labels associated with diffused features (raised indices) as the actual derivatives, and these invariants will be invariant to spatial coordinate transformations throughout time. Theorem 1 in combination with the Einstein notation convention allows one to construct differential invariants by placing indices interchangeably in both the raised and lowered positions, where the difference in position indicates whether the derivatives are taken before or after the diffusion process.

7.2. Higher-order geometry

Experiments show that this framework extends to higher-order geometric descriptions. Diffusion processes which incorporate first- and second- order geometry (for $I \subseteq \mathbb{R}^2$, a five-dimensional feature space, see e.g. [360]) are capable of breaking images into patches which have as boundaries both the middles and edges of objects [382, 379]. In this framework we have used as intermediate features a normalized product of first- and second- order derivatives. Such systems are invariant by Theorem 1.

There are alternatives to using sets of derivatives as geometric descriptions, and the proper choice is likely to depend on the problem at hand. The next section applies vector-valued diffusion to a local frequency representation in order to segment images on the basis of texture.

8. Spectra-limited diffusion and texture

This section proceeds by establishing a notion of texture that will motivate a description of local image structure which is based on frequency decomposition rather than differential geometry. This particular geometric description, used in conjunction with the vector-valued diffusion framework

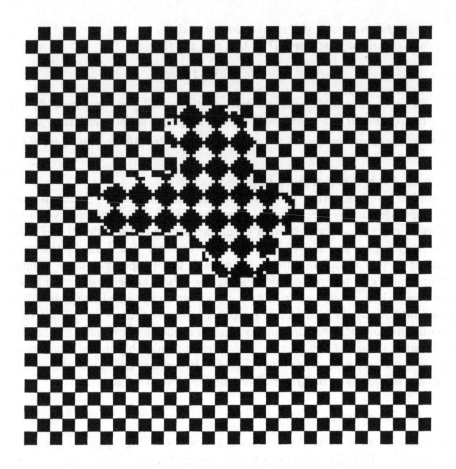

Figure 4.15. Textured patch on a textured background.

and an appropriately defined dissimilarity measure, gives rise to a diffusion process that produces a texture-based image decomposition.

8.1. Texture

The following definition is a relatively simple working definition of texture, which is not meant to capture all of the richness and variety of textures as they are perceived in the real world, but instead helps to establish a set of basic principles.

Definition 1 *A* texture *is a patch in an image that has a periodic or "nearly" periodic luminance profile.*

An image with such a patch is shown in figure 4.15.

Figure 4.16. Textures extended to a rectilinear grid show a great deal of regularity in their respective power spectra.

Several properties of such textures are worth noting. First, these patches are "extendible" in the sense that the periodic pattern could be repeated in order to create larger patches with the "same" texture. Second, the Fourier spectra of these textures have a great deal of structure when analyzed on regularly sized patches. Figure 4.16 shows the extended patches of the textures of figure 4.15 and the respective power spectra of those patches. Because of the periodic nature of these textures there are relatively few frequencies that have high energy. These power spectra suggest that there is some efficient encoding, based on the high-energy frequencies, that could be used to distinguish these textures. The difficulty is that textures are rarely analyzed in distinct rectilinear patches. The problem of segmenting images on the basis of texture typically involves distinguishing textures

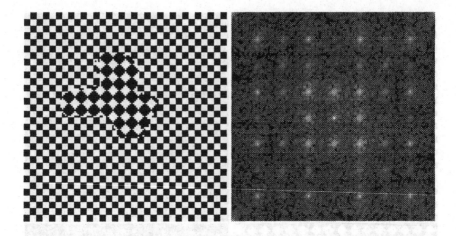

Figure 4.17. Irregular textured patches show the difficulty of using the Fourier transform as a means of segmenting textured patches. The regularly spaced bright areas indicate the presence of energy, but do not offer any immediate way of separating the effects of the two different textures or retrieving local information.

along irregular and often ambiguous boundaries. The spectral analyses of images that contain even a very few different textures with irregular boundaries are quite complex and typically lack the obvious structure of single texture spectra (e.g. see figure 4.17). The global nature of the Fourier transform makes it inadequate for segmenting textures.

A great deal of work in both natural and artificial vision has focused on the importance of filters that are local in both frequency and space. Gabor [121] shows that filters with optimal bandwidth in frequency and space are characterized in the spatial domain as sinusoids modulated by Gaussian envelopes. Indeed, the presence of frequency "channels" in natural vision has been argued on the basis of psychophysical data [138, 137]. In artificial vision such filters have been used to classify textures [207, 170, 65]. One approach to classify textures [65] is to compute at each pixel of a digital image an array (feature vector) of measurements from filters of various frequencies and orientations. Then apply techniques of statistical pattern recognition in order to classify pixels based on their proximity in the feature space. The difficulty with this kind of "local" spectral analysis is the inverse relationship between the extent of filters in the space and frequency domains. Any attempt to make a precise measurement of a particular frequency component at a particular position an in image results in an uncertainty about the exact location of the structures that gives rise to the measurement [377, 48-85]. The extreme case is the Fourier transform itself which is precise in frequency but global in space.

In the presence of noise and ambiguous region boundaries, decisions about the boundaries of textural patches can become difficult. In noisy conditions, measurements of digital images are often combined with low pass transforms in order to provide stable measurements. However, low-pass filters such as the Gaussian can have adverse effects on the characterization of objects or patches whose shapes depend on high frequency information.

Nonuniform or anisotropic diffusion has been proposed in earlier chapters in addition to stationary filtering with Gaussian kernels. However, Gaussian blurring (diffusion) and anisotropic diffusion are less adequate when trying to distinguish patches on the basis of high frequency information, particularly when such patches are similar in their low frequency characteristics. This is often the case with textures. In this section we develop a type of blurring which acts as a band-pass filter (rather than a low-pass filter) and a diffusion process which implements such blurring in a local, iterative fashion. We then introduce a conductance term analogous to the variable conductance term, $g(\mathcal{D}\vec{F})$, which controls geometry-limited diffusion. This conductance term allows characteristic wave forms to flow up to, but not across, textural boundaries.

8.2. Blurring the envelopes of functions

The difficulty with low-pass blurring is that it can undermine the periodic nature of textures and thereby destroy the information that distinguishes different textural regions. As an alternative, consider the representation of a luminance function (one dimension) as a periodic function, with a characteristic frequency ω, multiplied by an envelope that describes the magnitude. In one dimension

$$I(x) = \varphi(x,\omega)\psi(x,\omega), \qquad (4.11)$$

where $I : \mathbb{R} \mapsto \mathbb{R}$ is the luminance profile and $\psi : \mathbb{R} \times \mathbb{R} \mapsto \mathbb{C}$ is a periodic function, independent of $I(x)$, that has a characteristic frequency ω. The function $\varphi : \mathbb{R} \times \mathbb{R} \mapsto \mathbb{C}$ modulates the amplitude and phase of the periodic function $\psi(x,\omega)$ and describes the local presence of the characteristic frequency. Given a nonzero $\psi(x,\omega)$ and a luminance profile $I(x)$, solve for $\varphi(x,\omega)$:

$$\varphi(x,\omega) = \frac{I(x)}{\psi(x,\omega)}.$$

The function φ is the *envelope* of I for a frequency ω. If the characteristic wave form is defined for a continuous range of frequencies, there is an envelope $\varphi(x,\omega) = I(x)/\psi(x,\omega)$ for each frequency.

The function φ characterizes each point in the image over a range of characteristic frequencies. Note several things about φ, the envelope. First, it is generally complex valued. The pair of values for each frequency at every point encodes local spectral information, both magnitude and phase. Second, integrating $\varphi(x,\omega)$ over the entire domain produces a transform of $I(x)$. If the characteristic waveforms comprise a basis, then the integral

$$\int_I \varphi(x,\omega)\mathrm{d}x$$

is a complete representation of $I(x)$ over the image space $x \in I$. Third, if $\psi(x,\omega)$ is a sinusoid $e^{\iota\omega x}$, then the combination of blurring $\varphi(x,\omega)$ with Gaussian kernels and multiplying by $\psi(x,\omega)$ is precisely the same as a convolution of $I(x)$ with a Gaussian band-pass filter centered at frequency ω. Define a function $F(x,\omega,s)$

$$
\begin{aligned}
F(x,\omega,s) &= \psi(x,\omega)(G(s,x) \otimes \varphi(x,\omega)) \\
&= e^{\iota\omega x}(G(s,x) \otimes (I(x)e^{-\iota\omega x})) & (4.12) \\
&= (G(s,x)e^{-\iota\omega x}) \otimes I(x)) & (4.13)
\end{aligned}
$$

The resulting filter $Z(x,\omega,s) = G(s,x)e^{\iota\omega x}$ has both real and imaginary parts which form a pair of Gabor blobs (sine and cosine respectively). The magnitude of this kernel is a Gaussian $G(x,s)$, where the scale parameter s defines the standard deviation of the Gaussian and the effective width of the Gabor filter. For larger values of s these filters are wider in space and narrower in the frequency domain. The luminance function $I(x)$ can be described by measurements made with Gabor blobs over a range of scales and frequencies. Thus $F : \mathbb{R} \times \mathbb{R} \times \mathbb{R}^+ \mapsto \mathbb{C}$ is a frequency-tuned scale-space as described by Florack [107].

A degree of spatial coherence for the envelope $\varphi(x,\omega)$ is achieved by applying some low-pass filter. The more smoothing that is done to $\varphi(x,\omega)$ the narrower the bandwidth of the corresponding band pass filter. Consider Gaussian blurring as a process described by the heat equation,

$$\frac{\partial \varphi(x,\omega)}{\partial t} = \frac{\partial^2 \varphi(x,\omega)}{\partial x^2} \qquad (4.14)$$

and let the characteristic wave form be a sinusoid, $\psi(x,\omega) = e^{\iota\omega x}$. There is an equivalent process which modifies $I(x)$ with respect to a characteristic frequency ω in order to produce a smoother envelope for that frequency:

$$\frac{\partial F(x,\omega,t)}{\partial t} = \frac{\partial^2 F(x,\omega,t)}{\partial x^2} - 2\iota\omega\frac{\partial I(x)}{\partial x} - \omega^2 F(x,\omega,t), \qquad (4.15)$$

where $F(x,\omega,0) = I(x)$. This is a diffusion equation with fundamental solutions of the form

$$G(x,\omega,s) = \frac{1}{\sqrt{2\pi}s}e^{\frac{x^2}{2s^2}-\iota\omega x}. \qquad (4.16)$$

These are, once again, Gabor blobs. This process describes the "flow" of these Gabor blobs as they spread out in space. As the process evolves, it produces a progressively more narrow bandpass filter. The steady state solution of this equation (given the appropriate boundary conditions) is a sine and cosine wave pair.

8.3. Variable-conductance diffusion

Equation 4.15 describes a means of characterizing local frequency information. It specifies a scale-space for *each* characteristic frequency ω and provides a means of "smoothing" the function with respect to a particular frequency. The difficulty with this smoothing, as with Gaussian blurring, is the inherent trade-off between the size of the fundamental solutions in the spatial and Fourier domains. This trade-off has the same mathematical form as the uncertainty principle which dictates the fundamental limitations in quantum mechanical observations. We avoid this trade-off by employing the same principles used to construct the anisotropic diffusion process. The strategy is to introduce a variable-conductance term that limits the spread of the fundamental solution. Notice that Eq. (4.15) is the result of applying two first-order operators.

$$
\begin{aligned}
\frac{\partial F(x,\omega,t)}{\partial t} &= \frac{\partial^2 F(x,\omega,t)}{\partial x^2} - 2\imath\omega \frac{\partial F(x,\omega,t)}{\partial x} - \omega^2 F(x,\omega,t) \\
&= \left(\frac{\partial}{\partial x} - \imath\omega\right) \cdot \left(\frac{\partial}{\partial x} - \imath\omega\right) F(x,\omega,t) \qquad (4.17)
\end{aligned}
$$

Equation 4.17 has the same form, $D \cdot DF = \partial F/\partial t$, as the heat equation, where D is a linear differential operator. This similarity in form suggests a family of differential operators, analogous to the gradient operator, each sensitive to its associated characteristic frequency, ω. This family of operators $(\partial/\partial x - \imath\omega)$ operators has a somewhat intuitive interpretation when analyzed in the Fourier domain. Consider the Fourier representation of a directional derivative in a two-dimensional image space – it is a plane passing through the origin. The gradient operator is a set of planes (two planes if $I \subset \mathbb{R}^2$). Linear combinations of these planes produce a single plane – a directional derivative. The frequency-tuned derivative operators have the same interpretation except that they have zero response not at the origin, but at the characteristic frequency ω. The operator $(\partial/\partial x - \imath\omega)$ is a gradient that has undergone a translation (by an amount ω) in the Fourier domain as shown in figure 4.18.

The expressions (4.11-4.17) extend easily to image spaces of arbitrary dimension, and for this we use the vector notation to represent the space and frequency variables. Assuming an n-dimensional image, then $\vec{x}, \vec{\omega} \in \mathbb{R}^n$.

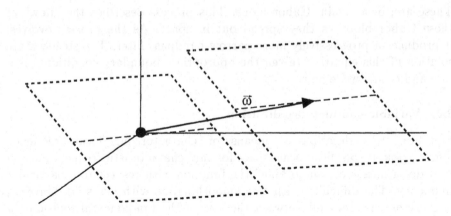

Figure 4.18. The operator $\nabla_{\vec{\omega}}$ is equivalent to an ordinary gradient ∇ that is shifted in the Fourier domain by the vector $\vec{\omega}$

Denote this "shifted" operator as $\nabla_{\vec{\omega}}$, and the corresponding Laplacian as $\Delta_{\vec{\omega}}$.

$$\nabla_{\vec{\omega}} = (\nabla - \imath\vec{\omega}) = e^{\imath\vec{\omega}\cdot\vec{x}}\nabla e^{-\imath\vec{\omega}\cdot\vec{x}}$$
$$\Delta_{\vec{\omega}} = (\nabla - \imath\vec{\omega}) \cdot (\nabla - \imath\vec{\omega}) = e^{\imath\vec{\omega}\cdot\vec{x}}\nabla \cdot \nabla e^{-\imath\vec{\omega}\cdot\vec{x}}$$

Recall that the gradient ∇ is the set of first-order derivatives in the form of a vector.

$$\nabla \equiv \frac{\partial}{\partial x_1} \cdots \frac{\partial}{\partial x_n}$$

Notice that the conventional gradient and Laplacian are a special cases of the operators $\nabla_{\vec{\omega}}$ and $\Delta_{\vec{\omega}}$ with $\vec{\omega} = 0$. Thus the diffusion Eq. (4.17) becomes

$$\frac{\partial F}{\partial t} = \nabla_{\vec{\omega}} \cdot \nabla_{\vec{\omega}} F \tag{4.18}$$

The form of (4.18) allows one to see more clearly the means of incorporating variable conductance into this frequency-sensitive diffusion process. Recall the physical analogy that the diffusion equation describes the "flow" of a substance for which F is a density function. The first derivative $\nabla_{\vec{\omega}} F$ is the difference in concentration of the substance and thereby the "driving force" of the quantity F from one location to another (infinitesimally close). The Laplacian $\nabla_{\vec{\omega}} \cdot \nabla_{\vec{\omega}} F(\vec{x})$ is the net flow into a location $\vec{x}$. Modulate the flow of F with a conductance term as follows:

$$\frac{\partial F}{\partial t} = \nabla_{\vec{\omega}} \cdot g(\vec{x})\nabla_{\vec{\omega}} F. \tag{4.19}$$

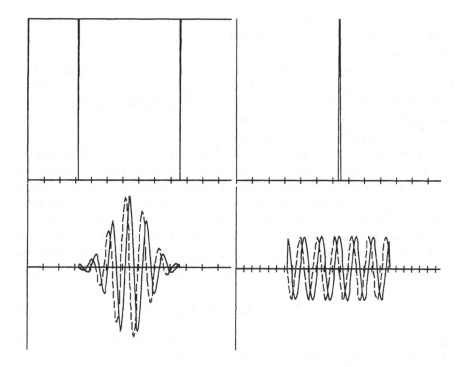

Figure 4.19. From top-left: A one-dimensional nonconducting box is constructed (a) and a single pulse (b) spreads at the characteristic frequency (c) finally filling the box (d).

Experiments in one dimension show that this equation behaves as expected. Solutions to (4.19) take the form of waves that "flow" but are impeded by regions of low conductivity. Figure 4.19 shows the evolution of a single pulse in a nonconducting "box". The pulse gradually spreads to fill the box forming a pair of waves whose combined magnitude is virtually constant.

Equation (4.19) provides the framework for a diffusion equation that is sensitive to frequency and also incorporates a conductance term which is sensitive to local structure.

$$\frac{\partial F}{\partial t} = \nabla_{\vec{\omega}} \cdot g(\mathcal{D}F)\nabla_{\vec{\omega}}F \qquad (4.20)$$

The dissimilarity $\mathcal{D}F$ depends on spatial derivatives of $F(\vec{x}, \vec{\omega}, t)$. The gradient measurement for the conductance term must incorporate some finite isotropic scale. The derivatives in the conductance term are measured with a Gaussian neighborhood function. This isotropic scale must reflect the same frequency sensitivity as the diffusion process, and gradient measurements should be conducted with kernels composed of Gabor blobs as in Eq. (4.16). This scale term slowly decreases as the process progresses in order to provide a large to small scale process that "narrows in" on relevant boundaries.

Experiments in one dimension, with a single nonzero frequency, confirm the desired behavior of (4.20). This process produces distinct regions with sinusoidal waves of nearly constant amplitude. The regions and the amplitude of the waves within regions reflect the presence of the characteristic frequency in the original data.

There is no need to implement this mixed-order partial differential equation as it is shown above, because the operator $\nabla_{\vec{\omega}}$ corresponds to a frequency shift followed by the application of the gradient operator ∇. To implement this alternative strategy, first shift the initial conditions (the luminance profile) in the frequency domain and then use the operator ∇ instead of $\nabla_{\vec{\omega}}$. Shifting the initial conditions and performing a variable-conductance diffusion is the same as blurring the envelope $\varphi(\vec{x}, \vec{\omega}, t)$. Assuming that the dissimilarity in (4.20) is some function $\mathcal{D}F \equiv \|\phi(F)\nabla_{\vec{\omega}}F\|$, then Eq. (4.20) becomes

$$\frac{\partial \varphi}{\partial t} = \nabla \cdot g(\|\phi(\varphi)e^{-i\vec{\omega}\cdot\vec{x}}\nabla\varphi\|)\nabla_{\vec{\omega}}\varphi. \qquad (4.21)$$

This form is easier than (4.20) to implement on a discrete grid, because it involves fewer derivatives and enables the conductance measurements be made with a Gaussian kernel instead of the Gabor pair.

8.4. Patchwise Fourier decomposition

In this section we make some simplifying assumptions about the conductance function and show that Eq. (4.21) implements a Fourier decomposition within patches. Consider an image space I with luminance function $I(\vec{x})$. Assume that I contains a patch P with a boundary ∂P as shown in figure 4.20. For the purposes of this example, assume that the conductance function is constant over time as in (4.19), identically 1 over the interior of P, and 0 along the boundary ∂P.

Consider the steady state solutions of Eq. (4.21) within the patch P. Notice that the nonconducting border around P recasts the problem of finding the steady state of (4.21) as a boundary value problem in the irregular patch with Neumann boundary conditions (adiabatic). This insures two properties in the solutions:

1. By the divergence theorem the integral

$$\int_P F(\vec{x}, \vec{\omega}, t)d\vec{x}$$

remains constant over time.

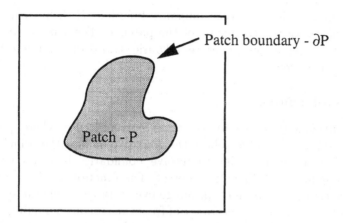

Figure 4.20. A patch with a nonconducting boundary is used to show that systems of frequency-sensitive variable-conductance diffusion equations are capable of computing a patchwise Fourier analysis.

2. The steady state solution is a boundary value problem (Laplacian) with constant boundary conditions, and solutions are constant over the patch.

Define a new function

$$f_P(\vec{x}) = \begin{cases} = I(\vec{x}) & \text{if } \vec{x} \in P \\ = 0 & \text{if } \vec{x} \notin P \end{cases}$$

and express the steady state solution in closed form.

$$A_P = \int_P \mathrm{d}\vec{x}$$

$$\lim_{t \to \infty} \varphi(\vec{x}, \vec{\omega}, t) = \frac{1}{A_P} \int_P \varphi(\vec{x}, \vec{\omega}, 0) \mathrm{d}\vec{x}$$

$$= \frac{1}{A_P} \int_P I(\vec{x}) e^{-\imath \vec{\omega} \cdot \vec{x}} \mathrm{d}\vec{x}$$

$$= \frac{1}{A_P} \int_I f_P(\vec{x}) e^{-\imath \vec{\omega} \cdot \vec{x}} \mathrm{d}\vec{x}$$

$$= \frac{1}{A_P} \mathcal{F} f_P(\vec{x}),$$

where $\mathcal{F}$ is the Fourier transform. The steady state solution within the patch is 1 over the area of the patch times the value of the Fourier transform of the luminance function defined only on the patch. *To the extent that patches are distinct and well defined, this process implements a patchwise Fourier transform.* The steady-state values within patches reflect not only the frequency characteristics of the luminance function, $I(\vec{x})$, but also the frequency characteristics of the shape of the patch P. For a perfectly flat luminance function this process generates a Fourier transform that describes only the *shapes* of patches.

8.5. Multi-valued diffusion

Although the argument above refers to an extreme case where patch boundaries are detectable and absolute, the goal is to employ the principles of multi-valued diffusion in order to resolve boundaries in cases where patches are ambiguous and difficult to detect. The function $\varphi : \mathbb{R}^n \times \mathbb{R}^n \times \mathbb{R}^+ \mapsto \mathbb{C}$ describes a scale-space of an image over a range of characteristic frequencies. Instead of a continuous range of frequencies, consider a finite set of "sampled" frequencies. Let $\vec{\Omega} = \vec{\omega}_1, \ldots, \vec{\omega}_m$, and let $\vec{F}_{\vec{\omega}_i}(\vec{x}, t) = \varphi(\vec{\omega}_i, \vec{x}, t)$. Then $\vec{F}_{\vec{\Omega}}(\vec{x}, t) = F_{\vec{\omega}_1}(\vec{x}, t), \ldots, F_{\vec{\omega}_m}(\vec{x}, t)$ is a multi-valued function $\vec{F}_{\vec{\Omega}} :$ $\mathbb{R}^n \times R^+ \mapsto \mathbb{C}^m$. Now treat this new function $\vec{F}_{\vec{\Omega}}$ with the multi-valued diffusion equation from (4.1),

$$\frac{\partial \vec{F}_{\vec{\Omega}}}{\partial t} = \nabla \cdot g(\mathcal{D}(G(s) \otimes \vec{F}_{\vec{\Omega}}))\nabla \vec{F}_{\vec{\Omega}}, \qquad (4.22)$$

so that

$$\frac{\partial F_{\vec{\omega}_1}}{\partial t} = \nabla \cdot g(\mathcal{D}_s \vec{F}_{\vec{\Omega}})\nabla F_{\vec{\omega}_1}$$

$$\vdots$$

$$\frac{\partial F_{\vec{\omega}_m}}{\partial t} = \nabla \cdot g(\mathcal{D}_s \vec{F}_{\vec{\Omega}})\nabla F_{\vec{\omega}_m}$$

where

$$\vec{F}_{\vec{\omega}_i}(\vec{x}, 0) = I(\vec{x})e^{\imath \vec{\omega}_i \cdot \vec{x}}.$$

Equation 4.22 is a system of equations in which each equation represents a distinct characteristic frequency. With perfectly defined boundaries, as described in Sect. 8.4, each equation computes a patchwise Fourier integral for a single frequency. This collection of equations produces a frequency-wise sampling of the Fourier transform within each patch.

Equation 4.23 is also vector-valued diffusion equation operating on a set of features that are derived from the local shape properties of an input image $I(\vec{x})$; *it is a special case of the geometry-limited diffusion framework described in previous sections* and it follows precisely the strategy depicted in figure 4.6. The process proceeds in three steps: the computation of features, the vector-valued diffusion, and the measurement and determination of patch boundaries. Unlike the examples in previous sections, the features $\vec{F}_{\vec{\Omega}}$ are not derived from I by a filter (i.e. convolution); they are obtained by a multiplication with characteristic wave forms of various frequencies. The values of the features depend on the absolute location in the image; they are not invariant to changes in the position of the origin in the image space. However, invariance is achieved, as in earlier sections, by ensuring that the dissimilarity operator is invariant to shifts in the image space coordinates. This vector-valued approach for segmenting images bears some similarity to conventional means for characterizing textures in images [377]. It utilizes a frequency-based representation of local structure and incorporates filters that are local in both frequency and space. However, it differs from conventional approaches in some important aspects. The narrow bandwidth in the frequency domain is obtained by blurring the *envelope* of an intensity function. The result is an incremental widening of these filters in the spatial domain. This blurring, however, is both nonstationary and nonlinear. The "spreading" of these frequency-tuned filters is limited by areas of low conductance. The result is a set of measurements that are *adapted* to local structure in images. Structure at different scales is combined by utilizing isotropic scale in the dissimilarity measurements and decreasing this scale as the process evolves.

8.6. Dissimilarity

The challenge is to construct a dissimilarity measure that describes the boundaries of textured patches. Patches in images are based not only on a single characteristic frequency but on the relationships between a number of frequencies. There are myriad options for this transformation ϕ associated with the dissimilarity measure. We choose a simple but effective dissimilarity operator based on the following principle: *Textural boundaries are distinguished by the relative changes in local spectral components.* That is, boundaries are affected not by the overall quantity of spectral energy but in the relative spectral composition at each point in the image.

Consider the range of $\vec{F}_{\vec{\Omega}}$, which is a $2m$-dimensional feature space D. The points in this feature space can be represented in hyper-spherical coordinates. Each point in the space D can be treated as a vector with both a magnitude and a direction. The principle stated above requires that the dissimilarity measure should reflect changes in the direction of vector $\vec{F}_{\vec{\Omega}}$, not its magnitude. In order to achieve a dissimilarity measure which does not depend on the magnitude of $\vec{F}_{\vec{\Omega}}$, use a set of normalized intermediate features $\vec{G}(\vec{F}_{\vec{\Omega}})$. This dissimilarity can also be expressed as a linear transformation $\phi(\vec{F}_{\vec{\Omega}})$ applied to Jacobian of $\vec{F}_{\vec{\Omega}}$, so that $\mathcal{D}\vec{F}_{\vec{\Omega}} =$

$\|\phi(\vec{F}_{\vec{\Omega}})D\vec{F}_{\vec{\Omega}}\|$. However, it is more convenient to write it as

$$\mathcal{D}\vec{F}_{\vec{\Omega}} = \left\|\nabla\vec{G}(\vec{F}_{\vec{\Omega}})\right\| = \left\|\nabla\frac{\vec{F}_{\vec{\Omega}}}{|\vec{F}_{\vec{\Omega}}|}\right\|, \tag{4.23}$$

where

$$|\vec{F}_{\vec{\Omega}}| = \left(\sum_{i=1}^{m} F_{\vec{\omega}_i}F_{\vec{\omega}_i^*}\right)^{\frac{1}{2}},$$

and F^* is the complex conjugate of F. This dissimilarity operator has the property that it is insensitive to linear intensity transformations in the original image and thus it does not favor areas in an image that have high contrast. Boundaries in low contrast images are detected on the basis of their frequency composition and are not penalized for having low energy across all spectral features. Lee et. al. [207] use a similar measure for calculating relative changes in frequency decomposition.

8.7. Results

The spectral features $F_{\vec{\omega}_1}, \ldots, F_{\vec{\omega}_m}$ are *samples* taken in the frequency domain that encode local frequency information in an image. If we insist on a finite number of equations in the system of diffusion equations, then choosing the number and placement of these samples could be difficult. Choosing too few samples results in processes that are unable to distinguish subtle textures, and yet each additional feature introduces a significant computational burden. For the examples in this section we have chosen a frequency sampling strategy that is a compromise between sample density and computation time. We sample the frequency domain on a rectilinear grid that includes the half plane minus the origin (which is the zero frequency), i.e., we explicitly ignore local intensity information, i.e. $\vec{\omega} = 0$.

The figures 4.21 to 4.23 show the results of this process using 40 features. In each example the process runs 100 iterations. The computational burden is addressed by utilizing the parallel nature of this diffusion process and using a highly parallel single-instruction-multiple-datastream (SIMD) computer. Because the diffusion requires only nearest neighbor differences, communication costs are low, making this application extremely efficient on this type of machine. These pictures show the original images and the results of computing a threshold on the dissimilarity measure, $\mathcal{D}\vec{F}_{\vec{\Omega}}$, after the final iteration. The dissimilarity measure controls the conductance throughout the process. A threshold of the dissimilarity measure provides a very simple way of locating boundaries with this technique.

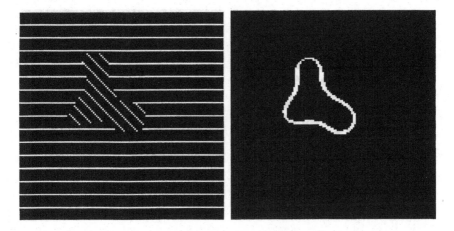

Figure 4.21. Left: Original image. Right: Dissimilarity after processing. A patch in an image is distinguished from the background by the orientation of the stripes.

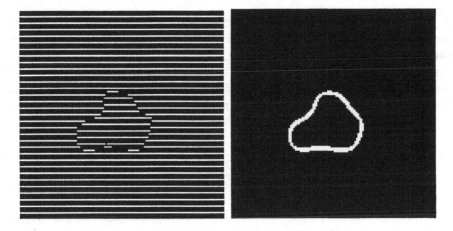

Figure 4.22. Left: Original image. Right: Dissimilarity after processing. A patch in an image is distinguished from the background by the relative phase of the stripes.

9. Conclusions

The vector-valued diffusion framework offers a powerful generalization of the anisotropic diffusion processes presented in previous chapters. It has applications to vector-valued images which result from imaging devices that produce multiple measurements as well as applications to sets of features that are derived from scalar input images. The dissimilarity measure is a metric that quantifies distance in the feature space and determines the kinds of structures that are preserved under the diffusion process.

The application of vector-valued diffusion to geometric descriptions offers

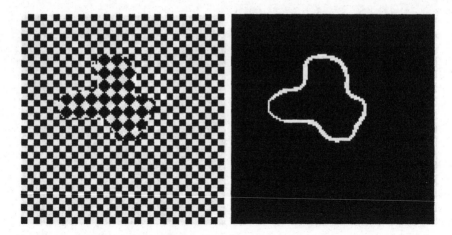

Figure 4.23.   Left: Original image. Right: Dissimilarity after processing. Textures are distinguished on the basis of orientation and size.

a means of characterizing higher-order structure in scalar images in the presence of noise. For geometric descriptions that consist of image derivatives, an invariant dissimilarity measure (orthogonal group) ensures that the process produces invariant results. There appears to be a wide range of options for geometric descriptions. One alternative is to use a local frequency decomposition in order distinguish patches on the basis of texture.

BAYESIAN RATIONALE FOR THE VARIATIONAL FORMULATION

David Mumford

Harvard University
Department of Mathematics
Cambridge, MA 02138, USA

1. Introduction

One of the primary goals of low-level vision is to segment the domain D of an image I into the parts D_i on which distinct surface patches, belonging to distinct objects in the scene, are visible. Although this sometimes requires high level knowledge about the shape and surface appearance of various classes of objects, there are many low-level clues about the appearance of the individual surface patches and the boundaries between them. For example, the surface patches usually have characteristic albedo patterns, textures, on them, and these textures often change sharply as you cross a boundary between two patches. Therefore, one approach to the segmentation problem has been to try to merge all the low-level clues for splitting and merging different parts of the domain D and come up with probability measures $p(\{D_i\})$ of how likely a given segmentation $\{D_i\}$ is on the basis of all available low-level information, and what is the most likely segmentation. Alternately, one sets $E(\{D_i\}) = -\log(p(\{D_i\}))$, which one calls the 'energy' of the segmentation, and seeks the segmentation with the minimum energy. In general, these models have two parts: a *prior model* of possible scene segmentations, possibly including variables to describe other scene structures that are relevant (e.g. depth relationships), and a *data model* of what images are consistent with this prior model of the scene. If we write w for the variables used to describe the scene, e.g. the subsets D_i or the set of all their boundary points Γ, then the prior model is some probability space (Ω_w, p), where Ω_w is the set of all possible values of w. The model is specified by giving the probability distribution $p(w)$ on all these values. The data model is a larger probability space $(\Omega_{w,d}, p)$, where $\Omega_{w,d}$ is the set of all possible values of w and of all possible observed images I. This model is completed by giving the conditional probabilities $p(I|w)$ of any image I given the scene variables w, resulting in the joint probability distribution:

$$p(I, w) = p(I|w) \cdot p(w)$$

The discussion above assumes implicitly that the spaces Ω_w and $\Omega_{w,d}$ are finite, although huge (e.g. the set of byte valued images I on a grid of size 256x256 has cardinality about 10^{150000}). In many situations, it is more convenient to consider images as real-valued functions of continuous variables, and to consider segmentations as sets of suitable measurable subsets of D: then more complicated probability spaces are needed, often using distributions as well as actual functions. We will not worry about this, as the expressions for the probability densities we use all look like $p = Z^{-1} \cdot p'$, where p' has a simple limiting expression as the exponential of an integral and Z is a normalizing constant introduced to make p into a probability measure. The only problem in the continuous limit is that $Z \to \infty$, hence $p \to 0$. But we can work with p' in the continuous limit, knowing that in finite approximations, the normalizing constant Z is finite. In terms of energy, this means that the E we work with should have an infinite constant added to it, which, in finite approximations, is finite.

These models are always used in conjunction with Bayes's theorem, which factors the joint probability distribution the other way:

$$
\begin{aligned}
p(I, w) &= p(w|I) \cdot p(I) \\
\text{hence} \quad p(w|I) &= \frac{p(I|w) \cdot p(w)}{p(I)} \\
&\propto p(I|w) \cdot p(w)
\end{aligned}
$$

The probability of w *given the data* I is called the *posterior probability* of w, and this what we want to calculate. In terms of energy, we can write:

$$
\begin{aligned}
E(w) &= -\log(p(w, I)) \\
&= -\log(p(I|w)) - \log(p(w)) \\
&= E_d(I, w) + E_p(w)
\end{aligned}
$$

where the goal is now to minimize $E(w)$. The log of the prior term, E_p, is sometimes called the 'regularizer' because it was initially conceived of as a way to make the variational problem of minimizing E_d well-posed. In what follows, it will play a much more central role of measuring how reasonable each scene model is, lower values being the more common scenes and higher values the less common ones.

What we want to do in the rest of this chapter is to present four such energy models for image analysis of increasing sophistication, which attempt to capture more and more of the subtleties of actual world scenes. It will be clear that none of these models captures all the important scene variables and that this is just the beginning of the exploration of probabilistic models for low-level vision. For instance, none of these models include explicit illumination variables, which are often essential to disambiguating scene structure. Also, we have not included any models of this type which build on *multiple* images, i.e. stereo pairs or motion sequences. Multiple images without a doubt make it infinitely easier to properly segment images (how many animals understand the content of photographs?). Notable models of this type are due to Belhumeur [29] and to Weber and Malik [375]. It is my belief that a robust solution to the general low-level vision problem can be found using this approach. The main obstacle is to find more effective and faster ways of estimating the w minimizing $E(w)$ than those presently available.

Although this energy approach seems on the surface to be totally different from the non-linear PDE's investigated in the rest of this book, they are in fact closely linked to each other. Some of these links can be seen in the contributions by Mitter and Richardson and by Nordstrom to this book. Others can be found in Geiger and Yuille's work [130].

2. Four Probabilistic Models

2.1. The Ising model

This is a model which comes directly from statistical physics, where it is used to describe a two-dimensional crystal of iron atoms subject to an external magnetic field. It was a very influential model in statistical physics, because it was the first mathematical model which was rigorously proven to model phase transitions (for discrete but infinite domains D) [283]. In vision, it models images which are made up of a set of white blobs against a dark background (or vice versa) and where one seeks to describe the white blobs by an auxiliary binary image, or, equivalently, by a subset $S \subset D$.

The prior model is very simple: we ask that S consists of a small number of compact blobs, i.e. that the length of the perimeter of S is as small as possible. More precisely, define ∂S as the boundary of S: in the discrete case, this is the set of pairs of horizontally or vertically adjacent pixels (α, β), one of which is in S and the other of which is not. In the continuous case, this is set of points (x, y) which are in the closure of S and $D - S$. Define $|\partial S|$ as the cardinality of ∂S in the discrete case and as the length (= 1-dimensional Hausdorff measure) of S in the continuous case. Then the prior model is determined by setting $E_p(S) = \nu|\partial S|$: this means that the

shorter the perimeter of S, the more likely S is to be the model of the scene. Let χ_S be the characteristic function of S, and let I be the observed image. The blobs are supposed to be characterized by being more or less bright compared to the background. However, we assume that the intensity of neither the blobs nor the background is uniform, but that it fluctuates randomly and is corrupted by many kinds of noise. If the image were simpler, we could recover S by simply thresholding I. However, if there is a substantial amount of noise present, no matter what threshold is chosen, we may find not S but S with lots of extra specks and hairs, minus small holes and cracks. This is exactly what the model seeks to correct. Assume for simplicity that on the blobs S the image I tends to have values bigger than 0.5 and that on the background, the image has values less than 0.5. (The data model can be modified for other thresholds: the original Ising model used 0). Then the data model is given by $E_d(I, S) = \mu \iint_D (I - \chi_S)^2 d\vec{x}$. This is equivalent to assuming that $I = \chi_S + n$, where n is white noise.
We may summarize the model by:

$$
\begin{aligned}
w &= S, S \subset D \\
E_p &= \nu |\partial S| \\
E_d &= \mu \int \int_D (I - \chi_S)^2 d\vec{x}, \qquad \chi_S = \text{char.function of } S
\end{aligned}
$$

In figure 5.1, we have illustrated this model by an image I consisting of a scene with a cow, tree and foreground in deep shade against a background of the sky and more distant parts of the scene. The figure shows both the original scene and the binary image given by the Ising model optimal S (for suitable ν, μ).

2.2. The Cartoon Model

The cartoon model is the model which has been most used in vision. It was invented in the discrete case, independently by S. and D. Geman in their influential paper [132] and by A. Blake and A. Zisserman [34]. J.Shah and I then investigated the corresponding continuous model in [264]. In fact, the variational problem of minimizing energy functionals of this continuous kind had also been independently invented by De Giorgi and his school at about the same time in modeling materials with 2 phases and a free interface [83].

Figure 5.1.   A scene with a cow and trees, and its Ising model results.

Instead of assuming that there exists a binary segmentation of the image into contrasting light and dark regions, in this model we assume that the real world scene consists of a set of *shaded* regions within which the intensity changes slowly, but across the boundaries between them, the intensity changes, in general abruptly. Thus what we want to infer is not the set S of light (or dark) foreground regions, but a *cartoon* consisting of a simplified noiseless version J of I. The cartoon has a curve Γ of discontinuities, but everywhere else is assumed to have small gradient $\|\nabla J\|$. The prior model can be built up by starting with the Ising model prior of Γ: $E_p^{(1)}(\Gamma) = \nu|\Gamma|$. We then put a prior on J by asking that its gradient be small: $E_p^{(2)}(J,\Gamma) = \iint_{D-\Gamma} \psi(\|\nabla J\|)d\vec{x}$, where ψ is some convex even function. The standard choice is $\psi(x) = x^2$, but $\psi(x) = |x|$ is also very interesting and should be more investigated. The full prior is $E_p = E_p^{(1)} + E_p^{(2)}$. The data model is again essentially the same as for the Ising model, except that instead of assuming $I = \chi_S + n$, where n is white noise, we assume $I = J + n$. This gives $E_d = \iint_D (I - J)^2 d\vec{x}$. We may summarize this model by saying that we seek to approximate an arbitrary function I by a piecewise smooth function J so that three things are kept as small as possible: i) the difference between I and J, ii) the gradient of J where it is smooth and iii) the length of the curve Γ where J has discontinuities.

In the case where D is discrete, it is always obvious that any energy functional like our E has a minimum, because we are minimizing over a finite set of possible Γ and the functional is continuous in the values of J, and goes to infinity if any of the values of J goes to infinity (i.e. it is 'proper' as a function of J). However, if D is continuous, it is not at all clear that E has a minimum in any sense. This is referred to as asking whether the variational problem associated to E is 'well-posed' or not. This was an open question in [264]. A so-called 'weak solution' was given by Ambrosio [10] where the cartoon J was allowed to be a very nasty sort of function, however: a so-called 'Special Bounded Variation' function. A first step in showing this weak solution was not horrendous consisted in proving that Γ was closed, see [83]. Shah and I conjecture that at minima of E, the curve Γ consists in a finite set of C^1 arcs (possibly with cusps at the ends of branches which terminate) but this is unproven: a survey of the theory of this functional is given in the chapter by Leaci and Solimini in this book. The computer scientist might be tempted to say: what do I care about the continuous case and the subtle estimates mathematicians require to prove that problems like min E are well-posed. The remarkable thing is that the estimates used in establishing the *existence* of well behaved solutions are exactly the same as the estimates which enable the engineer who wants to use the discrete model to be sure that his minima *behave predictably*: that this finite model doesn't produce artifacts or perceptually meaningless zigs and zags dependent on small details of how the problem is discretized.

In the discrete case, the functional E can be re-written in a suggestive way. The simplest form of the term $E_p^{(2)}$ in the discrete case is

$$E_p^{(2)} = \sum_{\text{adj. pixels } \alpha, \beta} \psi(J(\alpha) - J(\beta)).$$

Then it is easily seen that at a minimum of $E(J, \Gamma)$, Γ cuts an adjacent pair of pixels α, β if and only if $\psi(J(\alpha) - J(\beta)) > \nu$. Thus if we define

$$\psi'(x) = \min(\psi(x), \nu), \tag{5.1}$$

$$E_p'^{(2)}(J) = \sum_{\text{adj.pixels}} \psi'(J(\alpha) - J(\beta)) \quad \text{and} \tag{5.2}$$

$$E'(J) = E_d(J) + E_p'^{(2)}(J) \tag{5.3}$$

we find that

$$\min_{\Gamma} E(J, \Gamma) = E'(J).$$

This form enabled Blake and Zisserman [34] to analyze many properties of E, and to approximate E' by a third functional in which ψ' was replaced by a smooth ψ'', which, even though not convex itself, made E'' convex! They use this as a basis for a continuation method of getting good *local* minima of the original functional E.

To carry over this approach in the continuous case, however, requires that we modify the variable Γ, replacing it by a 'line process' $\ell(x, y)$ which is a smooth function with values in $[0, 1]$, mostly zero but climbing to one along Γ. One then replaces $E(J, \Gamma)$ with

$$E(J, \ell) = E_d(J) + \iint_D (1 - \ell)\psi(\|\nabla J\|)d\vec{x} + \iint_D \phi_c(\ell)d\vec{x}$$

for suitable ϕ_c. Such ϕ_c are described in the chapter by Mitter and Richardson in this book, where they also give theorems on when these functionals approximate the original $E(J, \Gamma)$. This approach seems very useful because it offers a way of taming the wildness inherent in having Γ itself as a variable.

In a nutshell, here is the cartoon model:

$$
\begin{aligned}
w &= (J(\vec{x}), \Gamma) \\
E_p &= \int\int_{D-\Gamma} \psi(\|\nabla J\|)d\vec{x} + \nu|\Gamma| \\
E_d &= \int\int_D (I - J)^2 d\vec{x}
\end{aligned}
$$

In figure 5.2, we give an illustration of this model applied to a close-up of Marilyn Monroe's eye. The original eye is shown on the left, then the cartoon J and finally the contours Γ.

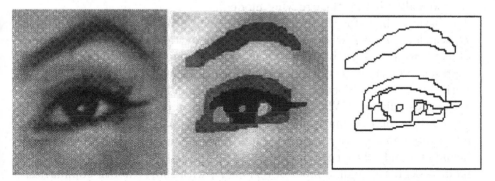

Figure 5.2. An image of the eye and its final cartoon and boundary, produced by graduated nonconvexity algorithm.

2.3. The Theater Wing Model

Although the model in the previous section is attractive and interesting from a mathematical point of view, it unfortunately is very crude as a model of image segmentations. One of its problems is that where 3 domains D_i meet at a point P, so that the boundary Γ has a singular point with 3 branches meeting at P, the branches will meet at $120°$ angles. This is because locally the effect of I is very weak and Γ behaves like soap bubbles do when 3 sheets meet: they form $120°$ angles. In images, on the contrary, 3 domains meeting usually means that the edge of a foreground object cuts across the edge of a more distant object. This gives instead a "T-junction" on Γ: Γ consists locally of a smooth curve Γ_1 through P with a second smooth branch Γ_2 ending at P. These singularities are not only typical of real world images, but they are very powerful psychological clues to depth relations, as the Gestalt school of psychology and especially Kanisza [172] discovered.

The problem is that we have not included even qualitative depth information in our model variables $\{w\}$ and occlusion edges in the real world are inherently asymmetrical, having a nearer and a farther side. (Nakayama refers to this by saying that an edge 'belongs' to one of its sides.) In the previous model, no region is considered foreground nor any background. Working with Mark Nitzberg [270], we sought a model which was a reasonable first step in modeling sets of regions occluding each other. This model goes back to the Ising model assumption that the individual regions have more or less uniform brightness, but it now assumes they are ordered in depth, and one region can vanish behind another, only to reappear elsewhere. The regions cannot change their depth relations however, interweaving like wicker chairs, nor can a region circle round and occlude itself (like your palm when you bend your thumb over it). This is why we called it the 'theatre wing model'.

The basic variable for this model is a sequence of regions $D_1, D_2, \cdots, D_k$ in D, which is *ordered*, where D_i represents the parts of the domain D where object i would be visible *if all closer objects were removed*. The ordering represents depth: object 1 is nearest, object 2 is behind it, object 3 behind both, etc., while object k, called the background and assumed equal to D, is most distant. What is the prior model here? Since we assume all singularities of Γ come from T-junctions where one occluding boundary interrupts another, we can now assume all D_i have smooth boundaries and include not only the length but the curvature of ∂D_i into the prior. Thus our model asks for a small number of regions D_i with short smooth boundaries. In some cases, it turned out that we also needed to ask that their areas not be too big either, so the final prior we chose was $E_p = \sum(\int_{\partial D_i} \psi(\kappa_i)ds + \epsilon \text{Area}(D_i))$. Here a typical choice is $\psi(x) = a + bx^2$.

The data model assumes that the intensity of each region is more or less uniform, but that its mean intensity may be anything. This leads us to $E_d = \sum \iint_{D_i'}(I - m_i)^2$, where D_i' is the *visible* part of D_i, i.e. D_i minus the parts $D_i \cap D_j$ occluded by nearer parts $j < i$, and where m_i is the mean value of I on this visible part (we have no way of knowing what is the brightness of the occluded parts of D_i!). In summary:

$$w = \{D_1, D_2, \cdots, D_k\} = D,$$
$$E_p = \sum \int_{\partial D_i} \psi(\kappa_i) ds + \epsilon \sum \text{Area}(D_i), \qquad \kappa_i = \text{curvature}(\partial D_i)$$
$$E_d = \sum \int \int_{D_i'} (I - \text{mean}_{D_i'}(I))^2 d\vec{x}, \qquad D_i' = D_i - \cup_{j<i} D_j$$

In figure 5.3 we show an example of this model. The image I depicts a beer bottle, an orange and a potato occluding each other. The figure shows how the theory correctly reconstructs their occlusion relations and gives its best shot at guessing how their contours continue behind each other. It might be thought that this kind of wild reconstruction of occluded contours is irrelevant to the interpretation of the scene. But curiously, extensive experiments by the Gestalt school of psychologists and more recently by Nakayama and his collaborators have shown that people very frequently make exactly such reconstructions, sometimes choosing one of several reasonable 'amodal' contours for seemingly unaccountable reasons. If people do it, it may not be absurd for computers to do it too.

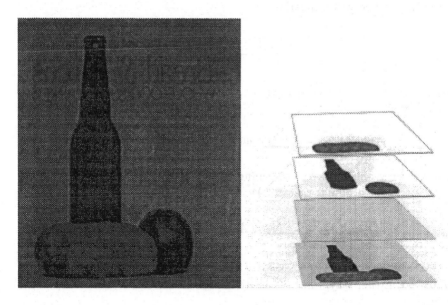

Figure 5.3. Still Life and its 2.1-D sketch.

2.4. The Spectrogram Model

So far our models have had one glaring omission: they have assumed that all visible surface patches have intensities which are slowly varying plus noise. In fact, this is wrong for a majority of surfaces: much more often, surfaces have a natural texture in which their intensity has strong systematic variation. To segment in real world scenes, we need to model not merely nearly uniform intensity but nearly uniform or slowly varying texture. There are many puzzles which are still unsolved about what constitutes a coherent, perceptually distinguishable texture, and we don't want to get involved with this. As in the previous section, working with Tai Sing Lee [207], we sought to model the simplest types of texture segmentation. Our approach here is to assume that each texture has, locally, a *spectral signature* which characterizes it.

Thus we start by taking a local Fourier analysis of the image I. There are two ways of doing this. The more traditional way is to form the windowed spatial (2-dimensional) Fourier transform of $I(x, y)$:

$$\mathcal{F}(I)(\vec{x_0}, \vec{\xi}) = \iint I(\vec{x} + \vec{x_0}) w(\vec{x}) e^{2\pi i \vec{x} \cdot \vec{\xi}} d\vec{x}$$

where w is the window function. If we note that we may rewrite this using the scaling and rotation matrix

$$A_\xi = \begin{pmatrix} \xi_1 & \xi_2 \\ -\xi_2 & \xi_1 \end{pmatrix}$$

noting that $A_\xi \begin{pmatrix} 1 \\ 0 \end{pmatrix} = \vec{\xi}$, we get:

$$\mathcal{F}(I)(\vec{x_0}, \vec{\xi}) = \iint I(\vec{x}) w(\vec{x} - \vec{x_0}) e^{2\pi i A_\xi (\vec{x} - \vec{x_0}) \cdot \vec{\xi}} d\vec{x}.$$

Then all we need to do is to change the windowing function with the frequency to get a new transform $\mathcal{W}$:

$$\mathcal{W}(I)(\vec{x_0}, \vec{\xi}) = \iint I(\vec{x}) w(A_\xi(\vec{x} - \vec{x_0})) e^{2\pi i A_\xi (\vec{x} - \vec{x_0}) \cdot \vec{\xi}} d\vec{x}.$$

This is exactly the *wavelet* transform of I associated to the wavelet $\psi(\vec{x}) = w(\vec{x}) e^{2\pi i \vec{x} \cdot \vec{\xi}}$. Either way, what we use to model the texture is its local spectral power:

$$\mathcal{P}(I) = |\mathcal{F}(I)|^2 \text{ or } |\mathcal{W}(I)|^2.$$

The model is based on the idea of finding a cartoon for $\mathcal{P}(I)$. More precisely, we seek firstly a set of boundary curves Γ in the spatial domain D (n.b. *not* boundaries in the 4-dimensional $D \times \hat{D}$ space of the variables $(\vec{x}, \vec{\xi})$ where $\hat{D}$ is a part of the dual two-dimensional vector space of the spatial frequency variables (ξ_1, ξ_2) with suitable high and low frequency cut-offs). Secondly, we seek a smooth spatial frequency description $J(\vec{x}, \vec{\xi})$ of the signal on $(D - \Gamma) \times \hat{D}$. The prior model is just like that of the cartoon: terms for the length of Γ and for the gradient (now 4-dimensional) of J. The data model has an important subtlety: one of the problems which bedevil texture segmentation is that spectral filters whose support overlaps the correct boundary between 2 quite distinct textures give erratic responses as a result of the partial fields visible in each texture. Thus J should not be compared with $\mathcal{P}(I)$ for such fields. For the windowed Fourier transform, this will be a strip around Γ of fixed size; for the wavelet transform, this is rather a shadow cast by Γ from high frequencies, i.e. at high frequencies, the window is smaller and the data can be modeled very close to Γ, while at low frequencies, the window grows and the strip of mixed responses grows. In either case, call $\mathcal{S}(\Gamma)$ the set of points of $D \times \hat{D}$ for which the corresponding filter overlaps Γ. Then the model can be summarized by:

$$
\begin{aligned}
w &= (J(\vec{x}, \vec{\xi}), \Gamma) \\
E_p &= \int\int\int\int_{(D-\Gamma)\times\hat{D}} \psi(\|\nabla_{\vec{x}}J\|, \|\nabla_{\vec{\xi}}J\|)\, d\vec{x}\, d\vec{\xi} + \nu|\Gamma| \\
E_d &= \int\int\int\int_{(D\times\hat{D})-\mathcal{S}(\Gamma)} (\mathcal{P}(I) - J)^2\, d\vec{x}\, d\vec{\xi}, \quad \mathcal{P}=\text{local spectral power}
\end{aligned}
$$

In figure 5.4, we show an example of this model for a lady with scarf: note how it finds most of the edges where the scarf has folds causing a break in texture statistics, but that it treats as uniform the slow changes where the scarf bends around her head. It should be noted that the figure probably does not present the minimizing Γ for this figure: rather it shows the best Γ found by either the Geman's annealing algorithm or Blake and Zisserman's continuation algorithm. I expect that the model will give even better results when a more effective optimizing technique is devised for energy functionals like this one.

I would also like to mention that this model should be quite interesting for *speech*. In speech, we have a function $I(t)$ of time, which gives us a time-frequency local power descriptor $\mathcal{P}(I)(t, \omega)$. Speech naturally breaks up into phonemes each with a typical power spectrum. This power spectrum changes slowly during a phoneme and changes rapidly when one phoneme succeeds another. Thus a model like that just given provides a method of segmenting speech without the detailed modeling of each individual phoneme required by the standard 'HMM' approach to speech.

Figure 5.4.   Lady with a scarf and the segmentation boundary.

VARIATIONAL PROBLEMS WITH A FREE DISCONTINUITY SET

Antonio Leaci

and

Sergio Solimini
Department of Mathematics
University of Lecce
I-73100 Lecce, Italy

1. Introduction

The present chapter is a short survey about the most recent contributions to the mathematical analysis of a variational approach to image segmentation proposed by D. Mumford and J. Shah.
Given a bounded open set $\Omega \subset \mathbb{R}^2$, we consider the functional

$$E(K, u) = \int_{\Omega \setminus K} |\nabla u|^2 dy + \mu \int_{\Omega \setminus K} |u - g|^2 dy + \nu \, \ell(K \cap \Omega)$$

where $K \subset \mathbb{R}^2$ is the union of a (*a priori* unknown) family of curves in $\bar{\Omega}$, $\ell(K \cap \Omega)$ denotes the sum of their lengths and $u \in C^1(\Omega \setminus K)$. By $|\nabla u|$ we denote the Euclidean norm of the gradient of u and g is a bounded measurable function defined on Ω with $|g(y)| \leq 1$. We consider the total length ℓ formally defined as the one dimensional **Hausdorff measure** $\mathcal{H}^1$, then the functional E makes sense for every closed set $K \subset \bar{\Omega}$.
We want to minimize the functional E over the class of all admissible pairs (K, u) defined as above, with K closed set.
Remark. For the sake of minimization the functional E can be seen as depending just on one variable. If the closed set K is prescribed, then the function u such that E attains its minimum is the unique solution $u(K)$ of the Neumann problem

$$\begin{aligned} -\Delta u + \mu(u - g) = 0 \quad &\text{in} \quad \Omega \setminus K \\ \tfrac{\partial u}{\partial n} = 0 \quad &\text{on} \quad \partial\Omega \cup K, \end{aligned}$$

hence we can set $E(K) = E(K, u(K))$. On the other hand, we may assume
$meas(K) = 0$, otherwise $E(K, u) = +\infty$. In this case the function u is
defined almost everywhere in Ω and K contains the discontinuity set $K(u)$
of u, hence u is the only meaningful variable and $E(u) = E(K(u), u)$. In
this approach the set $K(u)$ is not necessarily closed, so one actually extends
the meaning of the functional to the case of a nonclosed K. Therefore, in
this framework, it is not natural to *a priori* assume K closed.

The minimization problem for E makes sense even for $\Omega \subset \mathbb{R}^n$ with $n > 2$
if we replace ℓ by the $(n - 1)$ dimensional Hausdorff measure $\mathcal{H}^{n-1}$. The
case $n = 2$ is interesting in computer vision, namely in that framework the
functional E was introduced by Mumford and Shah [264]. In such a case,
the datum g is a digital image and the variable K which minimizes the
value of E will be the desired segmentation. In view of this application we
shall essentially focus on the properties of K. The multidimensional case
(n arbitrary) is more interesting from the mathematical viewpoint and can
have other applications (mechanics of fractures, interfaces, drops). We shall
subsequently call optimal segmentation any set K which corresponds to a
minimum point of E.

Remark. The utility of the above functional for the applied scientist in
vision is apparent: it gives an objective criterion in order to decide among
two sets K_1 and K_2 which one is a better segmentation. The only freedom
consists in the choice of two parameters ν and μ, such a choice corresponds
to fix respectively the scale and the contrast.

Other models have been considered which takes into account second order
derivatives of the function u and creases. On this subject we refer to
[34], [237] and to a paper by M. Carriero, A. Leaci and F. Tomarelli in
preparation.

2. Main questions about the functional

Some questions naturally arise about the functional E in the case of any
dimension n.

Questions.

1. Given g, does exist any optimal segmentation K?
2. Is such an optimal segmentation unique?
3. Is such an optimal segmentation regular in some suitable sense? In
   particular, when $n = 2$, is K a finite union of regular curves?

In the last years many authors have given contributions to answer the above
questions. We shall briefly outline what the main advances have been.

Answer to 1. Existence Theory.

The general method proposed by E. De Giorgi and L. Ambrosio in [82] to
get existence is a typical application of the direct methods in Calculus of

Variations. The first step is the weak formulation of the minimum problem in a suitable function space, called $SBV(\Omega)$, whose elements admit essential discontinuities along sets of codimension one. A general existence result has been established in every dimension n by L. Ambrosio in [10] by minimizing the functional depending on the function variable u in $SBV(\Omega)$.

As we already noticed, the singular set comes out as the set of discontinuity points of a minimizing function in $SBV(\Omega)$ and its closedness is not *a priori* guaranteed. The existence of a closed optimal segmentation was proved by De Giorgi, Carriero and Leaci in [83] arguing on a density property of the singular set of a minimizer presented below.

Answer to 2. Nonuniqueness.

One can easily observe that uniqueness cannot be expected by considering the dependence of optimal segmentation on the parameters μ and ν. When the scale parameter ν is very large, that is when we are looking at a very large scale with respect to the screen Ω, then the optimal segmentation comes out to be empty, as one can rigorously prove by simple computations (see [237]). On the other hand, if ν is very small, namely if we are looking to details in a very small scale, a lot of singular set K is going to appear, unless we are in a trivial case (g is constant). A simple continuity argument (see [237]) shows that we can find a limit value of ν which implies the existence of a nonempty segmentation K and at the same time keeps the empty set as an optimal segmentation.

Answer to 3. Regularity.

The answer to this question is still widely open, but the estimates which we are going to present in the following give some partial answers mainly in dimension $n = 2$. Nevertheless we point out that even if the purely qualitative regularity properties asked in Question 3 represent a beautiful problem for pure mathematics, in some cases the estimates in themselves are more interesting. This is particularly true from the point of view of the applications since the model is a continuous extrapolation of a discrete model in which purely qualitative regularity properties make no sense.

3. Density estimates

To present the known properties of an optimal segmentation K we use the following notation: for every $x \in \mathbb{R}^n$ and for every real number $r > 0$ we denote by $B(x,r)$ the ball of center x and radius r, namely

$$B(x,r) = \{y \in \mathbb{R}^n : |y - x| < r\}.$$

A simple property of an optimal singular set K is an upper bound on the "mean density" on every ball $B(x,r)$, i.e. the ratio between the measure of the trace of K in $B(x,r)$ and r^{n-1} (modulo a normalization constant). More precisely one can prove the following statement.

Density Bound (DB). There exists $\alpha > 0$ such that for every $x \in \Omega$, for every $r \leq 1$ with $B(x, r) \subset \Omega$ we have

$$\mathcal{H}^{n-1}(K \cap B(x, r)) \leq \alpha r^{n-1},$$

where α is given by $n + 1$ times the measure of the n-dimensional unit ball (e.g. $\alpha = 3\pi$ in the case $n = 2$).

The proof of (DB) is elementary. It just consists in replacing the part of K contained in $B(x, r)$ with $\partial B(x, r)$ and estimating the variation of E. The positiveness of this variation, guaranteed by the optimality of K, gives the bound in (DB).

A much more delicate task is to find lower density bounds. In this direction, one can prove the following two properties of an optimal segmentation K. The first one has been proved in any dimension n, the second one has been obtained only in the case $n = 2$.

Uniform Density Property (UD). There exists $\delta > 0$ such that for every $x \in K$, for every $r \leq 1$ with $B(x, r) \subset \Omega$ we have

$$\mathcal{H}^{n-1}(K \cap B(x, r)) \geq \delta r^{n-1}.$$

In the case $n = 2$ with $\mu = \nu = 1$ the above inequality holds with $\delta = \frac{1}{4}$ provided $r < \frac{1}{16\pi^{3/2}}$.

Uniform Concentration Property (UC). Let $n = 2$; for every $\epsilon > 0$ there exists $\delta(\epsilon) > 0$ such that for every $x \in K$, for every $R \leq 1$ with $B(x, R) \subset \Omega$, there is $B(y, r) \subset B(x, R)$ with $r \geq \delta(\epsilon)R$ and

$$\mathcal{H}^1(K \cap B(y, r)) \geq (2 - \epsilon)r.$$

In other words (UD) states that the trace of the singular set on a disk centered at one of its points is bounded from below in terms of the radius. Property (UC) requires furthermore that under a suitable choice of the constant δ, depending on a given positive number ϵ, the bound required in (UD) is satisfied by a portion K' of K "concentrated" in a sub-disk with a diameter close to the lenght of K' in terms of ϵ. Property (UC) is therefore stronger than (UD).

Theoretical consequences.
1. Property (UD) gives a positive lower bound on the density of K in every point of its closure. This in particular implies that such a density can never be equal to zero and therefore, by well known properties from geometric measure theory (see [237]), that K is closed up to a negligible set with

respect to the $(n-1)$ dimensional measure. This property was proved in [83] and [53] for every dimension n and for functionals more general than E. It was also proved in [79] in dimension two as a step in the proof of (UC).
2. Property (UC) implies the lower semicontinuity of the length with respect to the Hausdorff distance **d** along a sequence of sets K_h which satisfy it in correspondence of a function $\delta = \delta(\epsilon)$ which does not depend on h. The set of optimal segmentations is therefore closed with respect to **d**, so it is compact ([79]).
One more consequence of lower semicontinuity is the possibility of minimizing directly the functional E on the closed sets in two dimensions proving (UC) for the terms of a suitable minimizing sequence (see [79], [237]).
3. By (UD) also the following property of an optimal segmentation K has been obtained in [12], [53] for any dimension n:

$$\lim_{\rho \to 0} \frac{meas(\{x; dist(x, K) < \rho\})}{2\rho} = \mathcal{H}^{n-1}(K).$$

The above equality expresses the agreement between the Hausdorff measure and the **Minkowski content** of K. This is a useful information for the applications. It allows, for instance, the approximation scheme introduced in [12] and produces therefore the subsequently developed algorithms.

Applicable Consequences.
If x is a point of K "suitably isolated", namely such that in a spherical neighborhood $B(x, r)$ there is a trace of K of small length compared to r, it can be eliminated by (UD) (see Fig.1).
The same conclusion holds by (UC) if K has a consistent length on $B(x, R)$ but it is spread out in the ball (see Fig.2).
The two above assertions mean that a suitable algorithm can eliminate such essentially isolated parts of K without taking in this way any new decision, since the theory ensures that under this elimination the segmentation improves in the only way which makes formal sense, namely the value of the functional E decreases.

4. Rectifiability estimates

In the same line of the applicable consequences in the previous section, we want to examine other cases of "bad singular sets" which should be eliminated in the same sense as before even if they satisfy the qualitative regularity asked in Question 3 and the density estimates in the previous section. Let us consider the set in Fig. 3. It consists of a finite number of curves and it can be taken in such a way to satisfy (UC) (see [237]). This set is rectifiable because it is a finite union of curves but if we want to cover it by curves we have to use either many of them or to use curves of very big length.

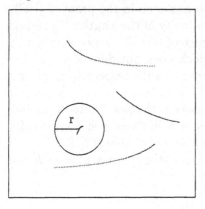

Fig. 1

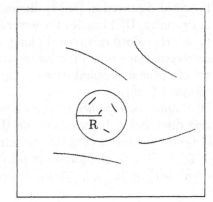

Fig. 2

A different situation is represented in Fig. 4 where the singular set may even consists of the union of infinitely many curves but it can be covered by a single curve of relatively small lenght. Note that, in spite of the possibly infinite number of the curves, it is conceivable that K_2 represents a more convincing segmentation than K_1.

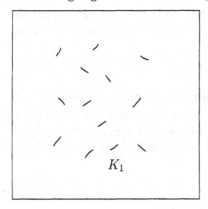

Fig. 3

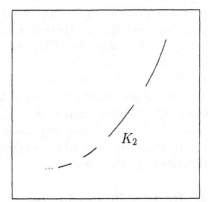

Fig. 4

The better shape of the set K_2 can be justified by the fact that, in some way, K_1 is "less rectifiable" than K_2. Rectifiability is traditionally a qualitative property. A nonrectifiable set is spread like K_1 over every scale (while K_1 looks like a curve in sufficiently local scale) and it can be seen like a cloud of points whose intersection with any curve has null length. A property which characterizes nonrectifiable sets says that they have negligible projection on almost every straight line.

For an optimal segmentation K we can give the following "rectifiability estimates", which in some sense quantify the validity of this property and allow us to discriminate the two sets in the above examples.

The first statement is concerned with the above characterization of rectifiable sets by projection and it consists in giving an *a priori* bound from below on the sum of the projections of the trace of K on the sides of a square centered at a singular point. Such a property is clearly analogous to (UD) and improves it in an obvious way, since the measure of the trace of K is bounded from below by the one of its projection.

Uniform Projection Property (UP). Let $n = 2$; then there exists $\delta > 0$ such that for every $x \in K$, for every square $S \subset \Omega$, centered at x and with side r, we have
$$\ell(p_1(K \cap S)) + \ell(p_2(K \cap S)) \geq \delta r,$$
where $p_i(K \cap S)$ $(i = 1, 2)$ denote the projections of $K \cap S$ on two adjacent sides of the square.

Property (UP) is already sufficient for the elimination of sets like K_1, as one can easily see in some particular cases as those in which all the curves are horizontal segments disposed in vertical rows.

Then we want to state a quantitative property connected to the original definition of rectifiability. Indeed we shall quantify the fact that the singular set has nonempty intersection with some curves and that on the other hand it can be consequently covered by curves. This situation has been investigated by G. David and S. Semmes in [81]. They have proved that the trace of K on a disk centered at one of its points almost contains a curve whose lenght is not too small compared to the radius. Such a property is clearly not enjoied by K_1.

Curve Filling Property (CF). Let $n = 2$; then for every $\epsilon > 0$, there exists $\delta > 0$ such that for every $x \in K$, for every $r \leq 1$, for every $B(x, r) \subset \Omega$, there exists a continuous curve γ with $\ell(\gamma) \geq \delta r$ and $\ell(\gamma \setminus K) \leq \epsilon \ell(\gamma)$.

A property like (CF) implies that with a simple argument (see [81], [237]) we can prove the existence of a single curve containing the set K and with length proportional to the one of K, giving in this way a third estimate on the rectifiability of K. Property (UP) can be derived from (CF) for optimal segmentations by using (DB). The relation between (CF) and (UC) is also evident. If we set $K' = K \cap \gamma$, we know that K' is contained in the curve γ, whose lenght is close to the one of K'. Since the curve is obviously contained in a ball with a diameter equal to its lenght, the concentration of K' is clearly achieved. Such an analogy is also reflected in the method of proof which is based on a variant of the elimination techniques introduced in [79] and a suitable use of the Coarea Formula. In [237] all the above estimates are derived from an abstract Elimination Theorem which may have other interesting concrete examples.

MINIMIZATION OF ENERGY FUNCTIONAL WITH CURVE-REPRESENTED EDGES

Niklas Nordström

Royal Institute of Technology (KTH)
Department of Numerical Analysis and Computing Science
S-100 44 Stockholm, Sweden

1. Introduction

Until the mid eighties almost all edge detection methods were based on local operators. Since the edges of interest usually exist at a coarser scale than the inner scale (corresponding to the pixel size,) of the image, these operators always incorporate some averaging or similar mechanism for suppressing fine scale variations. As a consequence the detected edges get blurred and may be displaced. These problems can be circumvented by applying some kind of nonlinear best fit technique. Such local methods exist in abundance. However, they typically take only a small number of possible local edge configurations into consideration. For all other configurations the blurring and displacement problems thus remain. Ideally one would of course like to consider all possible edge configurations. But then the best fit technique must be applied to image regions so large that the edges therein can be assumed to be sparse. Otherwise the desired fine detail suppression will be difficult, or even impossible, to achieve. Conceptually the simplest way to make sure that this is the case, is to use a global best fit technique, which is the essence of global edge detection or global segmentation.

Quite a few global edge detection methods have been proposed in the computer vision literature to this date. Many of these have more or less been formulated with particular analytical and/or computational techniques in mind. The method proposed in 1985 by Mumford and Shah [263, 264], on the other hand, is based more directly on the desired properties of the solution to the segmentation problem. In this section we outline a rigorous proof of existence of a solution to a slightly restricted version of this problem. Because of space limitations many details will unfortunately have to be left out. For the complete proof we refer to [274].

1.1. The Mumford and Shah Problem

The global edge detection problem posed by Mumford and Shah can be stated as follows. Given an image, represented by a function $g \in L_2(B)$, where the image domain B is some open rectangle in $\mathbb{R}^2$, find a closed set $D \subset \mathbb{R}^2$ and a function $f : B \setminus D \to \mathbb{R}$ that minimize the cost or energy functional

$$\mathcal{C}(D, f) \doteq \int_{B \setminus D}[(f - g)^2 + \mu\|\nabla f\|^2]\, dx + \lambda\, \text{length}(D) \qquad (7.1)$$

where μ and λ are strictly positive constants.[1] The set D is furthermore assumed to consist of a finite union of C^1-smooth curves meeting each other and the image boundary ∂B only at their endpoints. The *original image* (function) g is thereby estimated by a function f, which (if extended to B,) is smooth in some sense everywhere except on the set D. This set thus represents the edges. It will be referred to as the *discontinuity set*. Without further restrictions on the original image g, the *estimated image* (function) f will naturally be a member of the Sobolev space $\mathcal{H}^1(B \setminus D)$ consisting of all functions in $L_2(B \setminus D)$ whose first order distributional derivatives also belong to $L_2(B \setminus D)$. In the original problem formulation by Mumford and Shah the original image g was actually assumed to be continuous. With reference to the natural smoothing associated with any physical image acquisition system one could argue that this restriction is realistic. One goal of the imaging system design, however, is to reduce this smoothing to a minimum. Since the continuity assumption also implies that the noise must be continuous, we have simply omitted it in our analysis. This is of course not to say that any conclusions resting on this continuity assumption would be uninteresting.

1.2. The Fixed Discontinuity Set Problem

If the discontinuity set D is fixed, the last term in the cost (7.1) is constant. The minimization of the *total cost* $\mathcal{C}(D, f)$ then reduces to the elliptic problem of minimizing the *image cost* functional $\mathcal{C}_\Omega : \mathcal{H}^1(\Omega) \to \mathbb{R}$ defined by

$$\mathcal{C}_\Omega(f) \doteq \int_\Omega [(f - g)^2 + \mu\|\nabla f\|^2]\, dx$$

---

[1]Throughout this section the symbol ∇ will denote the distributional (row vector) gradient operator. The norm $\|\cdot\|$ (without subscript) is the Euclidean norm (on $\mathbb{R}^2$,) and $\doteq$ indicates equality *by definition*.

for open sets $\Omega \subset B$. This classical problem is well-known to have exactly one solution $f_\Omega \in \mathcal{H}^1(\Omega)$. Indeed, define for any $S \subset \Omega$ the bilinear forms

$$\langle f, \varphi \rangle_S \doteq \int_\Omega f\varphi \, dx \qquad f, \varphi \in L_2(\Omega)$$

and

$$a_S(f, \varphi) \doteq \int_\Omega (f\varphi + \mu \nabla f \nabla \varphi^T) \, dx \qquad f, \varphi \in \mathcal{H}^1(\Omega)$$

Then $\langle \cdot, \cdot \rangle_\Omega$ and a_Ω are inner products on $L_2(\Omega)$ and $\mathcal{H}^1(\Omega)$ respectively, which so equipped both become Hilbert spaces.[2] Moreover,

$$\mathcal{C}_\Omega(f) = \|g\|^2_{L_2(\Omega)} - 2\langle g, f \rangle_\Omega + \|f\|^2_{\mathcal{H}^1(\Omega)} \qquad \forall f \in \mathcal{H}^1(\Omega) \qquad (7.2)$$

where $\| \cdot \|_{\mathcal{H}^1(\Omega)}$ and $\| \cdot \|_{L_2(\Omega)}$ are the norms induced by a_Ω and $\langle \cdot, \cdot \rangle_{L_2(\Omega)}$ respectively.[3] Since $\mathcal{C}_\Omega$ is bounded below, there exists a sequence $(f_n)_{n \in \vec{N}}$ in $\mathcal{H}^1(\Omega)$ such that[4]

$$\mathcal{C}_\Omega(f_n) - \frac{1}{n} < \underline{\mathcal{C}_\Omega} \doteq \inf_{f \in \mathcal{H}^1(\Omega)} \mathcal{C}_\Omega(f) \qquad \forall n \in \vec{N}$$

By the parallellogram law

$$\begin{aligned}
\frac{1}{2}\|f_m - f_n\|^2_{\mathcal{H}^1(\Omega)} &= \|f_m\|^2_{\mathcal{H}^1(\Omega)} + \|f_n\|^2_{\mathcal{H}^1(\Omega)} - \frac{1}{2}\|f_m + f_n\|^2_{\mathcal{H}^1(\Omega)} \\
&= \mathcal{C}_\Omega(f_m) + \mathcal{C}_\Omega(f_n) - 2\mathcal{C}_\Omega\left(\frac{f_m + f_n}{2}\right) \\
&\leq \frac{1}{m} + \frac{1}{n} < \frac{2}{n} \qquad \forall m \geq n \qquad \forall n \in \vec{N}
\end{aligned}$$

Thus $(f_n)_{n \in \vec{N}}$ is a Cauchy sequence. By the completeness of $\mathcal{H}^1(\Omega)$ it then follows that

$$f_\Omega \doteq \lim_{n \to \infty} f_n \in \mathcal{H}^1(\Omega)$$

and $\mathcal{C}_\Omega(f_\Omega) = \underline{\mathcal{C}_\Omega}$, that is, f_Ω solves the minimization problem. For any $f \in \mathcal{H}^1(\Omega)$ such that $\mathcal{C}_\Omega(f) = \underline{\mathcal{C}_\Omega}$, we moreover have that

$$\frac{1}{2}\|f - f_\Omega\|^2_{\mathcal{H}^1(\Omega)} = \underline{\mathcal{C}_\Omega} + \underline{\mathcal{C}_\Omega} - 2\mathcal{C}_\Omega\left(\frac{f + f_\Omega}{2}\right) \leq 0 \Rightarrow f = f_\Omega$$

---

[2] While the inner product a_Ω depends on the constant μ, the topology that it generates does not, as different values of μ induce equivalent norms. For our purpose a_Ω is more convenient than the standard inner product on $\mathcal{H}^1(\Omega)$, which is defined as a_Ω, but with $\mu = 1$.

[3] In order to avoid lots of function restrictions, we will use the definitions $\|\varphi\|_{L_2(S)} \doteq \sqrt{\langle \varphi, \varphi \rangle_S}$ and $\|\varphi\|_{\mathcal{H}^1(S)} \doteq \sqrt{a_S(\varphi, \varphi)}$ whenever the right hand side expressions are well-defined. Thus $\|g\|_{L_2(\Omega)}$ is well-defined, even though $\| \cdot \|_{L_2(\Omega)}$ is not a norm (but rather a seminorm,) on the space $\mathcal{H}^1(B)$, to which g belongs.

[4] $\vec{N}$ denotes the set of natural numbers $\{1, 2, 3, \ldots\}$.

Hence the solution is unique. By straight forward calculation we furthermore find that

$$a_\Omega(f_\Omega, f) - \langle g, f \rangle_\Omega = \frac{C_\Omega(f_\Omega + f) - C_\Omega(f_\Omega) - \|f\|^2_{\mathcal{H}^1(\Omega)}}{2}$$

$$\geq -\frac{\|f\|^2_{\mathcal{H}^1(\Omega)}}{2} \qquad \forall f \in \mathcal{H}^1(\Omega)$$

Thus for any $f \in \mathcal{H}^1(\Omega)$

$$a_\Omega(f_\Omega, f) - \langle g, f \rangle_\Omega = \frac{1}{t}[a_\Omega(f_\Omega, tf) - \langle g, tf \rangle_\Omega]$$

$$\geq -\frac{\|tf\|^2_{\mathcal{H}^1(\Omega)}}{2t}$$

$$= -\frac{\|f\|^2_{\mathcal{H}^1(\Omega)}}{2}t \qquad \forall t > 0$$

Since the leftmost expression does not depend on t it must therefore be positive. It is, however, also linear in f, whence it must actually vanish. In other words,

$$a_\Omega(f_\Omega, f) = \langle g, f \rangle_\Omega \qquad \forall f \in \mathcal{H}^1(\Omega) \tag{7.3}$$

For the fixed domain Ω the estimated image f_Ω is thus bounded by

$$\|f_\Omega\|_{\mathcal{H}^1(\Omega)} = \sup_{\substack{f \in \mathcal{H}^1(\Omega) \\ \|f\|_{\mathcal{H}^1(\Omega)} \leq 1}} \langle g, f \rangle_\Omega \leq \sup_{\substack{f \in L_2(\Omega) \\ \|f\|_{L_2(\Omega)} \leq 1}} \langle g, f \rangle_\Omega = \|g\|_{L_2(\Omega)} \tag{7.4}$$

and in view of (7.2) the *optimal image cost* is given by

$$\underline{C_\Omega} = C_\Omega(f_\Omega) = \|g\|^2_{L_2(\Omega)} - \langle g, f_\Omega \rangle_\Omega \tag{7.5}$$

1.3. Outline of the Existence Proof

As a consequence of the discussion in the previous section the original minimization problem is reduced to that of minimizing

$$C(D, f_{B \setminus D}) = \underline{C_{B \setminus D}} + \lambda \, \text{length}(D) \tag{7.6}$$

with respect to the discontinuity set D. By earlier assumptions $D \in \mathcal{D} \doteq \bigcup_{N=0}^\infty \mathcal{D}_N$, where each $D \in \mathcal{D}_N$ is the union of N C^1-smooth curves—*edge segments*—in $\mathbb{R}^2$, (and $\mathcal{D}_0 \doteq \{\varnothing\}$.) Assume now that we restrict the discontinuity set collections $\mathcal{D}_N$, $N \in \vec{N}_0 \doteq \{0\} \cup \vec{N}$, so that

$$\inf_{D \in \mathcal{D}_N} \text{length}(D) \longrightarrow \infty \qquad \text{as } N \longrightarrow \infty \tag{7.7}$$

Then for some sufficiently large $\overline{N} \in \vec{N}$ we have

$$\inf_{D \in \mathcal{D}_N} \mathcal{C}(D, f_{B \smallsetminus D}) > \|g\|_{L_2(B)}^2 = \mathcal{C}(\varnothing, 0) \qquad \forall N \geq \overline{N}$$

which in turn means that there is an optimal number, $\check{N} \in \{0, \ldots, \overline{N} - 1\}$, of edge segments. In order to prove the existence of an optimal discontinuity set, it is thence sufficient to show that for each $N \in \vec{N}_0$, there exists some $\check{D}_N \in \mathcal{D}_N$ that minimizes the cost $\mathcal{C}(D, f_{B \smallsetminus D})$ with respect to $D \in \mathcal{D}_N$.

The smoothness constraints suggest that we represent the edge segments by N continuously differentiable parameterized curves $\gamma_n \doteq (\gamma_{n,1}, \gamma_{n,2}) :$ $[0,1] \rightarrow \mathbb{R}^2$, $n = 1, \ldots, N$. For convenience we also introduce the *image segmentation* $\gamma \doteq (\gamma_1, \ldots, \gamma_N)$ and define its associated discontinuity set $D_\gamma \doteq \bigcup_{n=1}^N \gamma_n([0,1])$. The image segmentation γ is thus an $\mathbb{R}^{2N}$-valued function with components $\gamma_{n,1}, \gamma_{n,2} \in C^1([0,1])$, $n = 1, \ldots, N$,[5] that is, it belongs to the product space $C^1([0,1])^{2N}$. If we give $C^1([0,1])$ the standard norm

$$\|\varphi\|_{C^1([0,1])} \doteq \bigvee_{l=0}^{1} \sup_{t \in [0,1]} |\varphi^{(l)}(t)| \qquad \varphi \in C^1([0,1])$$

where $\varphi^{(l)}$ denotes the lth derivative of φ and $\vee$ is the "max" operator, then for any $K \in \vec{N}$

$$\|(\varphi_1, \ldots, \varphi_K)\|_{C^1([0,1])^K} \doteq \bigvee_{k=1}^{K} \|\varphi_k\|_{C^1([0,1])} \qquad \varphi_1, \ldots, \varphi_K \in C^1([0,1])$$

naturally defines a norm generating the product topology on $C^1([0,1])^K$. The space $C^1([0,1])^{2N}$ of image segmentations is thereby topologized. We can now approach the existence problem by establishing lower semicontinuity of the *N-segment cost* $\mathcal{C}_N(\gamma) \doteq \mathcal{C}(D_\gamma, f_{B \smallsetminus D_\gamma})$ on a significant compact subset of $C^1([0,1])^{2N}$. A minimum must then exist on this subset. The fact that we must consider a subset of $C^1([0,1])^{2N}$, and hence a subset of $\mathcal{D}_N$, is theoretically disturbing. We have in particular been unable to include discontinuity sets with free endpoints or those that form cusps. The space of discontinuity sets that we allow is, however, rich enough to represent most cartoons. It is moreover quite possible that our techniques can be improved so that some restrictions on the discontinuity set can be lifted, without abolishing the explicit curve representation of the edges.

1.4. Admissible Image Segments

It is more or less trivial to arrange so that the *edge cost* $\lambda \operatorname{length}(D_\gamma)$ depends continuously on γ. In establishing the lower semicontinuity of the

---

[5]The space $C^1([0,1])$ consists of all continuously differentiable functions $\varphi : [0,1] \rightarrow \mathbb{R}$.

N-segment cost $\mathcal{C}_N$, it is thus the optimal image cost term $\mathcal{C}_{B \smallsetminus D_\gamma}$ that plays the critical role, and puts limitations on which image segmentations may be considered. To begin with we need a handle on the sensitivity of the optimal image cost $\mathcal{C}_\Omega$ with respect to the boundary $\partial\Omega$. This can be achieved if the domain Ω has certain global properties, in which case we shall refer to it as an *admissible image segment*.

Consider an open subset G of the image domain B. If F is an open set with closure $\overline{F} \subset G$, then the (Euclidean) distance between F and the boundary ∂G is strictly positive, and hence $F \subset G'$ for any set G' obtained from G by a sufficiently small perturbation of ∂G. It then follows that the optimal image cost $\mathcal{C}_{G'}$ over G' is bounded below by the optimal image cost $\mathcal{C}_F$ over the interior set F. Hence the difference $\mathcal{C}_{G'} - \mathcal{C}_G$ between the optimal image costs over G' and G respectively is bounded below by $\mathcal{C}_F - \mathcal{C}_G$. The desired lower semicontinuity then follows provided that F can be chosen so that $\mathcal{C}_F$ is an arbitrarily good approximation of $\mathcal{C}_G$.

It turns out that, if the optimal image function f_F can be extended to $\mathbb{R}^2$ without blowing up its norm, and the difference set $G \smallsetminus F$ is of sufficiently small measure, then $\mathcal{C}_F$ is indeed a good approximation of $\mathcal{C}_G$. For this reason we will consider the space $\mathcal{L}(\mathcal{H}^1(F), \mathcal{H}^1(\mathbb{R}^2))$, consisting of linear maps $P_F : \mathcal{H}^1(F) \to \mathcal{H}^1(\mathbb{R}^2)$ for which

$$\|P_F\|_{\mathcal{L}(\mathcal{H}^1(F),\mathcal{H}^1(\mathbb{R}^2))} \doteq \sup_{\substack{f \in \mathcal{H}^1(F) \\ \|f\|_{\mathcal{H}^1(F)} \le 1}} \|P_F(f)\|_{\mathcal{H}^1(\mathbb{R}^2)} < \infty$$

The maps in $\mathcal{L}(\mathcal{H}^1(F), \mathcal{H}^1(\mathbb{R}^2))$ of particular interest are those that extend functions (defined on F to $\mathbb{R}^2$.) We call such maps *extension operators*. Denoting the Lebesgue measure on $\mathbb{R}^2$ by m we then define admissible image segments as follows.

Definition 1.1 *We say that a bounded open set $G \subset \mathbb{R}^2$ is an* admissible *image segment if $\exists$ a collection $\mathcal{F}$ of open sets such that*

(i) $\overline{F} \subset G \qquad \forall F \in \mathcal{F}$
(ii) $\exists$ *an extension operator* $P_F \in \mathcal{L}(\mathcal{H}^1(F), \mathcal{H}^1(\mathbb{R}^2)) \qquad \forall F \in \mathcal{F}.$
(iii) $\sup_{F \in \mathcal{F}} \|P_F\|_{\mathcal{L}(\mathcal{H}^1(F),\mathcal{H}^1(\mathbb{R}^2))} < \infty$
(iv) $\inf_{F \in \mathcal{F}} m(G \smallsetminus F) = 0$

It is not immediately evident that the class of admissible image segments is rich enough to be of any interest. Luckily, however, there are relatively simple local criteria, by which the admissible image segment property can be verified.

Definition 1.2 *We say that a bounded open set $G \subset \mathbb{R}^2$ is a* Lipschitz domain[6] if $\forall x \in \partial G \; \exists$ an open interval Δ, a Lipschitz continuous function $\phi : \Delta \to \mathbb{R}$, a strictly positive number $d > 0$, and a rigid coordinate transformation $T : \mathbb{R}^2 \to \mathbb{R}^2$ such that the map $\Phi : \Delta \times] - d, d[\to \mathbb{R}^2 : y \mapsto T(y_1, \phi(y_1) + y_2)$ satisfies the conditions

  (i) $\Phi(\Delta \times]0, d[) \subset G$
  (ii) $\Phi(\Delta \times] - d, 0[) \subset \mathbb{R}^2 \smallsetminus G$
  (iii) $x \in \Phi(\Delta \times \{0\}) \subset \partial G$

Lipschitz domains are of big interest to us because of the following important result.

Theorem 1.3 *All Lipschitz domains are admissible image segments.*

A proof of this theorem, which is too involved to be presented here, can be found in [274]. Here we will merely describe its main steps. Let G be a Lipschitz domain. Since G is bounded and ∂G therefore compact, the map Φ in definition 1.2, which depends on the boundary point x, can actually be chosen from a finite collection $\{\Phi_m\}_{m=1}^M \; \forall x \in \partial G$. From this atlas on ∂G one then explicitly constructs a family $\{\{\Phi_{m,h}\}_{m=1}^M\}_{h>0}$ of modified atlases, by perturbing the functions $\phi_1, \ldots, \phi_M$ by an amount that gradually vanishes as $h \downarrow 0$. By careful design of these perturbations one can moreover arrange so that the sets $F_h \doteq G \smallsetminus \bigcup_{m=1}^M \Phi_{m,h}(\Delta \times] - d, d[)$, $h > 0$, are all Lipschitz domains such that $\partial F_h = \bigcup_{m=1}^M \Phi_{m,h}(\Delta \times \{0\})$, $U_0 \doteq F_{h_2} \subset \overline{F_h} \subset G \; \forall h \in H \doteq]0, h_1]$ for some $h_2 > h_1 > 0$, and $\lim_{h \downarrow 0} m(G \smallsetminus F_h) = 0$. Simultaneously it is possible to keep all the charts $\Phi_{1,h}, \ldots \Phi_{M,h}$, $h \in H$, and their inverses *uniformly* Lipschitz continuous, with a common Lipschitz constant $L < \infty$, while maintaining the chart ranges of the original atlas, so that $\Phi_{m,h}(\Delta \times] - d, d[) = U_m \doteq \Phi_m(\Delta \times] - d, d[) \; \forall h \in H, \, m = 1, \ldots, M$. The collection $\mathcal{U} \doteq \{U_m\}_{m=0}^M$, which is *independent of h*, can moreover be shown to be an open covering of G, and hence of $\overline{F_h} \; \forall h \in H$. For each such value of h one can then by standard methods construct an extension operator $P_{F_h} \in \mathcal{L}(\mathcal{H}^1(F_h), \mathcal{H}^1(\mathbb{R}^2))$, whose norm is bounded by a constant that only depends on L and some C^∞-partition of unity for G (and thus for F_h) subordinate to $\mathcal{U}$. If we let $\mathcal{F} \doteq \{F_h\}_{h \in H}$, the defining criteria for G being an admissible image segment are then satisfied.

We now continue with the cost sensitivity problem outlined above.

Lemma 1 *Let $G \subset B$ be an admissible image segment and let $\mathcal{F}$ be an collection of open sets with the properties (i)–(iv) in definition 1.1. Then the optimal image cost satisfies $\sup_{F \in \mathcal{F}} \mathcal{C}_F \geq \mathcal{C}_G$.*

Proof: Let $F \in \mathcal{F}$ and let $E \doteq G \smallsetminus F$. Then by (7.5) and (i)

$$\mathcal{C}_G - \mathcal{C}_F = \|g\|_{L_2(G)}^2 - \langle g, f_G \rangle_G - \|g\|_{L_2(F)}^2 + \langle g, f_F \rangle_F$$

---

[6]In the mathematical literature Lipschitz domains are also referred to as domains having the *Lipschitz property* or *domains of class $C^{0,1}$*.

$$= \langle g, g - f_G \rangle_E + \langle g, f_F - f_G \rangle_F$$

In order to bound the last term in this expression, we pick P_F according to (ii) and observe that $P_F(f_F)|G \in \mathcal{H}^1(G)$,[7] $P_F(f_F)|F = f_F$, and that

$$\|P_F(f_F)\|_{L_2(E)} \leq \|P_F(f_F)\|_{\mathcal{H}^1(E)} \leq \|P_F(f_F)\|_{\mathcal{H}^1(\mathbb{R}^2)} \leq \nu\|f_F\|_{\mathcal{H}^1(F)} \quad (7.8)$$

where $\nu \doteq \sup_{F \in \mathcal{F}} \|P_F\|_{\mathcal{L}(\mathcal{H}^1(F),\mathcal{H}^1(\mathbb{R}^2))} < \infty$ by (iii). Thus from (7.3) we have that

$$\begin{aligned}
a_E(f_G, P_F(f_F)) &= a_G(f_G, P_F(f_F)) - a_F(f_F, f_G)) \\
&= \langle g, P_F(f_F) \rangle_G - \langle g, f_G \rangle_F \\
&= \langle g, P_F(f_F) \rangle_E + \langle g, f_F - f_G \rangle_F
\end{aligned}$$

Consequently by the Cauchy-Schwartz inequality, (7.8) and (7.4)

$$\underline{\mathcal{C}_G} - \underline{\mathcal{C}_F} = \langle g, g - f_G - P_F(f_F) \rangle_E + a_E(f_G, P_F(f_F))$$
$$\leq \|g\|_{L_2(E)} \|g - f_G - P_F(f_F)\|_{L_2(E)} + \|f_G\|_{\mathcal{H}^1(E)} \|P_F(f_F)\|_{\mathcal{H}^1(E)} \leq$$
$$\|g\|_{L_2(E)}(\|g\|_{L_2(E)} + \|f_G\|_{\mathcal{H}^1(G)} + \nu\|f_F\|_{\mathcal{H}^1(F)}) + \|f_G\|_{\mathcal{H}^1(E)} \nu\|f_F\|_{\mathcal{H}^1(F)}$$
$$\leq [(2+\nu)\|g\|_{L_2(E)} + \nu\|f_G\|_{\mathcal{H}^1(E)}] \|g\|_{L_2(G)}$$

Since $g \in L_2(B)$ and $f_G \in \mathcal{H}^1(G)$ are independent of F, the dominated convergence theorem implies that the leading factor above vanishes as $m(E) \downarrow 0$. From (iv) it then follows that $\inf_{F \in \mathcal{F}}(\underline{\mathcal{C}_G} - \underline{\mathcal{C}_F}) \leq 0$. ∎

The primary value of the lemma above is that it captures the dependence of the optimal image cost on the boundary of the domain, without explicit reference to neither the properties of the boundary nor the optimal image function. It thereby effectively separates the issue of lower semicontinuity from that of characterizing the class of image segmentations, to which the existence proof here presented will apply. This is not merely a matter of convenience. It actually suggests a route for improving the existence proof by extending the class of known admissible image segments beyond the class of Lipschitz domains.

1.5. Admissible Image Segmentations

Up to this point we have only been concerned with the optimal image cost $\underline{\mathcal{C}_G}$ over some general subset G of the image domain B. The next natural step is to consider domains of the form $G = B \smallsetminus D$, where D is a discontinuity set.

---

[7]The restriction of a function φ to the set S is denoted by $\varphi|S$.

Definition 1.4 We say that $\gamma \in C^1([0,1])^{2N}$ is an admissible image segmentation of B if each connected component of $B \smallsetminus D_\gamma$ is an admissible image segment.

Theorem 1.5 The N-segment image cost C_N is lower semicontinuous on the set of admissible image segmentations.

Proof: Let $\gamma \in C^1([0,1])^{2N}$ be an admissible image segmentation of B, and let $\mathcal{G}$ be the collection of all the connected components of the set $B \smallsetminus D_\gamma$. Since $B \smallsetminus D_\gamma$ is an open set, $\mathcal{G}$ is countable. Thus, given $\epsilon > 0$, $\exists$ a finite collection $\{G_j\}_{j=1}^J \in \mathcal{G}$ such that

$$\sum_{G \in \mathcal{G}_0} C_G(0) = \sum_{Gin\mathcal{G}_0} \|g\|_{L_2(G)}^2 < \frac{\epsilon}{2}$$

where $\mathcal{G}_0 \doteq \mathcal{G} \smallsetminus \{G_j\}_{j=1}^J$. Since $G_1, \ldots, G_J$ are admissible image segments, $\exists$ corresponding collections $\mathcal{F}_1, \ldots, \mathcal{F}_J$ of open subsets of $G_1, \ldots, G_J$ respectively with the properties (i)–(iv) listed in definition 1.1. Furthermore, by lemma 1 we can pick $F_j \in \mathcal{F}_j$, $j = 1, \ldots, J$, so that

$$\sum_{j=1}^J C_{F_j} > \sum_{j=1}^J C_{G_j} - \frac{\epsilon}{2}$$

Now let $F \doteq \bigcup_{j=1}^J F_j$. Then by property (i) of the collections $\mathcal{F}_1, \ldots, \mathcal{F}_J$ we have that $\overline{F} = \bigcup_{j=1}^J \overline{F_j} \subset \bigcup_{j=1}^J G_j \subset B \smallsetminus D_\gamma$. Since D_γ and $\overline{F}$ are compact, this implies that D_γ and F are separated by a Euclidean distance $\delta > 0$. Thus for any image segmentation $\gamma' \in C^1([0,1])^{2N}$ such that $\|\gamma' - \gamma\|_{C^1([0,1])^{2N}} < \delta/\sqrt{2}$, we have that $F \cap D_{\gamma'} = \varnothing$, and hence $F \subset B \smallsetminus D_{\gamma'}$. It then follows that $f_{B \smallsetminus D_{\gamma'}}|F \in \mathcal{H}^1(F)$ and $f_{B \smallsetminus D_{\gamma'}}|F_j \in \mathcal{H}^1(F_j)$, $j = 1, \ldots, J$. Since $G_1, \ldots, G_J$ and hence $F_1, \ldots, F_J$ are pairwise disjoint, we then obtain

$$C_{B \smallsetminus D_{\gamma'}} = C_{B \smallsetminus D_{\gamma'}}(f_{B \smallsetminus D_{\gamma'}}) \geq C_F(f_{B \smallsetminus D_{\gamma'}}|F) = \sum_{j=1}^J C_{F_j}(f_{B \smallsetminus D_{\gamma'}}|F_j)$$

$$\geq \sum_{j=1}^J C_{F_j} > \sum_{j=1}^J C_{G_j} - \frac{\epsilon}{2}$$

For $j = 1, \ldots, J$ let $\widetilde{f_{G_j}}$ be the zero extension of f_{G_j} to $B \smallsetminus D_\gamma$. Then $f \doteq \sum_{j=1}^J \widetilde{f_{G_j}} \in \mathcal{H}^1(B \smallsetminus D_\gamma)$, and thus

$$C_{B \smallsetminus D_\gamma} \leq C_{B \smallsetminus D_\gamma}(f) = \sum_{G \in \mathcal{G}} C_G(f|G)$$

$$= \sum_{j=1}^J C_{G_j}(f_{G_j}|G) + \sum_{G \in \mathcal{G}_0} C_G(0) < \sum_{j=1}^J C_{G_j} + \frac{\epsilon}{2}$$

It now follows that $\mathcal{C}_{B \setminus D_{\gamma'}} > \mathcal{C}_{B \setminus D_{\gamma}} - \epsilon$ whenever $\|\gamma' - \gamma\|_{C^1([0,1])^{2N}}$ is sufficiently small. Since $\mathcal{C}_N(\gamma) = \mathcal{C}_{B \setminus D_{\gamma}} + \lambda \operatorname{length}(D_{\gamma})$ and

$$| \operatorname{length}(D_{\gamma'}) - \operatorname{length}(D_{\gamma})| = \left| \sum_{n=1}^{N} \int_0^1 [\,\|\dot{\gamma}'_n(t)\| - \|\dot{\gamma}_n(t)\|\,]\,dt \right|$$

$$\leq \sum_{n=1}^{N} \int_0^1 \|\dot{\gamma}'_n(t) - \dot{\gamma}_n(t)\|\,dt \leq N\sqrt{2}\|\gamma' - \gamma\|_{C^1([0,1])^{2N}} \to 0 \qquad \text{as } \gamma' \to \gamma$$

the proof is then complete.[8] ∎

Corollary 1.6 Let $\Gamma_N \subset C^1([0,1])^{2N}$ be compact and assume that each image segmentation $\gamma \in \Gamma_N$ is admissible. Then the restricted N-segment cost $\mathcal{C}_N|\Gamma_N$ attains its minimum.

1.6. Compact Image Segmentation Spaces

Our next goal is to specify a sequence $(\Gamma_N)_{N \in \vec{N}}$ of image segmentation spaces to which corollary 1.6 will apply. For each $N \in \vec{N}$ there will then exist an optimal discontinuity set $\check{D}_N \in \mathcal{D}_N \doteq \{D_{\gamma}\}_{\gamma \in \Gamma_N}$ such that $\mathcal{C}(D, f_{B \setminus D}) \geq \mathcal{C}(\check{D}_N, f_{B \setminus \check{D}_N})\ \forall D \in \mathcal{D}_N$. However, in order to arrive at an existence result for a variable number of edge segments, we must also ensure that (7.7) is valid, that is, that

$$\inf_{\gamma \in \Gamma_N} \sum_{n=1}^{N} \int_0^1 \|\dot{\gamma}_n(t)\|\,dt \longrightarrow \infty \qquad \text{as } N \longrightarrow \infty \qquad (7.9)$$

Let $\alpha \in]0,1]$. It is well known that the space $C^{1,\alpha}([0,1])$ consisting of Hölder continuously differentiable functions $\varphi \in C^1([0,1])$ with

$$\bigvee_{l=0}^{1} \sup_{\substack{s,t \in [0,1] \\ s \neq t}} \frac{|\varphi^{(l)}(s) - \varphi^{(l)}(t)|}{|s-t|^{\alpha}} < \infty$$

is a vector space with norm defined by

$$\|\varphi\|_{C^{1,\alpha}([0,1])} \doteq \|\varphi\|_{C^1([0,1])} + \bigvee_{l=0}^{1} \sup_{\substack{s,t \in [0,1] \\ s \neq t}} \frac{|\varphi^{(l)}(s) - \varphi^{(l)}(t)|}{|s-t|^{\alpha}} \qquad \varphi \in C^{1,\alpha}([0,1])$$

---

[8] We use dot notation, such as in $\dot{\gamma}_n$, for derivatives of curve parameterization functions.

It is also well known [3], that $C^{1,\alpha}([0,1])$ can be compactly embedded in $C^1([0,1])$. In other words, any bounded subset of $C^{1,\alpha}([0,1])$ is precompact in $C^1([0,1])$. From these facts it is easy to show that

$$K_{\alpha,\rho} \doteq \left\{ \varphi \in C^{1,\alpha}([0,1]) : \|\varphi\|_{C^1([0,1])} \vee \sup_{\substack{s,t\in[0,1] \\ s\neq t}} \frac{|\dot\varphi(s) - \dot\varphi(t)|}{|s-t|^\alpha} \leq \rho \right\}$$

is compact in $C^1([0,1])$. By Tychonoff's theorem, the product space $K_{\alpha,\rho}^{2N}$ is then compact in $C^1([0,1])^{2N}$. In order to enforce compactness of Γ_N it is therefore sufficient to define Γ_N as a closed subset of $K_{\alpha,\rho}^{2N}$ relative to the $C^1([0,1])^{2N}$-topology. The limit condition (7.9) as well as the admissibility conditions can then be taken care of by intersecting $K_{\alpha,\rho}^{2N}$ with some *closed* subset of $C^1([0,1])^{2N}$ that imposes the appropriate restrictions. If we define the image segmentation spaces

$$C_N(\alpha,\rho,\omega) \doteq \{\gamma \in K_{\alpha,\rho}^{2N} : \|\dot\gamma_n(t)\| \geq \omega \quad \forall\, t \in [0,1], \quad n = 1,\dots,N\}$$

we can summarize our results as follows.

Theorem 1.7 Let $\alpha \in\,]0,1]$ and $\rho,\omega > 0$ be constants. In addition, for each $N \in \vec{N}$, let A_N be a closed subset of $C^1([0,1])^{2N}$ with the property that each image segmentation $\gamma \in \Gamma_N \doteq C_N(\alpha,\rho,\omega) \cap A_N$ is admissible, and define $\mathcal{D} \doteq \{\varnothing\} \cup \bigcup_{N=1}^\infty \{D_\gamma\}_{\gamma\in\Gamma_N}$. Then $\exists\, \check D \in \mathcal{D}$ and $\check f \in \mathcal{H}^1(B \smallsetminus \check D)$ such that the cost $\mathcal{C}$ defined in (7.1) satisfies $\mathcal{C}(D,f) \geq \mathcal{C}(\check D, \check f) \; \forall\, f \in \mathcal{H}^1(B \smallsetminus D)$ $\forall\, D \in \mathcal{D}$.

Proof: It is trivial to verify that $C_N(\alpha,\rho,\omega)$ is closed in $C^1([0,1])^{2N}$. Hence Γ_N is compact in $C^1([0,1])^{2N}$. Since each $\gamma \in \Gamma_N$ is admissible and moreover

$$\sum_{n=1}^N \int_0^1 \|\dot\gamma_n(t)\|\, dt \geq N\omega \longrightarrow \infty \qquad \text{as } N \longrightarrow \infty$$

the theorem follows from corollary 1.6. ∎

It might seem somewhat artificial and irritating that the lengths of the edge segments are bounded below by $\omega > 0$. The purpose with the regularization, imposed by the smoothness term $\mu \int_{B\smallsetminus D} \|\nabla f\|^2\, dx$ in the cost (7.1), however, is to suppress the fine scale structure of the original image. Indeed, variations corresponding to spatial frequencies $\gg 1/(2\pi\sqrt{\mu})$ are in effect annihilated. In this perspective it is not too unreasonable to demand that the length of each edge segment be no less than some constant fraction of $\sqrt{\mu}$.

With the constants ρ and ω properly chosen, it is clear that the set

$$\bigcup_{N=1}^\infty C_N(\alpha,\rho,\omega)$$

of image segmentations is rich enough to account for almost any discontinuity set of interest. The question is then whether the set A_N can be specified without imposing too severe restrictions on the image segmentations in $C^1([0,1])^{2N}$. One such specification is presented in detail in [274]. Basically, the conditions listed there are as follows.

1. No edge segment can self-intersect except possibly by beginning and ending at the same point.
2. Distinct edge segments can only intersect at endpoints.
3. Wherever edge segments self-intersect or intersect with each other or the image boundary ∂B, the tangents are different, so that no cusps are formed.
4. The discontinuity set D must have the property that each neighborhood of each $x \in D$ intersects at least two distinct connected components of $B \smallsetminus D$, (otherwise some component of $B \smallsetminus D$ will not be a Lipschitz domain.)

This last condition might seem hard to express in terms of the image segmentation γ. However, the three preceding conditions render only a finite number of edge segment interconnections, each of which can be represented by a directed planar graph. Condition 4 an moreover be shown to depend only on this graph. It is thereby reduced to the exclusion of certain edge segment interconnection graphs, which are all easily expressed in terms of γ.

Because of the requirement that A_N be closed, the conditions 1–4 in the above above must be expressed interms of a few arbitrary small constants, which make them appear rather technical. The proof of admissibility is moreover quite involved. Furthermore, this particular construction of A_N is important foremost because it shows (by example) that the minimization of the cost $\mathcal{C}(D, f)$ makes sense over a family of discontinuity sets that is large enough to be of practical interest. The details are both somewhat arbitrary and probably easy to improve and therefore of less interest. For these reasons we will not elaborate on them any further, but refer the interested reader to [274].

1.7. Concluding Remarks

We have demonstrated the existence of a solution to the cost or energy functional minimization problem posed by Mumford and Shah as a method for global edge detection or image segmentation. As in the original problem formulation, we have considered edges represented by finite unions of smooth curves. While this particular representation might fail to lend itself to the most elegant mathematical proofs, it restricts the space of possible discontinuity sets –edges– to consist only of those corresponding to line drawings.

Viewed as human descriptions, it could be argued, that the line drawings capture exactly those discontinuities that are conceptually relevant, that is, those corresponding to desirable solutions. The restriction is therefore completely in line with the regularization technique used for selecting the optimal image function within the space $\mathcal{H}^1(\Omega)$ of desirable such functions, (even though the data—the original image function—might not be in that space.)

The basic strategy has been to prove lower semicontinuity of the cost functional with respect to some appropriate topology on the space of image segmentations, and then restrict the functional to a compact subset of this space. A proof based on completeness and convergence of a minimizing sequence would of course be more pleasing, but is ruled out because of lack of uniqueness of the solution. There is no reason why the solution should be unique, although one might expect it to be generically. The restriction approach will, in principle, always succeed. One can, for example, restrict the functional to a finite space. In this extreme case the existence problem becomes trivial, (and lower semicontinuity is not even an issue.) However, if the restriction is sufficiently mild, so that the space of possible discontinuity sets contains (descent approximations of) most line drawings, then the modified problem is still of theoretical and practical interest. In order to vindicate our approach, we have constructed a specific example of such a restriction. While the resulting image segmentation space certainly is rich enough to yield an interesting optimization problem, there are some details left to be desired. Most importantly, one would like to consider discontinuity sets with free endpoints. All that this really requires, is an extension of theorem 1.3 saying that: certain simple modifications of Lipschitz domains are also admissible image segments.

Because of the restriction technique, it seems likely that the solution to the problem will actually be on the boundary of the domain. It will thus in general not be critical, that is, it will not satisfy an Euler type equation. This is unpleasant in that the optimal solution then depends on the particular restriction, which in part is quite artificial. However, given the high dimensio-nality of the image segmentation space, this flaw is more or less cosmetic; the solution might still be in the interior of a vast majority of the one-dimensional projections of the restricted domain.

An alternative way to think of our conclusions is as follows. Existence of a solution to the general problem has *not* been proven. Rather we have found, that if a cost minimizing sequence does not have a convergent subsequence, then all but at most a finite number of its points are confined to a "boundary region" of the image segmentation space. The benefit of this assertion is that we know where this "boundary region" is, and can moreover make it arbitrarily "narrow", so that it is no longer crucial for adequate representation of the discontinuity sets of interest.

Finally we should comment on the assumption (7.7). It ought to be justified on its own merits rather than simply being a necessary condition for our main conclusions to hold. Indeed, smooth surface patches in the scene generically intersect along smooth curves, whose projections onto the image plane generically are smooth. Hence the edges in the "true" image should be expected to be piecewise smooth for just about the same reasons as the "true" image function, which we are trying to estimate, is expected to be piecewise smooth. This calls for mechanisms analog to those governed by the smoothness and edge cost terms in the cost (7.1). In this context condition (7.7) is merely a straight forward analog of the edge cost. The action of the smoothness term is roughly taken care of by imposing the bound, ($\rho < \infty$,) on the Hölder modulus of the derivatives of the components of the image segmentation components γ.

APPROXIMATION, COMPUTATION, AND DISTORTION IN THE VARIATIONAL FORMULATION

Thomas Richardson

AT&T Bell Laboratories
Murray Hill, NJ 07974, USA

and

Sanjoy Mitter [1]

Center for Intelligent Control Systems
Laboratory for Information and Decision Systems
Massachusetts Institute of Technology
Cambridge, MA 02139, USA

1. Introduction

Mumford and Shah [263][264] suggested performing edge detection by minimizing functionals of the form

$$E(f, K) = \beta \int_{\Omega} (f - g)^2 dx + \int_{\Omega - K} |\nabla f|^2 \, dx + \alpha \, |K| \, ,$$

where Ω is the image domain (a rectangle), dx denotes Lebesgue measure, g is the observed grey level image, i.e., a real valued function, f approximates g, K denotes the set of edges (a closed set), $|K|$ is the total length[2] of K, and β and α are real positive scalars. This approach is a modification of one due to Geman and Geman [132] that uses Markov random fields, which

---

[1]Research supported by the US Army Research Office under grant ARO DAAL03-92-G-0115 (Center for Intelligent Control Systems).

[2]In general K cannot be represented as a union of curves so one-dimensional Hausdorff measure is used to define 'total length.' See the contribution of Leaci and Solimini in this book.

was developed by Marroquin [249] and by Blake and Zisserman [34]. It is referred to as the variational formulation of edge detection.

The three terms of E 'compete' to determine the set K and the function f. The first term penalizes infidelity of f to the data while the second term forces smoothness of the approximation f, except on the edge set K. Thus, f will be a piecewise smooth approximation to g. The third term forces some conservativeness in the use of edges by penalizing their total length.

The primary difficulty in constructing an efficient algorithm to minimize E is appropriately representing the edges. One approach is to absorb the edges into the interaction between neighboring pixels. This idea appears in the anisotropic diffusion approach [293] [273], GNC (Graduated non-convexity) type algorithms [34], and in mean field annealing [33][129]. Another approach has emerged from the cross fertilization between computer vision and mathematical physics. The variational formulation, besides being a model for edge detection, now serves as a prototypical example of a 'free-discontinuity' problem. (See the contribution of Leaci and Solimini for an elaboration of this.) Within the framework developed for such problems the theory of Γ-convergence, approximation of one functional by another, has also been developed. Ambrosio and Tortorelli [12, 13] applied that theory to E. The approximation is achieved primarily by replacing the edge set with a function that modulates the smoothing of the image. Additional terms in the functional force this function to behave as if it were a smeared version of the corresponding edge set. The degree of smearing, i.e., the width of the effective edges, is controlled by a parameter. Convergence of the approximation to E occurs by taking the parameter to the appropriate limit, causing the effective edge width to tend to zero.

A benefit of replacing the edge set with a smooth function is that an obvious algorithmic approach suggests itself: discretize the functions and minimize the functional using descent methods. The product is an edge detector that can be represented as a coupled pair of non-linear partial differential equations (see Section 3). The idea of using coupled partial differential equations of this type is now being applied to many problems in computer vision. The chapter in this book by Proesmans, Pauwels, and Van Gool contains several examples. March [244] applied this approach to the stereo matching problem. In Section 2, we will outline the application of Γ-convergence to the variational formulation of edge detection and show how it leads to a coupled pair of partial differential equations.

The variational formulation was motivated in part by the desire to combine the processes of edge placement and image smoothing. Earlier edge detection techniques such as the Marr-Hildreth edge detector, the Canny edge detector, and their variants separated these processes; the image is first smoothed to suppress noise and control the scale, and edges are detected subsequently, as gradient maxima, for example. One consequence of this two step approach is pronounced distortion of the edges, especially at high curvature locations. Corners tend to retract and be smoothed out; the connectedness of the edges at T-junctions is lost. By introducing interaction between the edge placement and the smoothing, it was expected that this effect could be abated. There is evidence, both theoretical, in one dimension (see [34]), and experimental, in two dimensions, that this is indeed the case. However, the model is known to place undesirable restrictions on admissible edge geometries (see Section 5). Nevertheless, certain limit theorems that have been proven by one of us in [306] and are discused in Section 5 indicate that the restrictions may not be too serious and that, asymptotically, any edge geometry is possible. Moreover, heuristic arguments suggest that the approximate formulation indicated above behaves better with respect to these distortions/restrictions than the original model. The relaxation of the distortion is achieved at the cost of the smearing of the edges. Hence, there is trade-off here between the resolution of the edges and the systematic distortion of the model.

Localization of edges cannot be reasonably discussed without also making reference to scale. The notion of 'scale', scale of features and scale of representation, is widely held to be of fundamental importance in vision. (This book contains two substantial chapters devoted to 'scale-space.') One reason for this is that hierarchal descriptions of scenes offer potential reductions in complexity of various visual processing tasks. Coarse scale segmentation of an image, for example, can be used to identify regions of interest for further processing, thereby reducing the computational load. It is important, therefore, that coarse scale descriptions retain those features of the data that are required for effective decision making. In the case of edge detection, T-junctions and corners play important roles in estimating the depth and shape of objects in a scene [126]. It is desirable, therefore, to accurately represent these features even at coarse scales. The 'finger-print' images of gradient maxima of one dimensional images in scale-space [386][395] are well known; the localization of edges degrades badly as scale increases. Many two dimensional examples can be found in the literature.

By embedding the ideas implicit in the limit theorems mentioned above into the approximation scheme also mentioned above, one can develop an edge focusing scheme that essentially removes the restrictions on the edge geometry present in the original model and, at the same time, circumvents the smearing/geometry trade-off to produce well localized, sharp edges. We indicate how this is done in Section 6. A complete description can be found in [307]. The resulting algorithm is described by a coupled set of non-linear second order parabolic partial differential equations (eqns. (8.5)–(8.8)) with explicit parameters β and c which are adjusted in an appropriate way (see eqns. (8.9)–(8.11)). The adjustment induces focusing of the edges. The global coarse scale nature of the edges is retained by introducing scale stabilizing feedback mechanisms. The adjustment process commences after the non-linear parabolic equations have nearly converged to their equilibrium. The set of equations (8.5)–(8.8) and (8.9)–(8.11) should be viewed as an adaptive non-linear filter which performs edge detection via focusing. Indeed, the equations are the fundamental objects in this theory and, apparently, are far more well behaved (for example, convergence to global minima) than the original variational problem. It is apparent that the form and the properties of the Γ-convergent approximation mesh well with the parameter adjustment proposed for the edge focusing algorithm. In particular, the relaxation of the edge geometries due to the smearing of the edges is retained, by adjusting the parameters, while the edges are sharpened

2. Approximation via Γ–Convergence

To compute minimizers of E the critical question is how to represent the set K. This issue was raised in the section of this book by D. Mumford. A natural approach is to discretize K into "edge elements" and treat them combinatorially, adding or removing elements in an attempt to minimize E. Appending a stochastic component, based on a Markov random field model, leads to the simulated annealing approach first suggested in [132]. This tends to produce computationally impractical algorithms. Modifications which incorporate the edge elements into the interaction between image pixels have been proposed. One of these is based on mean field approximations of the Markov random field [128] [33] and another, GNC [34], is based on a homotopy of the interactions. Both these approaches have their strong points, and are in fact quite similar [33] [129]. A novel and powerful approach has appeared from the mathematical theory of approximation of functionals via Γ-convergence.

The concept of Γ-convergence is due, independently, to E. De Giorgi [84] and H. Attouch [23]. The idea is to approximate one functional, E for example, by more regular ones, E_c, so that minimizers of E_c approximate minimizers of E while enjoying greater regularity. For the variational formulation of edge detection, one would like to replace the edges, which are singular objects (in the context of 2-dimensional measure), with something more manageable.

In this section, we provide a definition of Γ-convergence, state some of its basic properties, and present the application to the variational formulation of edge detection.

Let (S, d) be a separable metric space and let $F_n : S \to [0, +\infty], n = 1, 2, \dots$ be a sequence of functions. We say this sequence $\Gamma(S)$ – converges to F :

$S \to [0, +\infty]$ if the following two conditions hold for all $f \in S$,

$$\forall f_n \to f \quad \liminf_{n \to \infty} F_n(f_n) \geq F(f)$$
$$\text{and } \exists f_n \to f \quad \liminf_{n \to \infty} F_n(f_n) \leq F(f).$$

The limit F, when it exists, is unique and lower–semicontinuous. The following theorem characterizes the main properties of Γ–convergence.

Theorem (Γ–Convergence.) Assume that $\{F_n\}$ $\Gamma(S)$–converges to F. Then the following statements hold.
(i) Let $t_n \downarrow 0$. Then, every cluster point of the sequence of sets

$$\{f \in S : F_n(f) \leq \inf_S F_n + t_n\}$$

minimizes F.
(ii) Assume that the functions F_n are lower semicontinuous and, for every $t \in [0, \infty)$, there exists a compact set $C_t \subset S$ such that for all n $\{f \in S : F_n(f) \leq t\} \subset C_t$. Then the functions F_n have minimizers in S, and any sequence of minimizers of F_n admits subsequences converging to some minimizer of F.

The point of (i) is that approximate minimizers of F_n approximate minimizers of F. Condition (ii) is useful for proving that minimizers of F_n exist and are well behaved.
To properly formulate the Γ–convergence results it is convenient to define E in terms of f alone, letting K be implicitly defined as the closure of the discontinuity set of f. This implicit definition is described in the chapter by Leaci and Solimini in this book.
We consider functionals of the form

$$E_c = \int_\Omega \left(\beta(f - g)^2 + \Phi(v) \mid \nabla f \mid^2 + \alpha(c\Psi(v) \mid \nabla v \mid^2 + (1 - v)^2/4c) \right) \ dx.$$
$$(8.1)$$

Here $\Phi(v)$ is playing the role of the K in E, i.e., it is modulating the smoothness constraint on f. The other terms involving v force $\Phi(v)$ to simulate the effect that K has in E. Implicitly, we have $0 \leq v \leq 1$. The algorithmic intention is to minimize E_c with respect to f and v. An obvious advantage the approximation offers over the original formulation is that v, since it is a function on Ω, can be discretized in a straightforward way and (local) minimizers of E_c can be computed using descent methods. Ambrosio and Tortorelli [13] proved that if one sets

$$\Phi(v) = v^2 \quad \text{and} \quad \Psi(v) = 1, \qquad (8.2)$$

then E_c Γ-converges to E as $c \to 0$, i.e., for any sequence $c_n \to 0$. Some computational results based on this functional have already appeared in March [245], see also Shah [341].

The choice for Φ and Ψ given above may be one of the simplest possible, but it is far from unique. (See [12] for an example which is not of the form 8.1.) When one considers algorithms based on these functionals there are trade-offs to be made between speed and performance. For example, the choice reflected in equation (8.2) leads to simple equations and fast computation. However, other choices may produce sharper singularities in Φ and hence less smearing of f near the edges. With slight modifications, the proof of Γ-convergence found in [12] and [13] can be made to go through for a large class of Ψ and Φ. In particular, one can choose Ψ to be any C^1 function satisfying

$$\Psi(x) > 0 \text{ for } x \in (0, 1],$$

$$2 \int_0^1 (1 - u)\Psi^{1/2}(u)du = 1.$$

Note that any C^1 function satisfying the first property can be made to satisfy the second property by suitable normalization. Given such a Ψ, one can choose Φ to be any C^1 function satisfying

$$\Phi(1) = 1,$$
$$\Phi(0) = 0,$$
$$\Phi(x) \in (0, 1) \text{ for } x \in (0, 1).$$

Although the conditions given above are sufficient for the proof of Γ-convergence, for algorithms based on 'gradient' descent on E_c one should also impose the condition that Ψ be monotonically non-decreasing and Φ be monotonically increasing on $(0, 1)$. Furthermore, for our implementation, which is discussed in Section 7, the condition $\lim_{x \to 0} \dot{\Phi}(x)/x < \infty$ should be imposed.

Even more general Ψ and Φ than defined above are possible. For example, setting

$$\Psi(v) = \Phi(v) = \frac{1}{2}e^{-(1-v)^2} \tag{8.3}$$

also produces a Γ-convergent set of functionals. Examples in the class defined above are

$$\Phi(v) = v^{2n} \quad \text{and} \quad \Psi(v) = \frac{(m+1)^2(m+2)^2}{4}v^{2m}, \tag{8.4}$$

where $m \geq 0$ and $n > 0$. Equation (8.2) is a special case of this with $(n, m) = (1, 0)$.

To formulate the Γ-convergence of E_c to E, one must find an appropriate metric space for f and v and extend the definition of E and E_c to this space. In [13], the choice of S is

$$(f, v) \in L^\infty(\Omega) \times \{v \in L^\infty(\Omega) : 0 \leq v \leq 1\},$$

and the metric is that induced by the $L^2(\Omega)$ norm. (In [12], slightly different choices were made.) First, it is possible to mathematically recast E (in a weak setting with $f \in \mathrm{SBV}(\Omega)$, see the chapter by Leaci and Solimini), whereby one defines K in terms of the discontinuities in f. In this way one can write $E(f, K) = E(f)$. The functional E can then be extended to S by setting $E(f, v) = E(f)$ if $v = 1$ (in $L^2(\Omega)$) and $f \in \mathrm{SBV}(\Omega)$ and $E(f, v) = \infty$ otherwise. Some additional care must be taken to define E_c on all of S. When ∇v and ∇f are not appropriately well defined, one should set $E_c(f, v) = \infty$; we refer the reader to [12] or [13] for technical details. Once the metric space is specified, the proof of Γ-convergence involves basically two steps. The first is to show that for any sequence $\{f_{c_i}, v_{c_i}\}$ where $c_i \to 0$, $f_{c_i} \to f$, and $v_{c_i} \to 1$ (in the appropriate sense, not pointwise) that $\liminf_{i \to \infty} E_{c_i}(f_{c_i}, v_{c_i}) \geq E(f, K)$. The second is to construct a sequence such that $\limsup_{i \to \infty} E_{c_i}(f_{c_i}, v_{c_i}) \leq E(f, K)$. If (f, K) minimizes E, then this second step requires constructing near minimizers of E_c. If

$$\liminf_{i \to \infty} E_{c_i}(f_{c_i}, v_{c_i}) < \infty,$$

then, roughly speaking, $x \in K$ implies $\lim_{i \to \infty} \Phi(v_{c_i}(x)) = 0$. (Thus at these points we do not have $v_{c_i}(x) \to 1$; however, K is a set of Lebesgue measure 0.) On the other hand, the last term in equation (8.1) forces $v_c(x)$ to converge to 1 for almost all $x \in \Omega$ (in the sense of Lebesgue measure), hence one has $\lim_{c \to 0} \Phi(v_c(x)) = 1$ for almost all $x \in \Omega$. The near minimizers of E_c are constructed by setting $\Phi(v_c(x)) \simeq 0$ on K and $\Phi(v_c(x)) \simeq 1$ outside some neighborhood of K, with a smooth transition in between. The approximations indicated here become equalities in the limit as $c \to 0$. The width of the transition depends on Ψ and on c. We give a brief heuristic description of how this occurs.

In the transition region of $\Phi \circ v$, we expect f to be relatively smooth, so only the terms not involving f in E_c will have a significant effect on the form of v there. In the following inequality,

$$c\Psi(v) \mid \nabla v \mid^2 + \frac{(1 - v)^2}{4c} \geq \Psi^{1/2}(v) \mid \nabla v \mid (1 - v),$$

equality holds only if $\mid \nabla v \mid = \Psi^{-1/2}(v) \frac{1-v}{2c}$. This suggests that (in one dimension) if $u_c(t)$ satisfies $\frac{\partial u_c(t)}{\partial t} = \frac{1-u_c(t)}{2c} \Psi^{-1/2}(u_c(t))$ with $\Phi(u(0)) \simeq 0$, then setting $v(x) = u_c(\text{dist}\,(x,\,K))$, for dist $(x,\,K) \leq \tau_c$ where $\Phi(u_c(\tau_c)) \simeq 1$ (with $u_c(\tau_c) \to 1$ as $c \to 0$), will produce near optimal transitions. This is how the near optimal v_c are constructed in [12] and [13]. Note that assuming that $u_c(0)$ does not depend on c, we obtain $u_c(t) = u_1(t/c)$. Thus the edge width is proportional to c. Let $G(s) = \int_0^s (1-r)\Psi^{1/2}(r)dr$. We now compute

$$\int_0^{\tau_c} \left(c\Psi(u(t))\mid \frac{\partial u_c(t)}{\partial t}\mid^2 + \frac{(1-u(t))^2}{4c} \right) dt \;\; =$$

$$\int_0^{\tau_c} \left(\Psi^{1/2}(u(t))\mid \frac{\partial u_c(t)}{\partial t} \mid (1-u(t)) \right) dt \;\; =$$

$$\mid \int_0^{\tau_c} \frac{\partial}{\partial t} G(u(t))dt \mid = G(\tau_c) \;\; \simeq \;\; \frac{1}{2}$$

with the last approximate equality becoming equality in the limit as $c \to 0$. In the one dimensional case, we now see that the last term in 8.1 will contribute approximately α times the number of discontinuity points of f. In two dimensions, one obtains approximately α times the length of K. Thus, we see that E_c approximates E.

3. Minimizing E_c

Local minimers of E_c can be found by gradient descent. Simple, practical algorithms can be obtained from finite element discretizations. In this section, we formulate the main ideas in the continuum setting.

The Euler-Lagrange equations associated with E_c are given by $\partial_v E_c = 0$ and $\partial_f E_c = 0$ with Neumann boundary conditions, where

$$\partial_f E_c \overset{\Delta}{=} \beta(f - g) - \nabla \cdot (\Phi(v)\nabla f)\,, \tag{8.5}$$

$$\partial_v E_c \overset{\Delta}{=} \dot{\Phi}(v)\,\alpha^{-1}\mid \nabla f \mid^2 \; - c\nabla \cdot (\Psi(v)\nabla v) + 2c\dot{\Psi}(v)\mid \nabla v \mid^2 +$$
$$(1 - v)/2c \tag{8.6}$$

are the functional derivatives of E_c.

Allowing f and v to depend on t, we can write a 'gradient' descent on E_c in the form

$$\frac{\partial}{\partial t} f(x, t) = -c_f \partial_f E \tag{8.7}$$

$$\frac{\partial}{\partial t} v(x, t) = -c_v \partial_f E \tag{8.8}$$

with Neumann boundary conditions, where c_f and c_v control the rates of descent; they would be constant for a strict gradient descent, but may not be in general. It turns out that better implementations (faster and with more stable convergence) can be obtained with c_v not held constant (see Section 7).

Since the functional E_c is not jointly convex in v and f, we do not expect to always reach a global minimum by a descent method. Thus the solution obtained will depend on the initial conditions and also on the parameters c_f and c_v.

Equation (8.7) (after substituting from equation (8.5)) strongly resembles the anisotropic diffusion scheme for image enhancement introduced by Perona and Malik [293] and, even more strongly resembles, the 'biased' anisotropic diffusion scheme of Norström [273]. The solution to the diffusion equation

$$\frac{\partial}{\partial t} f(x,t) = \Delta_x f(x,t), \quad f(x,0) = g(x)$$

(with Neumann boundary conditions) is identified with 'scale-space' smoothings of g, parametrized by t. The solution $f(x,t)$ is obtained by convolving $g(x)$ with a Gaussian kernel whose variance is linear in t. Perona and Malik [293] suggested performing image enhancement by controlling the diffusion coefficient to prevent smoothing across edges. Thus they were led to consider equations of the form

$$\frac{\partial}{\partial t} f(x,t) = \nabla_x \cdot (h(|\nabla_x f(x,t)|)\nabla_x f(x,t)), \quad f(x,0) = g(x).$$

They experimented with $h(s) = \frac{J}{1+(\frac{s}{K})^2}$ and $h(s) = e^{-(\frac{s}{K})^2}$, where J and K are constants. Equation (8.7) resembles this equation in that it is a diffusion with controlled conductivity. The control of the conductivity depends on $|\nabla f|$ indirectly through equation (8.8). The term $\beta(f - g)$ in equation (8.7) stabilizes the solution at some particular scale. Such a term also appears in the 'biased' anisotropic diffusion scheme studied in [273]. In [293], the authors analyze their scheme to show that the maximum principle holds, i.e., that the solution's extrema never exceed those of the original image[3]. They argue that this implies that no new 'features' (blobs) are introduced into the solution. Here, as in [273], this property is a trivial consequence of the formulation. The functionals E_c would *increase* if such new features appeared. (Truncating such a new feature would decrease E_c.) An advantage of the scheme represented by equations (8.7) and (8.8) is that it yields an explicit representation of the edges (via the function $\Phi(v)$). The resulting system of equations admits a particularly simple implementation in digital mesh connected parallel machines with simple processors or,

---

[3]It is unclear that the maximum principle can be invoked since an appropriate existence theorem for the Perona-Malik equation has not been proved.

potentially, in an analog network, such as discussed in [149]. Although a direct study of the existence, uniqueness, and well-posedness of (8.7) and (8.8) has not been carried out, local stability of these equations can be proved [305].

4. Remarks on Energy Functionals, Associated Non-Linear Diffusions and Stochastic Quantization

The methodology of minimizing energy functionals involving an "elliptic" form and a "geometric" term (leading to free discontinuity problems) for filtering and boundary reconstruction of images is quite general. For example, a modified energy functional which does not have some of the disadvantages of the functional considered in this chapter is

$$E(f, K) = \int_\Omega (f - g)^2 dx + \int_{\Omega/K} |\nabla f|^2 dx + \int_K |H^2| d\mathcal{H}^1$$
$$+ \text{ No. of singular points in } K \text{ with multiplicities,}$$

where H is the curvature of K (appropriately defined), and $\mathcal{H}^1$-represents the one-dimensional Hausdorff measure. This class of problems has been considered in the recent work of Ambrosio and Mantegazza (cf. thesis of Mantegazza at the University of Pisa in 1993). If an elliptic approximation to this class of free-discontinuity problems can be found, then the method of computation of minimizers via non-linear parabolic equations can in principle be constructed.

There is a connection between the ideas presented in this chapter and the ideas of stochastic quantization and the Bayesian formulation of Image Analysis (cf. Mitter [255] and the chapter of Mumford in this book). For this purpose consider

$$\exp(-E_c) = \exp(-\beta \int_\Omega (f - g)^2 dx) \cdot$$
$$\exp(-\int_\Omega [\Phi(v)|\nabla f|^2 + \alpha(c\psi(v)|\nabla v|^2 + \frac{(1 - v)^2}{4c})] dx)$$
$$\triangleq \exp(-\beta L(f, g)) \exp(-\Lambda_c(f, v)).$$

Then it is necessary to give meaning to the expression

$$d\mu_c = \exp(-\Lambda_c(f, v)) \prod_{x \in \mathbb{R}^2} d(f(x), v(x))$$

as a probability measure on an appropriate distribution space. This probability measure is formally interpreted as a prior measure on (f, v)

while the expression $\exp(-\beta L(f, g))$ is formally interpretated as a likelihood function.

The minimization of the energy functional E_c corresponds to computing the maximum a posteriori probability estimate of (f, v), but the probabilistic interpretation via the measure μ_c and the likelihood function allows one to compute other estimates such as the conditional mean estimate. Finally, the connection to stochastic quantization can be made by considering the coupled infinite dimensional stochastic differential equations

$$df(t, x) = -\partial_f E_c dt + d\xi(t, x)$$

$$dv(t, x) = -\partial_v E_c dt + d\eta(t, x),$$

where

$$\partial_f E_c \text{ and } \partial_v E_c$$

denote functional derivatives as above and ξ and η are infinite dimensional Brownian motions. Formally, μ_c is the invariant measure of this coupled pair of stochastic differential equations.

5. Scale, Noise, and Accuracy

The notion of scale-space representation of an image is a central one in this book. With regard to edge detection one of the central problems arising from the scale-space concept is the correspondence problem: which of the fine scale edges correspond to coarse scale edges? The problem is aggravated by the fact that the distortion of the edges, mentioned earlier, depends on scale. Typically 'coarse scale' implies more smoothing and, hence, more distortion. This is undesirable in many situations because salient features, corners and T-junctions for example, tend to be obscured. The correspondence problem is therefore of great importance.

Since the variational approach combats the distortion caused by smoothing, one hopes that the correspondence problem will be alleviated by using it. Although this appears to be the case, problems remain; there still are distortions, depending on scale, and the model intrinsically restricts the geometry of possible edge sets in an unnatural way. The analysis of Mumford and Shah [263] showed that edge sets produced by the variational approach have the following properties, which are illustrated in figure 8.1.

If K is composed of $C^{1,1}$ arcs, then

- at most three arcs can meet at a single point and they do so at 120°,
- they meet $\partial\Omega$ only at an angle of 90°,
- it never occurs that exactly two arcs meet at a point (other than the degenerate case of two arcs meeting at 180°), i.e., there are no corners.

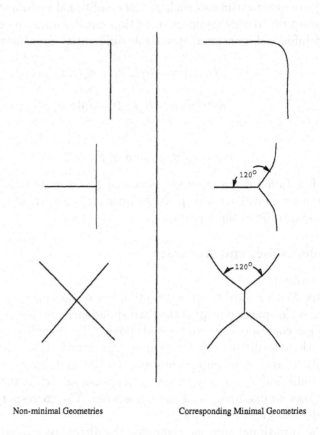

Non-minimal Geometries   Corresponding Minimal Geometries

Figure 8.1. Calculus of Variations Results

These results are a consequence of the fact that the term $|K|$ in E locally dominates the behavior of singularities in K. Hence, the types of singularities observed are identical to those of minimal surfaces. Of course 'real' edges are not restricted to these geometries. The dependence on scale derives from the interaction between the singularities of f and those of K. Roughly speaking, smaller values of β produce greater distortion. This becomes more clear in light of the theorem quoted below and the analysis of Mumford and Shah [264]. It turns out that under some mild assumptions on the data, it is possible to prove that the edge set produced by the variational formulation can approximate arbitrarily well, depending on the parameters, any edge geometry.

To quantify the disparity between one edge set and another, we introduce the Hausdorff metric. For $A \subset \mathbb{R}^2$, the ϵ-neighborhood of A will be denoted by $[A]_\epsilon$ and is defined by $[A]_\epsilon = \{x \in \mathbb{R}^2 : \inf_{y \in A} |x - y| < \epsilon\}$ where $|\cdot|$ denotes the Euclidean norm. Denoted by $d_H(\cdot,\cdot)$, the Hausdorff metric is evaluated by

$$d_H(A, B) = \inf\{\epsilon : A \subset [B]_\epsilon \text{ and } B \subset [A]_\epsilon\}.$$

Elementary analysis shows that d_H is a metric on the space of non-empty compact sets in R^2.

Suppose for the moment that we have ideal data: g is a piecewise smooth function. To make this clear we denote it by g_I. We assume that there exists a set K_g, a union of curves, satisfying length $(K_g) < \infty$ such that g_I is discontinuous on K_g and smooth elsewhere. More precisely, we require that $\int_{\Omega - K_g} |\nabla g_I|^2 \, d\mu < \infty$, that there exists a constant L such that g_I, restricted to any straight line segment lying in $\Omega - K_g$, is a Lipschitz function with Lipschitz constant L, and that g actually has a discontinuity everywhere on K_g except, possibly, for a set having zero total length. Under these conditions it can be proved that minimizers of E have the property that the edge sets converge to Γ_g in Hausdorff metric as $\beta \to \infty$. This means, in particular, that edge sets can assume arbitrary geometries asymptotically. This result can be interpreted as an asymptotic fidelity result for the variatio- nal approach. It implies that the distortions resulting from ad hoc functions of this type are local, small scale effects.

From a practical point of view this convergence is inadequate because $\beta \to \infty$ forces f to match g_I exactly; noise in g_I will result in the appearance of many spurious boundaries. However, noise and smearing effects can be incorporated into the result if they are scaled appropriately. Roughly speaking, if the admissible noise magnitude scales as $o(\beta^{-\frac{1}{2}})$ and the admissible smearing acts over a radius of $o(\beta^{-1})$, then their presence can be tolerated and we have the following,

Theorem (Approximation.) *For any fixed $\alpha > 0$ and $\epsilon > 0$ there exists $\beta^* < \infty$ such that if $\beta \geq \beta^*$ and K_β is minimal for E for some $g \in \Psi(\beta)$, then*

$$d_H(K_g, K_\beta) < \epsilon \, .$$

The proof may be found in [306] and [305]. The relevant mathematical framework is outlined in [11]. This theorem indicates how noise and localization defects should scale with the parameter β to maintain fidelity.

6. Edge Focusing via Scaling

In [307] the theory outlined above is applied to the Γ-convergent approximation to the variational formulation to produce an edge-focusing algorithm. The basic idea of edge-focusing is to start with coarse scale edges and then adjust them to get better localization without introducing finer scale edges. It is an attempt to circumvent the noise/accuracy/scale tradeoffs inherent in the underlying edge detection model.

Bergholm [31] developed an edge focusing algorithm based on the Marr-Hildreth edge detector. Since the variational model has inherently better localization, the correction required by the focusing algorithm is smaller. Prior to the existence of the theory outlined in previous sections, the main barrier to edge focusing based on the variational formulation was the computational difficulty of implementing such a scheme.

The edge focusing algorithm developed in [307] is based on the scaling conditions of the Approximation Theorem and on the Γ-convergent approximation to the variational formulation. Two problems immediately suggest themselves with regard to applying the Approximation Theorem. First, a real image has fixed noise which cannot be scaled since it cannot be identified, and second, smearing is fixed and cannot (in general) be removed. The algorithm resolves these objections by making a heuristic identification of noise with error and of distortion of the edges with smearing. A minimizer of the variational formulation provides a piecewise smooth approximation to the data and a nominal set of edges. If we assume that the edges are essentially correct but imprecisely localized, then, according to the Approximation Theorem, if we increase β, then the localization should improve. To prevent the introduction of smaller scale edges, we smooth the data g. This is accomplished by introducing feedback from f into g. This smoothing is suppressed in a neighborhood of the coarse scale edges to prevent smearing of the true edges in the data. These various mechanisms are balanced in accordance with the scaling conditions of the Approximation Theorem to achieve a stable, convergent algorithm. The edge focusing algorithm adheres to this paradigm with the additional feature that the variational formulation is replace by Γ-convergent approximation, where the degree of approximation is refined as the algorithm proceeds.

Since we intend to use an approximation to the variational formulation, it is prudent to consider whether the approximation deviates in a significant way from the original formulation with regard to distortion of edges. Although analysis is prohibitively difficult, we expect (and simulations have borne this out) that the spreading of the edges in the Γ-convergent approximation actually ameliorates some of the geometric distortion. We recall that the primary reason for the geometric distortion is that the term $|K|$ in E determines the structure of the singularities. Roughly speaking, this arises because the length term is one dimensional and scales linearly while the other terms are two dimensional integrals and hence scale quadratically in the size of the domain. (Actually this is not precisely true because singularities arise in f, but the dominance of the length term still occurs at singularities in K.) When the edges are smeared and length is replaced by a two-dimensional integral the concentration of cost in the length term is alleviated and, hence, we expect the distortion to be relaxed. The price paid for this is the lack of resolution of the edges. The edge focusing algorithm begins with thick edges, thus relaxing distortion, but ends by sharpening the edges while scaling the parameters in accordance with the Approximation Theorem. Thus, the edges are focussed as resolution increases.

Edge focusing is achieved by perturbing equations (8.7) and (8.8), introducing dynamics into β, c, and g. The additional dynamics take the following form,

$$\frac{\partial}{\partial t} g(x,t) = \epsilon \rho(v(x,t))(f(x,t) - g(x,t)) , \tag{8.9}$$

$$\frac{\partial}{\partial t} \beta(t) = \epsilon \beta(t) , \tag{8.10}$$

$$\frac{\partial}{\partial t} c(t) = -\epsilon c(t) , \tag{8.11}$$

where ϵ is a small positive constant included to reflect the fact that these equations are perturbations of equations (8.7) and (8.8).

The dynamics in β, c, and g are intended to come into effect only after the basic descent equations (8.7) and (8.8) have essentially converged. Thus, we assume $g(x,0)$ is the initial data and $f(x,0)$ and $v(x,0)$ satisfy their respective Euler-Lagrange equations with $\beta = \beta(0)$ and $g(x) = g(x,0)$. This implies the presence of a nominal set of edges, i.e., a function $v(x,0)$. We will be guided by the heuristic that the subsequent focusing should only focus the edges already found and not introduce new ones.

To understand the effect of these equations, it is best to first assume $\rho \equiv 1$ and eliminate equation (8.11), i.e., fix $c(t) = c(0)$. When this is done, one obtains the solution

$$\begin{aligned}
v(x,t) &= v(x,0) , \\
f(x,t) &= f(x,0) , \\
\beta(t) &= \beta(0)e^{\epsilon t} , \\
g(x,t) &= g(x,0)e^{-\epsilon t} + f(x,0)(1 - e^{-\epsilon t}) .
\end{aligned}$$

It can be checked that $\partial_f E = 0$ and $\partial_v E = 0$ for all t and that $f(x,t)$ (locally) minimizes E for all t. From the perspective of the Approximation Theorem, if we define $g_I = f(x,0)$ then by appropriately interpreting $g(x,t)$ as a smeared, noisy version of g_I we see that these equations effectively implement a continuous version of the limit process described in the Approximation Theorem. However, with the simplifying assumptions made here, there is no change in the edge function with time, i.e., no edge focusing. This is the role played by the dynamics in c and the function ρ. Consider now the full set of equations (8.9)-(8.11). Equation (8.11) reduces c. This heuristic mimics the Γ-convergence to the variational formulation. Thus, edges are sharpened as t increases. Equation (8.9) introduces feedback from f into g, effectively smoothing g. We choose ρ to suppress the smoothing of g in a neighborhood of the edges, i.e., we choose ρ so that $\rho(v(x,t))$ will be approximately zero inside some neighborhood of the edges and approximately one outside some larger neighborhood. We make a correspondence with the conditions of the Approximation Theorem by interpreting the effect of ρ on $f(x,t)$ as a smearing of the edges in the ideal data which is iterpreted as $\lim_{n \to \infty} f(x,t)$.[4] Hence, the width of the larger neighborhood should shrink as $\beta^{-1}(t)$. A simple and reasonable choice, for example, is $\rho = \Phi$ since in this case the neighborhood width is proportional to $c(t) = (c(0)/\beta(0))\beta^{-1}(t)$. Since the edges in the data are not smoothed and $\beta(t)$ is becoming large, the singularities of the edge function should converge to the 'true' edge locations.

7. Discretization and Parameter Choices

In this section, we address some of the issues which arise as a consequence of discretization of E_c. In particular, appropriate step sizes for the discrete versions of the descent algorithm are given, and the relative rates of the gradient descent and the scaling dynamics are considered. A more detailed version of this section can be found in [307].

We assume a discretization in which f and g are defined on a rectangular subset of $\mathbb{Z}^2$, i.e., the lattice generated by the vectors $(0,1), (1,0)$. The discrete version of v is defined on the inter-leaving subset of a square

---

[4]See [307] for a detailed description of the correspondences.

lattice which is twice as dense and rotated 45°, i.e., the lattice generated by $(1/\sqrt{2}, 1/\sqrt{2})$, $(1/\sqrt{2}, -1/\sqrt{2})$ and translated by $(1/2, 0)$. Presumably one can consider different lattice spacings, but it turns out that, by scaling, the lattice spacing can be absorbed into the other parameters.

Each pixel y in the discretization of v is uniquely associated with the two nearest pixels x and x' in the discretization of f. For each such y let $df(y)$ denote $(f(x) - f(x'))^2$. The derivatives of Ψ and Φ as real functions will be denoted $\dot{\Psi}$ and $\dot{\Phi}$ respectively. Note that in equations (8.7) and (8.8), time scaling can be absorbed into c_f and c_v. Thus, time is discretized simply by substituting

$$\frac{\partial}{\partial t} f(x, t) \to f(x, t+1) - f(x, t) ,$$

$$\frac{\partial}{\partial t} v(y, t) \to v(y, t+1) - v(y, t).$$

We now address the question of choosing c_f and c_v. A standard gradient descent would have both c_f and c_v constant. If we try to set c_v constant then it must be chosen small since $|\nabla f|^2/\alpha$ can be quite large, hence convergence will be slow. A computationally efficient choice that gives much faster convergence is to approximate a Newton type descent. By appropriately choosing $c_v(y)$ and making certain mild approximations[5], we obtain

$$v(y, t+1) = \frac{1}{2} \left(v(y) + \frac{\frac{1}{4c} + 2c\dot{\Psi}(v(y)) \sum_{y' \in \mathcal{N}_v(y)} v(y')}{\dot{\Phi}(v(y)) df(y)/(\alpha v(y)) + \frac{1}{4c} + 8c\dot{\Psi}(v(y))} \right), \quad (8.12)$$

where all quantities on the right hand side are evaluated at time t. This update formula enjoys many desirable properties. Note, in particular, that $v(y, t+1)$ is an average of $v(y, t)$ and a well behaved quantity which lies between 0 and 1, so $v(y, t+1) \in [0, 1]$ is guaranteed. Setting c_f constant is much less problematic, and for our simulations we have set it to $(2\beta + 8)^{-1}$ since this gives a good rate of convergence without allowing overshoot in f. The initialization of f and v will effect which local minimum is reached by the initial gradient descent. We expect this will have little effect on the edge focusing part of the algorithm. We have experimented with $f(0) = g(0)$ and also letting $f(0)$ be the solution to $f = \beta(0)\Delta(f - g)$ with Neumann boundary conditions. The second choice is better when more smoothing is desirable, i.e., in noisy or textured images. In general, we set $v(0) = 1$. With these choices, we observe that the basic descent on f and v converges in about 30 iterations for the range of parameters we have experimented with. (Larger values of c and smaller values of β will reduce the rate of convergence.)

---

[5] An approximation is only required when $\dot{\Psi} \not\equiv 1$; the details may be found in [307].

The scaling dynamics associated with edge focusing can be discretized in a straightforward way.

$$
\begin{aligned}
g(x, t+1) &= g(x,t) + \epsilon \rho(x,t)(g(x,t) - f(x,t)) , \\
\beta(t+1) &= (1-\epsilon)^{-1}\beta(t) , \\
c(t+1) &= (1-\epsilon)c(t) ,
\end{aligned}
$$

where ρ and ϵ are to be specified in each case.

Termination of the computation is best controlled through the value of $c(t)$. As $c(t)$ becomes very small, the discretization error becomes more significant. For the choices of Ψ and Φ used for the simulations presented in this chapter, we allowed $c(t)$ to become small enough so that the effective edge width is one pixel. (Effective edge width can be defined as the width of the set $\{\Phi(y) < 1/2\}$, for example.)

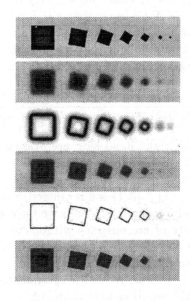

Figure 8.2. Square images, 120 × 490. From top to bottom: Data g, no edge smoothed data f_I, pre-focusing $\Phi(v)$, pre-focusing f, final $\Phi(v)$, and final f. Initial parameters are $\beta = 0.01, \alpha = 0.008, c(0) = 2.0$ and final parameters are $\beta = 0.1, c = 0.2$.

8. Simulation Results

In this section, we present the results of some simulations of the edge focusing algorithm. Image intensity is linearly scaled to lie in the range $[0, 1]$. In the two dimensional plots and in the images, we plotted $\Phi(v)$ on the same mesh as f, i.e., the plotting mesh corresponds to $\mathcal{L}_f$. For each $x \in \mathcal{L}_f$, we plot the minimum value of $\Phi(v)$ among the four nearest neighbors.

Figure 2 serves as a demonstration of the behavior of the edge focusing algorithm and its dependence on the parameters α and β. The image is a synthetic image of several squares of various sizes and rotation. The squares have side lengths of $60, 50, 40, 30, 20, 10$, and 6 pixels. We chose squares to illustrate the effects observed at high curvature edges, and rotated them to demonstrate the rotational invariance of the algorithm. The functions Φ and Ψ are as in equation (8.2) and we set $\rho(v) = \Phi(v)$. The initial value f_I is given by the no-edge smoothing of g associated with $\beta(0)$. We have carefully chosen the parameters to make the detection of some of the squares marginal. For Figure 2, the initial values are $\beta(0) = 0.01$, $\alpha = 0.008$, and $c(0) = 2.0$. This value of $\beta(0)$ corresponds to smoothing of the image over a radius of approximately 10 pixels, relatively large compared to the sizes of the squares. Such smoothing would not be required for good quality 'real' images. The final values of the parameters are $\beta(T) = 0.1$ and $c(T) = 0.2$ (where T is the time of termination), and ϵ was chosen so that 200 iterations with scaling are required to reach T. Figure 2 presents the data $g, f_I, \Phi(v(0)), f(0), \Phi(v(T))$, and $f(T)$ from top to bottom. Note the accurate localization of those edges detected. There is little visible distortion in the edges even of the smallest square whose edges were detected at all. Figure 3a and 3b illustrate the effects of variation in α and β. Both are images of $\Phi(v(T))$. The image consists of the original with two more copies of the series of squares with intensities chosen so that the difference in intensity between the background and square has been reduced by a factor of 0.7 and 0.5 respectively. By scaling, one can see that this has the same effect as increasing α by a factor of 2 and 4, respectively. To generate Figure 3a, we used parameters $\beta(0) = 0.01, \alpha(0) = 0.003, c(0) = 2.0$ and $\beta(T) = 0.1, \alpha(T) = 0.003, c(T) = 0.2$. The parameters used to generate Figure 3b are the same except $\beta(0) = 0.006$ and $\beta(T) = 0.06$.

Figure 4 demonstrate the algorithm on 'real' images. Figure 4 is 512×512 pixels. In general, ϵ is chosen so that 200 iterations with scaling are required. The data is in Figure 4a. The image has been processed for two different sets of parameters to indicate the stability of the edges under a change in scale. In both cases, the displayed images are the following. Figures b,c, and d are $\Phi(v(0)), f(T)$, and $\Phi(v(T))$, respectively. Figures e,f, and g reiterate b,c, and d for the second set of parameters. The first set of parameters is given by

$$\beta(0) = 0.01, \quad \alpha = 0.001, \quad c(0) = 2.0, \quad \beta(T) = 0.1, \quad c(T) = 0.2$$

and the second by

$$\beta(0) = 0.2, \quad \alpha = 0.005, \quad c(0) = 2.0, \quad \beta(T) = 2.0, \quad c(T) = 0.2 \ .$$

The value $\beta = 0.01$ roughly corresponds with smoothing by averaging over a 10×10 window and hence represents a high degree of smoothing. The value $\beta = 0.2$ roughly corresponds with smoothing by averaging over a 2×2 window. The degree of smoothing in the two examples is widely different and yet the edges found at the coarse scale are essentially a subset of those found at the finer scale. Virtually no scale dependent distortion is visible.

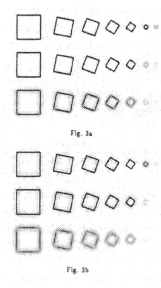

Fig. 3a

Fig. 3b

Figure 8.3. Square images of varying intensities, 340×490. Final $\Phi(v)$. For Figure 3a the initial parameters are $\beta = 0.01, \alpha = 0.003, c(0) = 2.0$ and the final parameters are $\beta = 0.1, c = 0.2$. For Figure 3b the initial parameters are $\beta = 0.006, \alpha = 0.003, c(0) = 2.0$ and the final parameters are $\beta = 0.06, c = 0.2$.

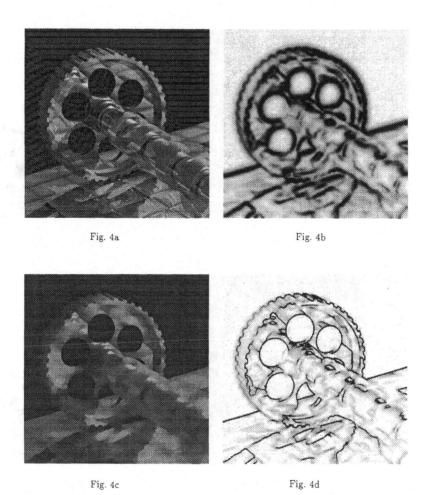

Fig. 4a   Fig. 4b

Fig. 4c   Fig. 4d

Figure 8.4a, b, c, d.   Cam image 510×512. Original g, Prescaling $\Phi(v)$, final f, and final $\Phi(v)$, respectively, with initial parameters $\beta = 0.01$, $\alpha = 0.001$, $c(0) = 2.0$ and final parameters $\beta = 0.1$, $c = 0.2$.

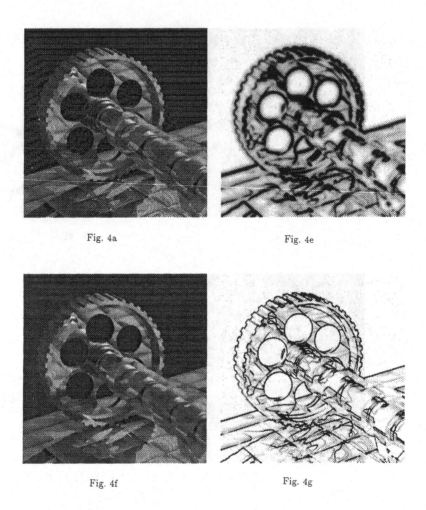

Fig. 4a

Fig. 4e

Fig. 4f

Fig. 4g

Figure 8.4a, e, f, g.   Original g, Prescaling $\Phi(v)$, final f, and final $\Phi(v)$, respectively, with initial parameters $\beta = 0.2$, $\alpha = 0.005$, $c(0) = 2.0$ and final parameters $\beta = 2.0$, $c = 0.2$.

Acknowledgements: We are indepted to Pietro Perona and to Stefano Casadei for help with images and image processing software.

COUPLED GEOMETRY-DRIVEN DIFFUSION EQUATIONS FOR LOW-LEVEL VISION

Marc Proesmans

and

Eric Pauwels[1]

and

Luc van Gool

ESAT-MI2, Catholic University Leuven
Mercierlaan 94, B-3001 Leuven, Belgium

Abstract. This chapter introduces a number of systems of coupled, non-linear diffusion equations and investigates their role in noise suppression and edge-preserving smoothing. The basic idea is that several maps describing the image, undergo coupled development towards an equilibrium state, repre- senting the enhanced image. These maps could e.g. contain intensity, local edge strength, range, or another quantity. All these maps, including the edge map, contain continuous rather than all-or-nothing information, following a strategy of least commitment. Each of the approaches has been developed and tested on a parallel transputer network.

1. Introduction and basic philosophy

1.1. Energy minimization and systems of coupled diffusion equations

Optimization of energy-functionals provides a clear-cut and, from a conceptual point of view, very attractive framework for the regularisation and processing of images. It is not surprising therefore that it has been the inspiration and point of departure for many investigators (cfr. [362, 34, 264, 273] to name just a few). The underlying basic idea is that a given input-signal g is transformed into an output-signal f in such a way that the result *minimizes* a predefined cost- or energy-functional. In its simplest form both

---

[1]Post-doctoral Research Fellow of the Belgian National Fund for Scientific Research (NFWO).

in- and output can be thought of as grey-level functions, defined on some open set $\Omega \subset \mathbb{R}^2$. Moreover, f is assumed to be piecewise smooth with a discontinuity set K, which represents the sharp grey-level transitions (i.e. edges) in the (processed) image.

Minimizing the cost-functional will amount to a delicate balancing act, in effect trying to accommodate two conflicting requirements: the output f should, on the one hand, be as *smooth* as possible, but at the same time, it should show a high degree of fidelity to the input-signal g, in particular to the salient features present in g. Consequently, the functional will penalize a noisy output for its lack of smoothness, but at the same time assign a high cost to an output signal f that deviates significantly from the original g. Furthermore, in view of the fact that the output is expected to contain sharp edges, discontinuities in f are acceptable, but these too come at a cost: failing to tax the occurrence of discontinuities would result in an overfragmentation of the output. Typically the functional will accrue a cost proportional to the measure of the discontinuity set (denoted $|K|$).

These considerations led Mumford and Shah [264] (see also others chapters in this book) to study the following regularizing functional for an input-signal g:

$$E_g(f, K) = \int_{\Omega \setminus K} \left[\alpha \|\nabla f\|^2 + \beta(f - g)^2 \right] \, dx dy + \nu |K| \qquad (9.1)$$

(α, β and ν are adjustable parameters). Minimizing this functional (over the set of piecewise smooth functions) forces f to be smooth (i.e. $\|\nabla f\|^2$ small) and "close" to g (i.e. $(f - g)^2$ small), except on the set of discontinuities K (the size of which is measured using some appropriate measure $|K|$). Hence this formulation incorporates two simultaneous regularizations: the term $\|\nabla f\|^2$ regularizes f in $\Omega \setminus K$, while the term $|K|$ regularizes the boundary. Unfortunately, the functional is non-convex which makes the minimization-problem extremely difficult to solve. Moreover, experimental evidence and theoretical considerations indicate that this particular approach runs into all sorts of practical problems: oversegmentation, shape-sensitivity, very rigid constraints on the geometry of the boundary,...; for more detail, see [264] and other chapters in this book. These limitations prompted Shah [341] to propose a different set of minimizing functionals which incorporate an auxiliary function v introduced by Ambrosio and Tortorelli [12]. The basic idea is to insist on the continuity and smoothness of f but to use v as a sort of "fuzzy" but smooth edge-indicator (for the grey-value f) which, at each point indicates the "likelihood" of the existence of an edge. Hence, v is assumed to be close to 1 in the vicinity of an edge in the grey-level f, and close to 0 everywhere else.

Shah's original strategy was to determine v (for a given grey-value function f) by minimizing the functional

$$V_f(v) := \int_\Omega [\alpha(1-v)^2 \|\nabla f\| + \frac{\rho}{2}\|\nabla v\|^2 + \frac{v^2}{2\rho}] \, dx\, dy, \qquad (9.2)$$

(again, α and ρ are adjustable parameters). Indeed, minimization of V_f (for given f) will force both v and $\|\nabla v\|$ to be small except at places where $\|\nabla f\|$ is large. This captures in accurate mathematical language what a fuzzy edge-indicator is supposed to do.

Conversely, when the edge-indicator v is given, the grey-value function f is determined (in Shah's scheme) by minimizing

$$F_v(f) := \int_\Omega [\|\nabla f\|^2 + \frac{1}{v^2 \sigma^2}(f-g)^2] \, dx\, dy. \qquad (9.3)$$

Notice how away from the edges (i.e. $v \approx 0$) the functional penalizes infidelity with respect to the initial data g, whereas in the region near edges (i.e. $v \approx 1$), smoothness gains importance.

Since the functionals F and V are convex, Shah then suggests to use the method of steepest descent in an attempt to find (at least approximately) a simultaneous minimum for both functionals. This is accomplished via a dynamic approach: make f and v additionally dependent on the time t and introduce a time-evolution which forces both functions to evolve proportional (but opposite) to the first variation:

$$\frac{\partial f}{\partial t} = -\delta_f(F) \quad \text{and} \quad \frac{\partial v}{\partial t} = -\delta_v(V).$$

This yields the following set of coupled nonlinear diffusion equations (after multiplying out the factor v^2 in the denominator of $\delta_f(F)$):

Shah:
$$\begin{cases} \dfrac{\partial f}{\partial t} = v^2 \nabla^2 f - \dfrac{1}{\sigma^2}(f-g) \\[2mm] \dfrac{\partial v}{\partial t} = \rho \nabla^2 v - \dfrac{v}{\rho} + 2\alpha(1-v)\|\nabla f\|. \end{cases} \qquad (9.4)$$

This PDE-system tends to converge rather efficiently to a stable solution, albeit one that may look noticeably blurred. Notice that the two coupled equations are indeed smoothing processes (due to the terms $\nabla^2 f$ and $\nabla^2 v$), but through their coupling, there is a constant exchange of information: the grey-level f steers v via its gradient, and conversely, v modulates the diffusion coefficient of f. As a consequence of this, away from the discontinuity regions, the diffusion term for f is strongly suppressed by the conduction coefficient $v^2 \approx 0$. This results in an output image for which regions with slowly varying intensity retain their noisy appearance. For a more detailed discussion and applications to real images we refer to Shah's original paper [341].

To overcome this drawback we modify the functional F in eq.(9.3) slightly to

$$F_v(f) = \int_\Omega [\|\nabla f\|^2 + \frac{(1-v)^2}{\sigma^2}(f-g)^2] \, dx dy. \tag{9.5}$$

From a qualitative point of view there still is a similar shift of relative importance from (data-)fidelity to smoothness when when approaching edges, but the corresponding evolution equation will never switch off the diffusion term:

Modified Shah:
$$\begin{cases} \dfrac{\partial f}{\partial t} &= \nabla^2 f - \dfrac{(1-v)^2}{\sigma^2}(f-g) \\[2mm] \dfrac{\partial v}{\partial t} &= \rho \nabla^2 v - \dfrac{v}{\rho} + 2\alpha(1-v)\|\nabla f\|. \end{cases} \tag{9.6}$$

This set of equations will do a more convincing job as far as the noise-removal is concerned, but the smoothing will inescapably blur the edges, once again resulting in poor localisation. So, at this point, the essential factor that seems to be hampering further progress is the averaging and smearing out intrinsic to smoothing. We need diffusion to suppress the noise, but that very same diffusion seriously foils our attempts at localising important and salient structures such as edges. A possible way out of this conundrum is provided by Perona & Malik's "anisotropic diffusion" which we will discuss briefly in the next section.

1.2. Harnessing Perona & Malik's anisotropic diffusion

In their seminal paper [293] (see also other chapters in this book), Perona and Malik succeeded in creating a "discerning" (or selective) diffusion process which exhibits a dichotomous behaviour: small fluctuations are ironed out while large gradients are sharpened and enhanced (the Matthew effect again!).

Their basic idea was to return to the interpretation of the heat equation as a transport equation: if the diffusion coefficient c is a constant, the heat equation can be recast in the following form

$$\frac{\partial f}{\partial t} = c\Delta f = c \, \text{div}(\nabla f) = -\text{div} \, J,$$

where $J = -c\nabla f$ is the *flux* (proportional but opposite to the gradient). The diffusion coefficient c tells us how effective the gradient is in driving this flux. It is clear that, if the boundaries in an image are known, choosing $c = 0$ at the boundaries and $c = 1$ elsewhere, will block smoothing across a region's boundaries but allow diffusion to take place within the region.

Of course, in practice, the boundaries are not known in advance, but since we can safely assume that large gradient values are indicative of boundaries, the most obvious solution is to choose c as a decreasing function of $\|\nabla f\|$. This results in the Perona-Malik equation

$$\frac{\partial f}{\partial t} = \text{div}(c(\nabla f)\nabla f),$$ (9.7)

where the most frequently used functional forms for c are

$$c(\nabla f) = e^{-(\|\nabla f\|/K)^2} \quad \text{and} \quad c(\nabla f) = \frac{1}{1 + (\|\nabla f\|/K)^2}.$$

Simple theoretical considerations show that this algorithm does indeed succeed in sharpening edges: gradients above a certain threshold are enhanced, whereas below this threshold, diffusion will tend to have a smoothing effect.

Comparing eqs.(9.6) and (9.7) it seems natural from an intuitive point of view, to try and counteract the smearing out of f by exchanging the Laplacian $\nabla^2 f$ for Perona & Malik's non-linear diffusion operator. This yields the set of coupled PDEs

$$\begin{cases} \dfrac{\partial f}{\partial t} &= \text{div}(c(\nabla f)\nabla f) - \dfrac{(1-v)^2}{\sigma^2}(f-g) \\ \dfrac{\partial v}{\partial t} &= \rho\nabla^2 v - \dfrac{v}{\rho} + 2\alpha(1-v)\|\nabla f\|. \end{cases}$$ (9.8)

(The first equation is now almost identical to Nordström's diffusion process which is derived through optimization of a different, though related, functional [273]). This system of PDEs does indeed converge for a wide range of σ's: after a reasonably short time (i.e. after about 20 iterations of the corresponding finite difference scheme), the process practically always seems to have stabilized and no change is observed if the process is continued (see Fig. 9.1).

There is no reason why we can't apply a similar boosting strategy to enhance the v-map which, due to the diffusive Laplacian $\nabla^2 v$, performs rather poorly at localizing the discontinuity. Introducing the Perona-Malik non-linearity in this diffusion operator will result in sharper and crisper edges. We thus obtain the following system of coupled PDEs

$$\begin{cases} \dfrac{\partial f}{\partial t} &= \text{div}(c_1(\nabla f)\nabla f) - \dfrac{(1-v)^2}{\sigma^2}(f-g) \\ \dfrac{\partial v}{\partial t} &= \rho\,\text{div}(c_2(\nabla v)\nabla v) - \dfrac{1}{\rho}v + \dfrac{1}{2}\alpha v(1-v)\|\nabla f\| \end{cases}$$ (9.9)

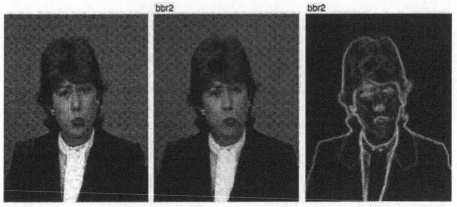

Figure 9.1. Original (noisy) image g of "Claire" (left) together with processed image f and edge-map v (white pixels indicate $v \approx 1$, while dark areas indicate $v \approx 0$) using the combined Shah-Perona-Malik algorithm (9.8). The result shown was obtained after about 100 iterations of the finite-difference scheme based on eq.(9.8).

which we applied to an image of an arterial network (see Fig. 9.2). The improvement in comparison to the gradient obtained with Sobel convolution masks is obvious.

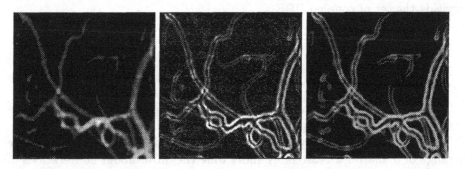

Figure 9.2. Original image g of an arterial network (left), together with the edge map extracted using a Sobel operator (middle, enhanced using clipping and rescaling) and the edge-indicator v based on eq.(9.9).

To get an idea of the performance of these processing methods we compare them to the median filter. To this end we took the "Claire" image, added noise (see Fig. 9.1) and applied the diffusion system (9.8), as well as (a number of iterations of) a 3×3 median filter (the result is shown in Fig. 9.4). It is immediately obvious that it takes at least 10 iterations of the median filter to get a noise reduction comparable to the one obtained with the diffusion approach. However, that number of median iterations results in a noticeable deterioration of the image quality.

1.3. Systems of coupled geometry-driven diffusions

Comparison of eqs.(9.4), (9.8) and (9.9) at this point, shows that in each case we have a system of coupled diffusion (or evolution) equations which evolve towards a limit configuration representing the processed image and edges. Although functionals provided the original inspiration and starting point, we have turned our attention to the corresponding evolution equations for the pragmatic reason that they are more amenable to application-specific tailoring and far more easy to implement than a direct minimization of the corresponding functionals.

So, instead of trying to find a functional which embodies the desired global characteristics of a solution, we prefer to concentrate on *simultaneous evolution- or diffusion equations for a number of important image-features* (such as grey-level, edge-indicator, range, motion, etc...), *which at each incremental time-step do the best they can, given the actual local configuration.* To compensate for this "short-sighted" (local) approach, we make the response of each feature conditional on the state of the other features. Put differently, the evolution equation for an individual feature is coupled to all the other evolution equations and this continuous exchange of information between the evolving features compensates for the local nature of the evolution or diffusion equations, (hopefully!) steering the evolution to a non-trivial and meaningful limit. Loosely speaking, one can say that optimality over space and time has been traded in for optimality with respect to the *locally available* (feature) information.

It is instructive to highlight the characteristics which make such systems especially attractive:

- The evolution equations are local and proceed to their stationary limit without the need for a difficult and time-consuming global search; iterations of these local processes allow non-local influences to accumulate so that global interactions become possible.
- Moreover, these evolution equations are easily parallelizable. This is particularly important for efficient implementations, both in software (parallel computing) and in hardware (VLSI-implementations).
- The non-linearity of the equations allows for non-trivial and non-propor- tional responses to the input and can offer filtering with a level of sophistication needed in e.g. medical and remote sensing applications.
- Coupling the different equations provides a way of information transfer between the different features (cue-integration) and ensures the internal consistency of the final result.
- Finally (but by no means the least important observation), although functionals are very attractive from a theoretical point of view, they represent difficult problems. The PDEs on the other hand always

carry within them a straightforward (numerical) solution strategy. Admitted- ly, such a strategy might require extra attention to be paid to problems of stability, convergence, ... but these are in a sense problems intrinsic to the numerical approach. Because PDEs dictate local behaviour ("what will happen in this neighbourhood in the next short time-interval?") they are far easier to deal with than functionals that determine the global behaviour.

Browsing through the above list of rationales, one comes across several of the basic architectural principles underpinning biological vision. These include fine-grained parallelism, retinotopically organized and functionally specialized areas (in casu equations), an abundance of bi-directional couplings between them, non-linearities from the earliest stages onwards and relatively localized connections within each area (e.g. only 10^{12} synapses versus 10^9 neurons in a pigeon brain, a far cry from the all-to-all connectivity in traditional neural nets). As we hope the sequel will corroborate, the proposed **CO**upled **DI**ffusion e**Q**uations (CODIQ for short) framework holds good promise for the robust extraction of a number of important "low-level" features, including other than intensity and edges. It is our aim in this chapter to explore some of the possibilities of this paradigm and to show that it is a flexible yet powerful tool for low-level vision. We remind the reader that a more fundamental and theoretically oriented discussion of coupled diffusion equations can be found in the chapter by Richardson and Mitter in this book.

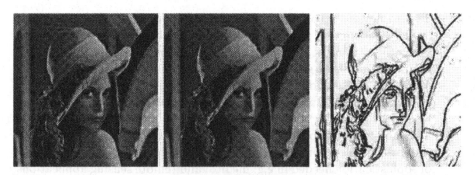

Figure 9.3. Original "Lena"-image g (left), its f-map (middle) and its (reversed) edge map $1 - v$ (right) using CODIQ system (9.8). To get beter localisation of the v-map, the ρ-parameter was decreased over time.

2. Diffusion based on second order smoothing

2.1. Introduction

Experimentation shows that for a number of applications, the systems of coupled (non-linear) diffusion equations introduced in section 1 have the disadvantage that

- the results show a rather poor response in the neighbourhood of ramp edges (oversegmentation);
- edges can be sharpened, but contrast (measured as the difference in intensity across the edge) is not enhanced.

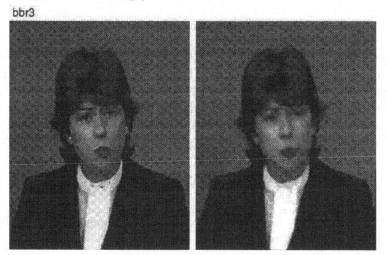

Figure 9.4. "Claire"-image processed using geometry-driven diffusion eq.(9.8) (left), and image processed using 10 iterations of 3 × 3 median filter (right).

These drawbacks can be eliminated by turning to what we will refer to as "second order smoothing". (To keep the analysis that follows as transparent as possible we will, for the moment, restrict ourselves to the one-dimensional case $\Omega = \mathbb{R}$.)

Recall that the original inspiration for the work by Mumford and Shah was furnished by the functional

$$F(f) = \int_{\mathbb{R}} \left[\left(\frac{\partial f}{\partial x} \right)^2 + \sigma^2 (f - g)^2 \right] dx$$

which was modified in such a way that it makes sense for piecewise continuous functions. In the same vein, minimization of a functional involving second rather than first order derivatives of the image-function provides the starting point for *second order smoothing*. More precisely, the objective is to minimize the integral

$$F(f) = \int_{\mathbb{R}} \left[\left(\frac{\partial^2 f}{\partial x^2} \right)^2 + \sigma^2 (f - g)^2 \right] dx$$

where the first derivative has been exchanged for the second. Since the gradient no longer features in the functional, linear segments no longer contribute to the cost-functional and continuous piecewise-linear solutions (e.g. ramps) become a lot "cheaper". The Euler-Lagrange equation corresponding to this functional is the fourth order ODE

$$\frac{d^4 f}{dx^4} + \sigma^2 (f - g) = 0. \tag{9.10}$$

Unfortunately, as this equation involves a fourth-order differential, the evolution equation based on this ODE will be very noise-sensitive. For this reason we introduce a new function β which is meant to be an approximation of $\partial f / \partial x$. Of course, this entails changing the cost-functional to one in which the first derivative of β plays the role of the second derivative of f and which involves an extra term penalizing large discrepancies between β and $\partial f / \partial x$:

$$F_\lambda(f, \beta) = \int_{\mathbb{R}} \left[\left(\frac{\partial \beta}{\partial x} \right)^2 + \sigma^2 (f - g)^2 + \lambda^2 \left(\beta - \frac{\partial f}{\partial x} \right)^2 \right] dx \tag{9.11}$$

The Euler-equations for this approximating functional are given by the system of coupled second differential equations

$$\begin{cases} \dfrac{\partial^2 f}{\partial x^2} - \dfrac{\sigma^2}{\lambda^2}(f - g) - \dfrac{\partial \beta}{\partial x} & = 0 \\[2mm] \dfrac{\partial^2 \beta}{\partial x^2} - \lambda^2 \left(\beta - \dfrac{\partial f}{\partial x} \right) & = 0, \end{cases} \tag{9.12}$$

which means that we have exchanged one fourth order equation for two second order ones.

2.2. Diffusion and edge-enhancing using second-order smoothing

The point of departure for geometry-driven second order smoothing is the set of coupled diffusion equations based on the Euler-Lagrange equations (9.12):

$$\begin{cases} \dfrac{\partial f}{\partial t} & = d_1 \dfrac{\partial^2 f}{\partial x^2} - \sigma^2 (f - g) - s_h \dfrac{\partial \beta}{\partial x} \\[2mm] \dfrac{\partial \beta}{\partial t} & = d_2 \dfrac{\partial^2 \beta}{\partial x^2} - \xi^2 \left(\beta - \dfrac{\partial f}{\partial x} \right) \end{cases} \tag{9.13}$$

(some additional parameters d_1, d_2, s_h and ξ have been introduced which can be used to fine-tune the problem at hand.) True to our philosophy we will concentrate directly on the PDEs and change them in such a way that they exhibit the required behaviour, rather than try and tinker with the functionals in an attempt to generate interesting global behaviour.

Notice how once again we arrive at a system of coupled non-linear diffusion equations for features (in this case β and f). Typical for the behaviour of this set of coupled equations is that in the neighbourhood of a discontinuity, it will generate under- and overshoots, the relative size of which is controlled by s_h (the so-called "shooting-parameter"). As a result of this discontinuities will be smoothed much less than in an ordinary smoothing approach.

This model can be improved when edges in the first order derivative (i.e. β) are sharpened. Such a process will result in jumps for the β-function which correspond to sharp transitions in the slope of f. In that case a piecewise linear profile can be recovered from a noisy original one. The relevant equations are

$$\begin{cases} \dfrac{\partial f}{\partial t} &= d_1 \dfrac{\partial^2 f}{\partial x^2} - \sigma^2(f-g) - s_h \dfrac{\partial \beta}{\partial x} \\[2mm] \dfrac{\partial \beta}{\partial t} &= d_2 \operatorname{div}\left(c(\dfrac{\partial \beta}{\partial x}) \dfrac{\partial \beta}{\partial x} \right) - \xi^2(\beta - \dfrac{\partial f}{\partial x}) \end{cases} \tag{9.14}$$

The diffusion process exhibits an acceptable behaviour on ramp edges (cfr. Fig. 9.5), even if the noise amplitude in β is larger than the discontinuity steps: this is due to the coupling with f.

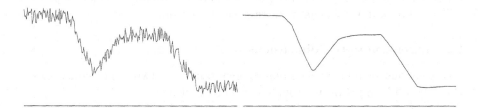

Figure 9.5. The system (9.14) driven by second-order information can handle noisy ramp edges.

Contrast enhancement can be obtained by sharpening f, while retaining the shooting parameter:

$$\begin{cases} \dfrac{\partial f}{\partial t} &= d_1 \operatorname{div}(c(\nabla f)\nabla f) - \sigma^2(f-g) - s_h \dfrac{\partial \beta}{\partial x} \\[2mm] \dfrac{\partial \beta}{\partial t} &= d_2 \nabla^2 \beta - \xi^2(\beta - \dfrac{\partial f}{\partial x}) \end{cases} \tag{9.15}$$

The diffusion process on a noisy step edge clearly results into a Mach-band profile (see Fig. 9.6). The contrast enhancement can easily be adapted depending on the applications. However, as in the case of the Perona-Malik algorithm, the response on ramp edges is poor. Somewhat better results

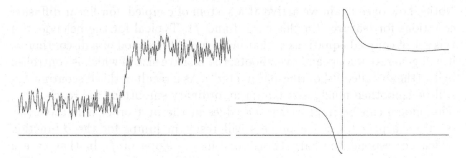

Figure 9.6. Systems driven by second-order dynamics in eq.(9.15) can generate over- and undershoots (Mach band effect).

can be obtained when using β instead of ∇f to drive the f-diffusion:

$$\begin{cases} \dfrac{\partial f}{\partial t} &= d_1 \mathrm{div}(c(\beta)\nabla f) - \sigma^2(f-g) - s_h \dfrac{\partial \beta}{\partial x} \\[2mm] \dfrac{\partial \beta}{\partial t} &= d_2 \nabla^2 \beta - \xi^2 \left(\beta - \dfrac{\partial f}{\partial x}\right) \end{cases} \qquad (9.16)$$

Since β is smoother than $\partial f / \partial x$, the edge enhancement will not be as spectacular as for the previous equations, but by the same token, the process will not break down as easily on ramp edges or slopes.

2.3. Extension to two dimensions

The previous systems can be readily generalized to two dimensions. Again, the original inspiration is provided by the functional

$$F_g(f) = \iint_\Omega \left[\left(\frac{\partial^2 f}{\partial x^2} + 2\frac{\partial^2 f}{\partial x \partial y} + \frac{\partial^2 f}{\partial y^2} \right)^2 + \frac{\sigma^2}{2}(f-g)^2 \right] \, dx\,dy \qquad (9.17)$$

As before, direct minimization of this functional will result in fourth order Euler-Lagrange differential equations. To bypass such complications we introduce two new functions which are meant to be approximations of the first derivatives of f:

$$\beta_x \approx \frac{\partial f}{\partial x} \quad \text{and} \quad \beta_y \approx \frac{\partial f}{\partial y}. \qquad (9.18)$$

The contribution of second derivatives to the functional can then be exchang- ed for the following (symmetrised) approximation:

$$Q = \frac{\partial \beta_x}{\partial x} + \frac{\partial \beta_y}{\partial y} + \left(\frac{\partial \beta_x}{\partial y} + \frac{\partial \beta_y}{\partial x} \right).$$

Adding a term penalizing deviation from the approximations (9.18), the functional (9.17) then becomes

$$F_\lambda(f) = \iint_\Omega \left[Q^2 + \lambda \left((\beta_x - \frac{\partial f}{\partial x})^2 + (\beta_y - \frac{\partial f}{\partial y})^2 \right) + \frac{\sigma^2}{2}(f - g)^2 \right] dx dy.$$

The Euler-Lagrange equations for this functional are

$$\frac{\partial f}{\partial t} = (\frac{\partial^2 f}{\partial x^2} + \frac{\partial^2 f}{\partial y^2}) - (\frac{\partial \beta_x}{\partial x} + \frac{\beta_y}{\partial y}) - \sigma^2(f - g),$$

$$\frac{\partial \beta_x}{\partial t} = 2\frac{\partial^2 \beta_x}{\partial x^2} + \frac{\partial^2 \beta_x}{\partial y^2} + \frac{\partial^2 \beta_y}{\partial x \partial y} - 2\lambda(\beta_x - \frac{\partial f}{\partial x}),$$

$$\frac{\partial \beta_y}{\partial t} = 2\frac{\partial^2 \beta_y}{\partial y^2} + \frac{\partial^2 \beta_y}{\partial x^2} + \frac{\partial^2 \beta_x}{\partial x \partial y} - 2\lambda(\beta_y - \frac{\partial f}{\partial y}).$$

Introducing some additional parameters, these equations have been generali- zed to

$$\frac{\partial f}{\partial t} = (\frac{\partial^2 f}{\partial x^2} + \frac{\partial^2 f}{\partial y^2}) - s_h(\frac{\partial \beta_x}{\partial x} + \frac{\beta_y}{\partial y}) - \sigma^2(f - g),$$

$$\frac{\partial \beta_x}{\partial t} = 2\frac{\partial^2 \beta_x}{\partial x^2} + \frac{\partial^2 \beta_x}{\partial y^2} + \ell\frac{\partial^2 \beta_y}{\partial x \partial y} - 2\lambda(\beta_x - \frac{\partial f}{\partial x}),$$

$$\frac{\partial \beta_y}{\partial t} = 2\frac{\partial^2 \beta_y}{\partial y^2} + \frac{\partial^2 \beta_y}{\partial x^2} + \ell\frac{\partial^2 \beta_x}{\partial x \partial y} - 2\lambda(\beta_y - \frac{\partial f}{\partial y}).$$

Notice that if f represents *range data*, then β_x and β_y yield estimates of the normal directions to the surfaces, i.e. information considered essential in its own right by Marr for his 2.5-D sketch.

Finally, to counteract the smearing out due to the diffusion terms, the Perona-Malik non-linearity is introduced. Adding the non-linearity to the first equation, one can easily create Mach-band effects:

$$\frac{\partial f}{\partial t} = (\text{div}(c(\nabla f)\nabla(f))) - s_h(\frac{\partial \beta_x}{\partial x} + \frac{\beta_y}{\partial y}) - \sigma^2(f - g). \qquad (9.19)$$

An example is given in Fig. 9.7. Furthermore, if one is interested in reconstructing ramp edges, one can introduce non-linearity in the β-equations:

$$\frac{\partial \beta_x}{\partial t} = 2\frac{\partial}{\partial x}\left(c(\frac{\beta_x}{\partial x})\frac{\beta_x}{\partial x}\right) + \frac{\partial}{\partial y}\left(c(\frac{\beta_x}{\partial y})\frac{\beta_x}{\partial y}\right) + \ell\frac{\partial^2 \beta_y}{\partial x \partial y} - 2\lambda(\beta_x - \frac{\partial f}{\partial x})$$

$$\frac{\partial \beta_y}{\partial t} = 2\frac{\partial}{\partial y}\left(c(\frac{\beta_y}{\partial y})\frac{\beta_y}{\partial y}\right) + \frac{\partial}{\partial x}\left(c(\frac{\beta_y}{\partial x})\frac{\beta_y}{\partial x}\right) + \ell\frac{\partial^2 \beta_x}{\partial x \partial y} - 2\lambda(\beta_y - \frac{\partial f}{\partial y}).$$

Figure 9.7.   Original image of license plate (left) and result after contrast enhancement using the Mach-effect in eqs.(9.19).

3. Application to multispectral images

Another area where it seems natural to exploit the idea of coupled diffusions are applications involving multispectral data. This could be very useful in remote sensing where sensor data typically comprise a number of spectral components (e.g. red, green and near-infrared, the so-called RGN-system). The same is true for a number of medical applications where, for example, a typical NMR will produce two components. (For an in-depth discussion of multispectral and multivalued diffusion we refer to the chapter by Withaker and Gerig in this book.)

It stands to reason that, in order to obtain the best possible result, it is of major importance to pool all the available information and to integrate clues over the different spectral bands. Put differently, when processing one spectral component one should bring to bear all the information that has come available via the other components. Formulated like this, it is obvious that multispectral analysis is an area where coupled diffusions come to full advantage. In order to illustrate their potential, we show in Fig. 9.8 a synthetic set of three noisy images g_R, g_G and g_N. Each component shows one bright and three dark patches, one of which has a slightly different intensity. Applying the coupled diffusions (9.8) to each individual spectral

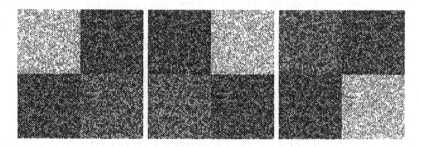

Figure 9.8. Synthetic image components g_R, g_G, g_N

component produces unsatisfactory output (see Fig. 9.9): The v-map captures only the boundaries of one quadrant (in this case, the upper left) and, as a consequence, the three other quadrants are lumped together. Obviously,

Figure 9.9. Result of separate diffusion on the g_R-component: f_R (left) and v_R (right).

the reason for this poor result is that separate diffusion on each individual component fails to take advantage of the fact that for each boundary, there is at least one spectral component for which this boundary is clearly demarcated. Ideally, this information should be used to steer the diffusion in the other spectral bands.

It is clear how this situation can be remedied: the different components should interchange local gradient information. This means that if one component shows a high gradient at some point, this gradient should also block diffusion at the corresponding points in the other component images. To fit this sort of behaviour in the diffusion framework, we introduce a new function γ which is driven by the *maximal* gradient value and it is this function that is used to modulate the diffusion coefficients for the separate spectral components and to drive the v-map. So we arrive at the following system of five coupled diffusions:

$$\frac{\partial f_R}{\partial t} = \operatorname{div}(c(\gamma)\nabla f_R) - \frac{(1-v)^2}{\sigma^2}(f_R - g_R)$$

$$\frac{\partial f_G}{\partial t} = \operatorname{div}(c(\gamma)\nabla f_G) - \frac{(1-v)^2}{\sigma^2}(f_G - g_G)$$

$$\frac{\partial f_N}{\partial t} = \operatorname{div}(c(\gamma)\nabla f_N) - \frac{(1-v)^2}{\sigma^2}(f_N - g_N) \qquad (9.20)$$

$$\frac{\partial v}{\partial t} = \rho\nabla^2 v - \frac{v}{\rho} + 2\alpha(1-v)\gamma$$

$$\frac{\partial \gamma}{\partial t} = \xi\nabla^2\gamma - (\gamma - max(\|\nabla f_R\|, \|\nabla f_G\|, \|\nabla f_N\|))$$

with $c(\gamma) = 1/(1 + (\gamma/\tau)^2)$ and τ a tunable threshold. The drastically improved results can be seen in Fig. 9.10. For applications on real satellite-images we refer the interested reader to [299].

Figure 9.10. Multi-spectral approach f_R, f_G, f_N, v ($\tau = 10.0$, $\sigma = 7.0$, $\xi = 0.0$, $\rho = 3.0$, $\alpha = 0.05$).

4. Application to optical flow

4.1. Introduction

Most methods for computing motion fields rely on spatial and temporal gradients of the image intensity. Since the optical flow problem is ill-posed, additional constraints are required. This is not the place to give an extensive discussion of the relevant literature, but we want to mention some key references. One of the simplest solutions is adding a quadratic smoothness constraint as was done by Horn and Schunck [157]. In an attempt to estimate displacement more accurately, other constraints have been introduced using higher order spatial and temporal derivatives [267]. Some methods use over- constrained systems of equations [50, 376], rather than functional descriptions. Still other methods try to look for specific features within the image — such as corners [266].

The major obstacle to accurate estimation is the fact that most strategies tend to enforce a degree of smoothness on the flow field, thereby blurring the sharp transitions along discontinuities in the flow. Part of the problem lies in the size of the operator masks required to estimate the spatio-temporal derivatives of the grey-value distribution. Indeed, large masks may be necessary to eliminate the effect of noise and to bridge relatively large interframe distan- ces, but unavoidably these large masks will blur the flow-discontinuities.

To cope with discontinuities, several methods have been proposed ranging from controlled smoothness of the resulting field [4], to non-linear variational methods [66, 289]. The method presented here is an elaboration of earlier work [289] and offers an alternative solution which comprises three parts.

- Basically, the method takes the variational approach to the optical flow problem as a starting point. It shares with most other methods a formulation based on the differential form of the optical flow constraint (expressing the assumption of a constant image irradiance), and a smoothness constraint to regularize the solution.
- Unlike other methods however, the above approach is supplemented with a mechanism to match each point in one frame to a corresponding point in another frame. The idea is very similar to the matching process of correlation based approaches [303]. The main advantage is that there is no need for large operator masks to overcome large interframe distances. Moreover, the mechanism allows for a dual implementation. Inconsistencies within the resulting scheme turn out to be concentrated nearby flow discontinuities.
- Finally, the discontinuity information is fed back to the original optical flow scheme in a non-linear way. This allows an accurate reconstruction of the optical flow field while preserving its discontinuities.

The rest of this section is organized around these three topics. Furthermore, an approach will be outlined to separate the real optical flow boundaries and occluding regions. The scheme will be illustrated on synthetic and real images sequences. Finally, we would also like to refer the reader to the discussion of optical flow in the chapter by Kimia, Tannenbaum and Zucker.

4.2. The optical flow equations

We will adhere to the standard notation used in the optical flow literature and use $I(x, y, t)$ to denote the time-dependent image intensity, and (u, v) for the velocity vector field. The basic assumption underlying the classical Horn & Schunck approach to optical flow is that the intensity values in the image, although shifting their position, remain essentially constant, whence it follows that intensity-variations over small time intervals are due to the optical flow. This observation gives rise to the so-called optical flow constraint equation:

$$I_x.u + I_y.v + I_t = 0 \tag{9.21}$$

(subscripts indicate corresponding partial derivatives).

Evidently, this single equation is insufficient to determine the two motion components and most techniques use some kind of regularity constraint to reduce the space of admissible functions. A very popular approach is the so-called *smoothness constraint* originally introduced by Horn & Schunck [157]: solve (9.21) subject to the condition that $\|\nabla u\|^2 + \|\nabla v\|^2$ should be as small as possible. In practice this amounts to the more robust minimisation of the following cost-functional:

$$E = \int\int_\Omega (\lambda(I_x.u + I_y.v + I_t)^2 + (u_x^2 + u_y^2 + v_x^2 + v_y^2))\ dxdy. \tag{9.22}$$

Its main advantage is its simplicity, since minimizing the functional leads to a set of linear diffusion equations. In fact, the corresponding Euler-Lagrange equations are

$$\begin{cases} \dfrac{\partial u}{\partial t} &= \nabla^2 u - \lambda I_x(I_x.u + I_y.v + I_t) \\[2mm] \dfrac{\partial v}{\partial t} &= \nabla^2 v - \lambda I_y(I_x.u + I_y.v + I_t) \end{cases} \tag{9.23}$$

which can be discretized using finite differences and solved by an iterative solution scheme. The smoothness constraint thus allows to estimate both velocity components, but at the same time has the unwelcome side-effect that it forces the estimated vector field to vary smoothly across boundaries and occluding regions.

A number of methods have been introduced [4, 267] that try and capture the discontinuities in the optical flow field by judiciously varying the λ-parameter. Typically, λ will increase for regions with high temporal derivatives, thus decreasing the relative importance of the smoothing terms in eqs.(9.23). Nevertheless, some degree of smoothing remains active at all times. Other approaches [66, 289] modulate the smoothing term by taking into account gradient information of the flow field, which is reminiscent of the Perona-Malik non-linearity.

However, to determine the existence of an optical flow boundary, it doesn't seem appropriate to restrict attention to gradients of the optical flow field. First of all, such gradients are not necessarily indicative of discontinuities. The smooth rotation of a circular object about its centre provides a case in point. Furthermore, the use of these gradients can cause over-fragmentation in the flow field during the iteration process. We therefore propose an alternative way of detecting discontinuities in the flow field which we will elaborate in section 4.4. Presently, however, we intend to address another problem.

4.3. Successive approximation of the velocity field

The optical flow constraint equation (9.21) is an idealized mathematical construction as it relates temporal and spatial *derivatives* at a single point. However, in practice, these derivatives must be estimated using *difference quotients*, and, unless the luminosity-profile is planar, the accuracy is limited by the size of the spatial and temporal increments ($\Delta x, \Delta y$ and Δt). The dimensions of spatial increments are determined by the pixel-size (or - density) and are typically sufficiently small to allow accurate determination of the spatial derivative. The estimation of the time derivative is another matter as it is tied up with the velocities. It is easy to see that for velocities which are comparable to the spatial wavelengths that occur in the grey-level function, the increment quotient can yield results that are totally wrong (see Fig. 9.11). This is bound to happen when e.g. fine-grained texture is present in the picture. Therefore, if accurate results are required for larger velocities, the gradient information has to be estimated with larger operator masks, resulting in a considerable degradation and smearing-out of the discontinuities in the flow field.

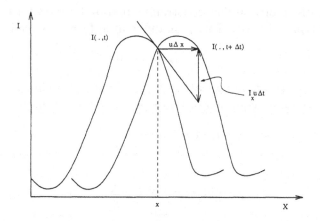

Figure 9.11. Whenever the velocity u is sufficiently large so that the displacement $u\Delta t$ is comparable to spatial wavelengths in the grey-value function I, the estimate for the time-derivative I_t risks to be seriously inaccurate. As an illustration, consider the point x, where the estimate for I_t will be zero.

To investigate how we can remedy this problem, let us consider the simplified case of a 1D-intensity profile that translates with a constant velocity u. Since the velocity is constant the optical flow constraint takes on a particular simple form:

$$I(x,t) = I(x - u\Delta t, t - \Delta t). \qquad (9.24)$$

Now let's assume that we have an initial estimate (u_0 say) for the (unknown) velocity u: hence

$$u = u_0 + u_r$$

where u_r represents the *residual* velocity. The initial estimate u_0 can be used to predict the intensity profile at time t starting from time $t - \Delta t$; the discrepancy between the actual and predicted profile then provides us with information to estimate the residual velocity u_r. The key point about this stepwise procedure is that in general u_r is smaller in magnitude and therefore easier to estimate accurately using only two frames separated by a fixed time-lag.

Since the residual velocity u_r will shift this new profile $I(x - u_0\Delta t, t)$ to $I(x,t)$ in the time-span Δt, it is instructive to reformulate the new set-up in terms of the old terminology:

$$\tilde{I}(x,t) := I(x,t) \quad \text{and} \quad \tilde{I}(x, t - \Delta t) := I(x - u_0\Delta t, t - \Delta t). \qquad (9.25)$$

In terms of this notation eq.(9.24) can be written as

$$\tilde{I}(x,t) = \tilde{I}(x - u_r\Delta t, t - \Delta t),$$

which is again an optical flow constraint equation but now for the intensity profile $\tilde{I}$ moving with velocity u_r. In differential form this reads:

$$\tilde{I}_x \cdot u_r + \tilde{I}_t = 0$$

Notice that the x-derivatives satisfy $\tilde{I}_x(x,t) = I_x(x,t)$. This means that we can recast the above equation in terms of the initial estimate

$$I_x \cdot (u - u_0) + \tilde{I}_t = 0 \qquad (9.26)$$

where

$$
\begin{aligned}
\tilde{I}_t &\approx \frac{\tilde{I}(x,t) - \tilde{I}(x, t - \Delta t)}{\Delta t} \\
&= \frac{I(x,t) - I(x - u_0\Delta t, t - \Delta t)}{\Delta t} \\
&= \frac{I(x,t) - I(x + u_r\Delta t, t)}{\Delta t}
\end{aligned}
$$

The second line in this equation array draws attention to the fact that the estimate can be expressed in terms of the original intensity profile I and the previously estimated u_0. But equally important, the third line underlines that an accurate estimate of $\tilde{I}_t$ should be easier to come by than an estimate for I_t, since the residual velocity u_r is hopefully small compared to u. Substituting this into eq.(9.26) we can conclude that the initial estimate u_0 provides us with the following acceptable approximation:

$$I_x \cdot (u - u_0) + \frac{I(x,t) - I(x - u_0 \Delta t, t - \Delta t)}{\Delta t} = 0.$$

Collecting all the terms that can be computed once the initial estimate u_0 is made in $J(u_0)$, we obtain

$$I_x \cdot u + J(u_0) = 0 \tag{9.27}$$

where

$$J(u_0) = -u_0 \cdot I_x(x,t) + \frac{I(x,t) - I(x - u_0 \Delta t, t - \Delta t)}{\Delta t}.$$

The extension to the more realistic 2D-case is straightforward: if we can make an initial estimate (u_0, v_0) for the velocity field such that

$$(u, v) = (u_0, v_0) + (u_r, v_r),$$

then we can rewrite the optical flow constraint equation (9.24) (approximately) as:

$$I_x u + I_y v + J(u_0, v_0) = 0,$$

where

$$J(u_0, v_0) = -I_x \cdot u_0 - I_y \cdot v_0 + \frac{I(x,t) - I(x - u_0 \Delta t, y - v_0 \Delta t, t - \Delta t)}{\Delta t}.$$

Substituting this approximation into the cost functional (9.22) yields corresponding Euler-Lagrange equation:

$$\begin{cases} \dfrac{\partial u}{\partial t} = \nabla^2 u - \lambda I_x (I_x u + I_y v + J(u_0, v_0)) \\[2mm] \dfrac{\partial v}{\partial t} = \nabla^2 v - \lambda I_y (I_x u + I_y v + J(u_0, v_0)) \end{cases} \tag{9.28}$$

In principle (u_0, v_0) can be any (initial) estimate of the velocity components, but in practice, the estimates of the previous iteration are used (the initial step assumes a zero velocity field all over). Within this scheme, the equations are updated each iteration step and in between the iterations, new estimates for the spatio-temporal gradients are computed at a shifted location $(x - u_0 \Delta t, y - v_0 \Delta t)$. The estimated displacement will try and stabilize onto a point in the other frame with the same intensity characteristics. The whole idea is quite similar to the principle of correlation-based approaches, since a matching point (based on equality of intensity) is assigned to each point in the reference frame, and the smoothness term favours continuity in this matching.

4.4. Discontinuities and non-linear diffusion

The issue of edges in motion fields is more intricate than for intensity images. Consider Fig. 9.12. A rectangle moves over a still background. All

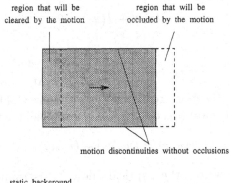

Figure 9.12.   Moving rectangle: edges with and without occlusion.

around the rectangle motion edges are found. However, these are of basically two types: edges with and without occlusion. Only the latter ones can be expected to be sharply delineated. Both types need to be detected and, if possible, distinguished.

Up to now the estimates (u_0, v_0) have been used to allocate a position in a previous frame. Imagine for a moment that the frames would be swapped, running the motion in reverse so to speak. In that case the estimates (u_0, v_0) could be used to recompute the spatio-temporal information at a position $(x + u_0 \Delta t, y + v_0 \Delta t)$ in a next frame at time $t + \Delta t$. From this point of view we have a forward and a backward scheme, where it is just a matter of switching the reference frame. At first glance both schemes seem to be equivalent, but they are not! In fact, nearby optical flow boundaries and occluding regions large inconsistencies between the two schemes may be found. Indeed, depending on whether the first or second frame is taken as reference, regions that are about to be occluded or to re-emerge, remain without matching points. This observation lies at the basis of our construction of an edge indicator within the flow field.

For most regions in the flow field, the forward and backward schemes are consistent, by which we mean that they yield the same result for both the forward and backward scheme. If we consider two successive frames a and b, this can be expressed mathematically as

$$\vec{v}_a(x,y) = -\vec{v}_b(x - u_a\Delta t, y - v_a\Delta t)$$

or

$$\vec{v}_b(x,y) = -\vec{v}_a(x - u_b\Delta t, y - v_b\Delta t)$$

with $\vec{v}_a = (u_a, v_a)$ and $\vec{v}_b = (u_b, v_b)$. Since the roles of the frames are reversed from one scheme to the other, the velocity estimates in both schemes will be opposite to each other (Fig. 9.13). This explains the minus sign in the above equations. In the neighbourhood of discontinuities, however, the schemes

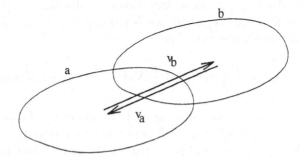

Figure 9.13. Opposite velocities in CODIQ scheme.

are very likely to be inconsistent. Indeed, for a number of points in frame a, one can not find matching points in frame b, and vice versa, due to occlusion by the moving object. The likelihood of there being an edge (by virtue of occlusion) can be expressed in relationship with the following quantities

$$\vec{C}_a = \vec{v}_a(x,y) + \vec{v}_b(x - u_a\Delta t, y - v_a\Delta t)$$

for frame a and

$$\vec{C}_b = \vec{v}_b(x,y) + \vec{v}_a(x - u_b\Delta t, y - v_b\Delta t)$$

for frame b. These vector quantities can be used to define a (scalar) *consistency measure c* which, in its turn, can be fed back to the optical flow scheme. For it to be useful, this consistency measure is constructed following some guidelines:

1. First of all, the consistency measure c should be normalized, to reduce its dependency on the magnitude of $\vec{C}$. As before, we will assume that c takes values between 0 and 1, corresponding to low, respectively high, consistency between the forward and backward schemes.

2. Since $\vec{C}$ is very likely to be scattered along the discontinuity line, some smoothing is advisable. Furthermore, changing the diffusion coefficient allows to control the region of interest along the discontinuity.
3. The most important role of c is to ensure that the smoothing process is halted nearby discontinuities.
4. Last but not least, it is possible that some velocity estimates are temporarily inconsistent during the solution process, even when they are not located on a flow discontinuity. Therefore consistent velocities should drive such transient non-consistent estimates to the right solution.

In keeping with the main theme of this chapter, it is only natural to try and realise these requirements via (extra) diffusion equations. The first two can be implemented elegantly via an additional Shah-diffusion process: given the vector field $\vec{C}$ on the region R, we determine the consistency measure c such that it minimizes the functional

$$V_{\vec{C}}(c) := \int_R \left(\alpha c^2 \|\vec{C}\| + \frac{\rho}{2} \|\nabla c\|^2 + \frac{(1-c)^2}{2\rho} \right) \, dx\, dy.$$

As before, this cost-functional will force c to be smooth and close to 1, except at places where $\vec{C}$ is significantly different from zero. Here the consistency c will approach zero, as it should to be true to its name. Since the functional V is convex, it can be minimized (dynamically) by solving the evolution equation based on the first variation (i.e. method of steepest descent):

$$\frac{\partial c}{\partial t} = \rho \nabla^2 c + \frac{1-c}{\rho} - 2\alpha c \|\vec{C}\| \tag{9.29}$$

The current value for c is then used to prevent non-consistent areas in the velocity field from corrupting consistent ones. As it can easily been seen by writing down the discretization, a diffusion process of the form

$$\frac{\partial u}{\partial t} = \text{div}(c\nabla u) + \ldots$$

will only allow u-information to dissipate along paths where the consistency measure c is high. In other words, inconsistent regions are prevented from propagating (confusing) information to consistent ones.

4.5. Integration

Integrating all of the above ideas, we obtain a CODIQ-scheme of six coupled diffusion equations, four of which describe the optical flow constraints, while the other two measure the consistency of the system. A flow chart of this algorithm is sketched in Fig. 9.14. More specifically, if we streamline the

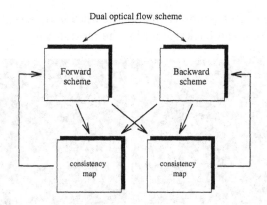

Figure 9.14. CODIQ optical flow scheme with feedback loops.

notation by introducing vector symbolism $\vec{w}_i = (u_i, v_i)$ (with $i = a$ or b to distinguish between back- and forward, and an additional 0-subscript indicates the initial estimate), we arrive at the following CODIQ-system which we propose to use when estimating a velocity field:

$$
\begin{cases}
\dfrac{\partial u_a}{\partial t} &= div(c_a \nabla u_a) - \lambda I_x (I_x u_a + I_y v_a + J(\vec{w}_{a0})) \\[2mm]
\dfrac{\partial v_a}{\partial t} &= div(c_a \nabla v_a) - \lambda I_y (I_x u_a + I_y v_a + J(\vec{w}_{a0})) \\[2mm]
\dfrac{\partial u_b}{\partial t} &= div(c_b \nabla u_b) - \lambda I_x (I_x u_b + I_y v_b + J(\vec{w}_{b0})) \\[2mm]
\dfrac{\partial v_b}{\partial t} &= div(c_b \nabla v_b) - \lambda I_y (I_x u_b + I_y v_b + J(\vec{w}_{b0})) \\[2mm]
\dfrac{\partial c_a}{\partial t} &= \rho \nabla^2 c_a + \dfrac{1 - c_a}{\rho} - 2\alpha c_a \| \vec{C}_a (\vec{w}_a, \vec{w}_b) \| \\[2mm]
\dfrac{\partial c_b}{\partial t} &= \rho \nabla^2 c_b + \dfrac{1 - c_b}{\rho} - 2\alpha c_b \| \vec{C}_b (\vec{w}_a, \vec{w}_b) \|
\end{cases}
\tag{9.30}
$$

This set of coupled diffusion equations has been discretized using finite differences. The evolution process itself proceeds by iteratively updating the equations, recomputing the spatio-temporal information for each iteration step: we always use newly obtained values as the $\vec{w}_{i0}$ for the next iteration.

4.6. Experimental Results

We applied the algorithm to a set of synthetic image sequences which consist of a textured object moving on an identically textured background. This means that the object is not visible in the stationary case and can only be discriminated from the background by the motion cue.

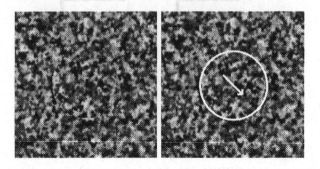

Figure 9.15. Translating circle

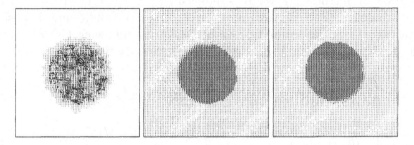

Figure 9.16.   Left: Horn protect& Schunck approach using 3×3 operator masks ($\lambda = 0.001$). Middle and right: Optic flow fields of the CODIQ scheme ($\lambda = 0.001$, $K = 0.2$, $\rho = 0.5$, $\alpha = 10.0$).

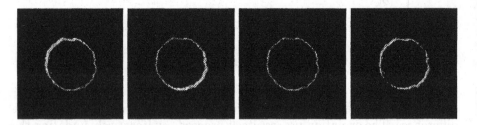

Figure 9.17. Left two images: discrepancies δ_1 and δ_2. Right two images: boundary ∂ and occluded regions ω.

Figure 9.15 shows a circle translating diagonally on an identically textured background. Note that these sequences are not easy to analyse, since at various locations, the gradient information is not consistent with the optical flow directions.

The operator mask is 3×3, which is small compared to a real velocity of about 3.5 pixels/frame. The classical Horn & Schunck approach, although able to extract the optical flow field with larger masks, shows clear distortions in the flow field for these small mask sizes (Figure 9.16). The CODIQ-approach on the other hand clearly succeeds in finding the correct displacement field. Also interesting are the consistency (or rather, the complementary *discrepancy*) images in Fig. 9.17: here we plotted the discrepancy $\delta := 1 - c$ between the forward and backward velocity field (white means $\delta \approx 1$, whereas black indicates $\delta \approx 0$). They clearly show the cleared and occluded regions as a thickening of the boundary contours. The left image primarily shows the region which is about to be cleared (i.e. become visible), while the second image contains the region which is on the verge of being occluded.

Obviously, both consistency images contain part of the information concerning flow boundaries or occluding regions. It is interesting to note that these data can be rearranged to exhibit even more clearly what part of the inconsistency is due to boundary- and what part to occlusion-effects. To this end we define a boundary indicator

$$\partial := min(\delta_1, \delta_2)$$

and an occlusion indicator:

$$\omega := max(\delta_1 - \partial, \delta_2 - \partial).$$

The operations are equivalent to the AND and EXOR operation on binary signals.

We point out that if the edges don't take part in occlusion, our edge indicator, which is based on inconsistencies due to occlusion, should, in principle, fail to detect these edges. However, due to the small but non-zero support of operator masks, the computation of the spatio-temporal gradients nearby flow discontinuities involves object points as well as background pixels. Thanks to this effect, non-occluding edges do show up in the discrepancy images.

Figure 9.18 shows a circle rotating at 5 degrees/frame, yielding a maximum velocity (at the outer perimeter) of about 4 pixels/frame. The method clearly succeeds in finding the correct flow field for the rotating circle. Figure 9.20 shows a three dimensional plot of the velocity magnitude. It increases linearly from the center to the object boundaries and then sharply drops to zero. (Note that for this example, one ought to be careful when using optical flow gradients for object-background discrimination.)

Obviously, there aren't any occluding regions, and for each point in one frame, there is an exact match in the other frame. This also means that

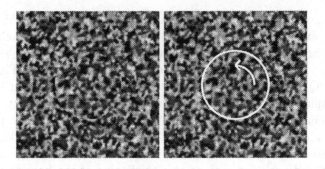

Figure 9.18. Rotating circle.

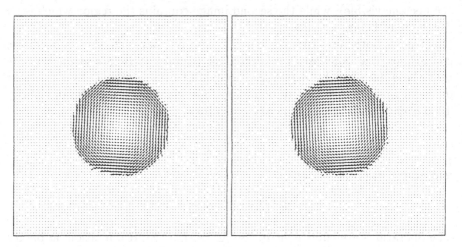

Figure   9.19.   Optic   flow   fields   of   the   CODIQ   scheme
$(\lambda = 0.001, K = 0.2, \rho = 0.5, \alpha = 10.0)$.

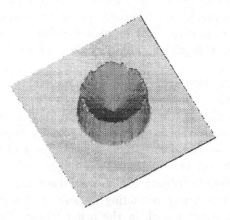

Figure 9.20. 3-dimensional plot of velocity magnitude.

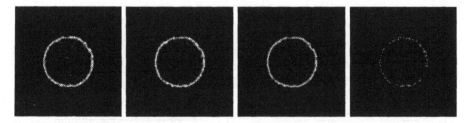

Figure 9.21. Left two images: discrepancies δ_1 and δ_2. Right two images: indicator for boundary ∂ and occluded regions ω.

in principle no inconsistencies can be found. But as mention above, the non-zero size of operator masks saves the day. That is why the discrepancy images clearly show a circular optical flow boundary. On the other hand, ω vanishes practically everywhere, and the sparse non-zero values can be accounted for by discretization errors and noise.

Figure 9.22. Three successive frames of Wanda moving to the left.

Figure 9.22 shows experiments on a real sequence: three successive frames of a fish (called Wanda, would you believe it?!) moving to the left. As Fig. 9.23 shows, the computed optical flow field is reasonably accurate. The magnitude plot of the velocity in Fig. 9.25 clearly shows a rather smooth flow field, whereas the classical Horn & Schunck approach suffers from much more internal distortions within the field. As for the optical flow boundaries, local thickening can be observed in the consistency images (Fig. 9.24), indicating the existence of occluding (and occluded) regions.

We draw attention to the fact that at some points the consistency images are noticeably blurred. There are several reasons for that. First of all, the object boundaries in the original (intensity) image are rather vague and fuzzy to begin with. But there is also another factor contributing to this effect. Consider for example a textured object on a uniform background. The method will very likely find the correct displacement field within the object. In the background however, the spatio-temporal gradients do not provide any information about the flow field whatsoever. In fact, one cannot tell

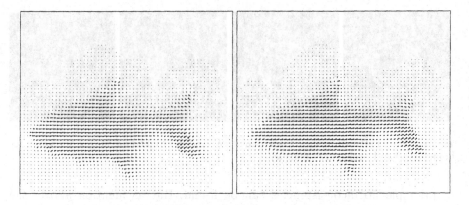

Figure 9.23. Optic flow fields of the CODIQ scheme.

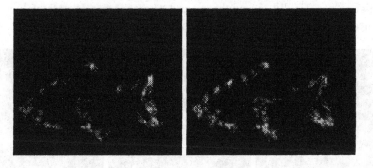

Figure 9.24. Discrepancies δ_1 and δ_2.

whether the background is moving or stationary. In the proposed scheme, such boundaries can be found to be consistent, so that limited smoothing across the boundary does occur. In the next section on stereo, an approach is described to counteract this effect.

As a last illustration, figure 9.26 shows a sequence taken by a camera which is following a moving train within a play school scene. The difficulty with this example is that the local spatio-temporal information is very low at various spots within the image. The algorithm however succeeds in filling in the missing information. The optical flow field clearly shows low velocity values on the train and wagons (the camera is panning to follow the train). In summary, the goal of this section was to improve optical flow computations without adding too many assumptions or complexities about the scene. The proposed CODIQ method presents a merger of correlation and differential techniques. Although the inclusion of correlation aspects does not really unify the analysis of short-range and long-range motion (but see also the multi-scale techniques introduced in the stereo section),

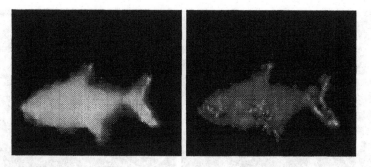

Figure 9.25. Left: velocity magnitude with CODIQ scheme. Right: velocity magnitude according to Horn & Schunck.

a wide range of velocities can be handled without the need to fall back on larger filters. Consequently, a more accurate determination of the optical flow field becomes possible. Discrepancy measures indicate the presence of flow boundaries of occluding or occluded regions. Another important aspect of the method is the introduction of a non-linear diffusion term which allows to reconstruct the optical flow field preserving its discontinuities.

5. Application to stereo

5.1. The simplified case of a calibrated stereo rig

Stereo matching can be considered as a special case of the Optical Flow Problem. For a single calibrated stereo setup, corresponding points lie on horizontal epipolar lines. Consequently, there is no *vertical* disparity between the two scenes, and the *horizontal* disparity is linked in a straightforward way to the depth of the points under consideration. Interpreting the (horizontal) disparity as the (x-component of a) velocity-field, we see that we can recast this problem in terms of optical flow terminology:

$$
\begin{aligned}
u_a &\equiv d_L \equiv \text{disparity for backward (i.e. left) scheme} \\
u_b &\equiv d_R \equiv \text{disparity for forward (i.e. right) scheme} \\
v_a &\equiv v_b \equiv 0
\end{aligned}
$$

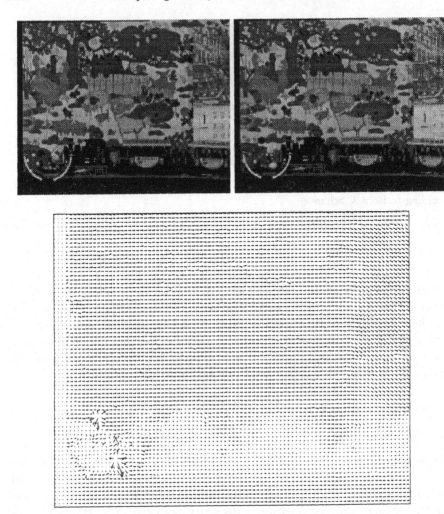

Figure 9.26. Play school scene: the camera is panning to follow the moving train.

Specifying the CODIQ-system (9.30) for this restricted configuration yields:

$$
\begin{cases}
\dfrac{\partial d_L}{\partial t} &= div(\gamma(c_L)\nabla d_L) - \lambda I_x(I_x d_L + J(d_{L0})) \\[2mm]
\dfrac{\partial d_R}{\partial t} &= div(\gamma(c_R)\nabla d_R) - \lambda I_x(I_x d_R + J(d_{R0})) \\[2mm]
\dfrac{\partial c_L}{\partial t} &= \rho\nabla^2 c_L + \dfrac{1-c_L}{\rho} - 2\alpha c_L \|\vec{C}_L(d_L, d_R)\| \\[2mm]
\dfrac{\partial c_R}{\partial t} &= \rho\nabla^2 c_R + \dfrac{1-c_R}{\rho} - 2\alpha c_R \|\vec{C}_R(d_L, d_R)\|
\end{cases}
\qquad (9.31)
$$

Now the dynamics of these processes is fixed, we can once again start the iterations and allow the system to converge to a stationary limit. Experiments with this system yield encouraging results (we refer to [299] for more details) as long as we make sure that the vertical disparity is negligible.

5.2. Extension to a general disparity field

5.2.1. *Motivation*

The results obtained for the calibrated stereo rig are reasonably accurate, even if the scheme has to bridge quite large disparities (a situation not uncommon when processing stereo views). However, it is to be expected that this approach will in general find it difficult to converge to the correct results when disparities are so large that local measurements are unable to pin them down. In such cases, successive approximation of the disparity field might provide (just as for optical flow) a welcome stepping stone. Finally, relaxing the strict calibration constraints on the stereo rig might result in a non-vanishing vertical disparity field which the simplified model cannot accommodate.

A more generally useful approach will therefore have to encompass the following items:

- both vertical and horizontal disparities, to deal with a non-calibrated stereo setup;
- a multi-scale analysis that can be used to generate successive approximations of the disparity field;
- time dependent parameters.

In what follows we will address these points and test our solution on a stereo view of two manikins (see Fig. 9.27). The stereo pair is not calibrated in the sense that the viewangle of the camera is slightly rotated from one view to the other. This induces vertical disparities which can be quite large. we point out the related work has been reported in [343, 245, 246].

5.2.2. *Multi-scale approach*

Recall that in the case of the optical flow problem we got round the problem of large velocity fields by using an iterative scheme that allowed us to make successively more accurate estimates of the velocity field. Iteration was based on a bootstrapping effect steered by neighbouring consistent velocities. A faster and more robust way of dealing with systematically large displacements is to exploit a multi-scale analysis.

This is exactly what we will do next: we will use a multi-scale scheme to iteratively finetune the disparities. The idea is basically simple: starting from a multi-scale representation of the input, we construct the disparity field for the data on the coarsest grid, where disparities are small. This result is used as the initial estimate for the next grid in the pyramid, a process that is repeated until we reach the original input scale.

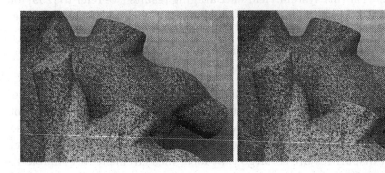

Figure 9.27. Stereo view of male and female manikin.

Let us be a bit more precise. The image domain Ω is approximated using a number of ever-finer grids $G_1 \ldots G_M$ with corresponding decreasing mesh sizes $h_1 > \ldots > h_M$, where $h_{k+1} = h_k/2$. As we mentioned earlier, the idea is to use the solution of the coarser grid G_k as the initial estimate for the solution on the finer grid G_{k+1}.

The original data are transferred to each successive coarser level by applying a 3×3 Gaussian (or binomial, if you like) averaging kernel as a restriction operator. Conversely, a finer grid is recovered from the coarser one by expanding the linear dimensions of the latter by a factor two (filling the new intermediate positions with zeros) and convolving this expanded grid with a 3×3 bilinear interpolator.

For this particular stereo view, the number of scales is $M = 3$. At the coarsest level, the algorithm is indeed able to find the correct disparities and their boundaries (Fig. 9.28). The task is considerably simplified by the fact that at the top of the pyramid, disparities have been reduced by a factor 4 (with respect to the original data). As an additional advantage of starting at the coarsest scale we point out that at that scale it suffices to use the simplified stereo scheme since the vertical disparities are hardly visible.

For the stereo images at the finer scales, we used an enlarged CODIQ-scheme of 6 equations: the four displayed in eq. (9.31) supplemented by two additional equations dictating the (completely similar) dynamics for the vertical disparity field. (It goes without saying that in this augmented set-up, the $\vec{C}$-process will depend on both the vertical and horizontal disparity.) Adding vertical disparity is really necessary, especially in the "outer" regions of the stereo pair where the rotation of the camera introduces a vertical disparity of about ±1.5 pixels. This might seem small but it is certainly not negligible as it will cause serious disruption if not taken into account. A

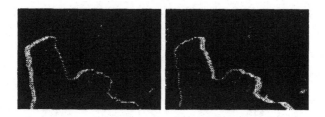

Figure 9.28.   Stereo consistencies at coarsest scale.

cursory glance at Fig. 9.33 shows that this extended algorithm is perfectly capable of extracting depth.

5.2.3. *Two further refinements*

We conclude this section by drawing attention to two further refinements which we introduced to improve the final results. First of all, in order to refine the stereo boundaries, we used a variable parameter setting for λ and K in the CODIQ-scheme used for the stereo. Initially, the smoothing factor λ and non-linearity threshold K are chosen to be quite high and are subsequently decreased in time. This will ensure that consistent regions, containing a lot of gradient information, will lock on to each other in both views, without interference from inconsistent regions. The initial estimate may for that reason look quite noisy, since neither the non-linearity nor the smoothing will drive inconsistent regions to the correct solution. Therefore, smoothing as well as non-linear behaviour are gradually enhanced, in order to correct inconsistencies within the disparity fields. Furthermore, possible stereo boundaries will be sharpened due to the increasing non-linear effect. Finally, as Fig. 9.28 shows, no boundaries are found between the large (male) manikin and the background. This is due to the fact that the background doesn't contain any gradient information at all, and consequently will not appear to be inconsistent. There is therefore a real danger that information from the background will seep into non-related manikin-regions. In order to avoid this sort of unwanted information-mixing, we introduce an additional map (denoted by w) to discriminate between textured and flat intensity regions. This w map (which takes values between 0 and 1 and can be thought of as a "fuzzy" texture-indicator) is akin to the v map described earlier and obeys a similar dynamics:

$$\frac{\partial w}{\partial t} = \rho \nabla^2 w - \beta w e^{-(w/\tau)^n} + \alpha(1-w)\|\nabla f\| \tag{9.32}$$

where $\|\nabla f\|$ is the gradient of the intensity image, and $\rho, \alpha, \beta, \tau$ and n are tunable parameters. Textured regions will tend to have high gradient values $\|\nabla f\|$ and this will drive w towards 1, whereas for relatively flat non-textured

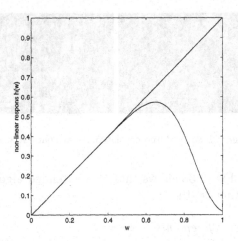

Figure 9.29.   Non-linearity $h(w) = we^{-(w/\tau)^n}$ in the dynamics of the w-map. Varying the parameters n and τ allows one to change the height and position of the maximum.

regions, w will tend to zero. The dynamics of the w-process becomes more transparent when we forget the exponential for a moment and consider the limit for small diffusion ($\rho \to 0$). Equation (9.32) then simplifies to

$$\frac{\partial w}{\partial t} = -\beta w + \alpha(1 - w)\|\nabla f\| \tag{9.33}$$

representing the relaxation over time of a "w-particle" on the interval $[0, 1]$ which is connected to the endpoints with (different) elastic springs. Adding the exponential $e^{-(w/\tau)^n}$ to the dynamics can be thought of as allowing the "β-spring" to "snap" when it is "overstretched" (i.e. when w becomes too large, see Fig. 9.29). As soon as w crosses a certain threshold, the elastic restraining force connecting the "particle" to the origin will drop dramatically, allowing w to take off towards one.

Switching diffusion back on will create the possibility that neighbouring regions may be dragged along by these high gradient patches, resulting in a very discriminating w-map. The effect is shown in Fig. 9.30 on one of the two views. The background is clearly discriminated from the texture surface, i.e. both manikin surfaces.

For each of the two views, a w_1 and w_2 map is generated, and integrated with the consistency maps c_1 and c_2. The final results are shown in Figs. 9.31, 9.32 and 9.33. Figure 9.31 shows clearly the occluding regions for this stereo view, and Fig. 9.33 gives two views on the reconstructed depth map. The results show a nice and acceptable fit to the actual manikin surfaces.

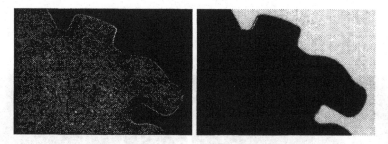

Figure 9.30.   Gradient map $\|\nabla f\|$ and texture map w (white: $w \approx 1$; black: $w \approx 0$) for the left image in Fig. 9.27.

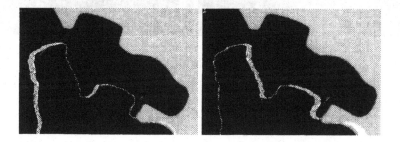

Figure 9.31.   Stereo consistencies for both schemes.

6. Conclusion

In this chapter we have shown how systems of coupled non-linear diffusion equations provide a powerful yet flexible conceptual framework for many aspects of low-level vision. Such systems exhibit many attractive characteristics: they are highly parallelizable and are therefore natural candidates for VLSI-implementation. Time-consuming global searches through high-dimensional spaces are no longer necessary. Their coupling ensures internal consistency of the final result and the non-linearity allows a rich and interesting dynamics. In short, they open up an interesting new avenue of research in low-level image processing.

Acknowledgements   Part of this work was carried out under Esprit BRA 3001 "Insight".

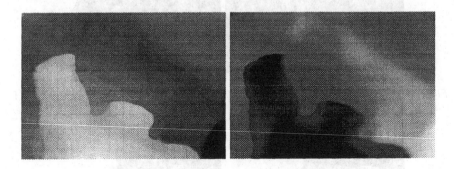

Figure 9.32. Stereo disparities for both schemes.

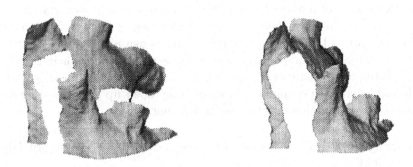

Figure 9.33. 3D reconstruction: two views.

MORPHOLOGICAL APPROACH TO MULTISCALE ANALYSIS: FROM PRINCIPLES TO EQUATIONS

Luis Alvarez

Departamento de Informatica y Sistemas
Universidad de Las Palmas
Campus de Tafira, 35017 Las Palmas, Spain

and

Jean-Michel Morel

Ceremade, Université Paris IX Dauphine
75775 Paris, Cedex 16 , France

1. Introduction

A numerical image can be modelized as a real function $I_0(x)$ defined in $\mathbb{R}^N$ (In practice, $N = 2$ or 3). The main concept of vision theory and image analysis is *multiscale analysis* (or *"scale space"*). Multiscale analysis associates with $I(0) = I_0$ is a sequence of simplified (smoothed) images $I(t, x)$ which depend upon an abstract parameter $t > 0$, the *scale*. The image $I(t, x)$ is called *analysis of the image I_0 at scale t*. The formalization of *scale-space* has received a lot of attention in the past ten years ; more than a dozen of theories for image, shape or "texture" multiscale analysis have been proposed and recent mathematical work has permitted a formalization of the whole field. We shall see that a few formal principles (or axioms) are enough to characterize and unify these theories and algorithms and show that some of them simply are equivalent. Those principles are *causality* (a concept in vision theory which can be led back to a maximum principle), the *Euclidean (and/or affine) invariance*, which means that image analysis does not depend upon the distance and orientation in space of the analysed image, and the *morphological invariance* which means that image analysis does not depend upon a contrast change.

The characterization and the classification of the numerous theories of image and shape analysis will be obtained by identifying the underlying partial differential equations (which have been pretty more than implicit in many theories !) The axiomatic characterization leads, as we shall see, to a significant improvement of most proposed algorithms as well as to new ones, with more invariance properties. Among the theories which will be axiomatically here, we shall comment

- the *Raw Primal Sketch* by Hildreth and Marr [248],
- the *Scale-Space* by Witkin [386], Koenderink [187], ...
- the *Intrinsic Heat Equation* by Gage, Hamilton [125], Grayson [139], Angenent [17, 18], ...
- the *Motion by Mean Curvature* by Osher [285], Sethian [339], Evans and Spruck [97], Giga, Goto [60, 61], Barles [26, 27], Souganidis [28], ...
- the *Entropy Scale-Space* by Kimia, Tannenbaum and Zucker [180, 182],
- the *Dynamic Shape* by Koenderink and Van Doorn [195],
- the *Curvature Primal Sketch* by Mackworth and Mocktarian [256, 257, 256], Asada and Brady [22], ...
- the *Morphologie Mathématique* by Matheron, Serra [336] and the "Fontainebleau school",
- the *Anisotropic Diffusion* by Perona and Malik [293],
- the *Affine Scale-Space of Curves* by Sapiro and Tannenbaum [323],
- the *Affine Morphological Scale-Space* of images by Alvarez, Guichard, Lions and Morel [6],
- the *Affine Morphological Galilean Scale-Space* of movies by the same authors [144].

The classification of these multiscale theories will lead us to focus on the only one of them (in $N = 2$) matching simultaneously all invariance and stability requirements partially satisfied by others: the Affine Morphological Scale-Space (AMSS). This multiscale analysis can be defined by a simple Partial Differential Equation,

$$\frac{\partial I}{\partial t} = |\nabla I|(t.div(\nabla I/|\nabla I|))^{1/3}, \quad I(0,x) = I_0(x) \qquad (10.1)$$

where $I(t,x)$ denotes the image analysed at scale t and point x. (This parabolic equation admits a unique "viscosity solution" in the Crandall-Ishii-Lions (1990 [72]) sense). As we shall see, the equation of the Affine Morphological Scale-Space handles independently all *level sets* of the analysed image and therefore is compatible with the Morphologie Mathématique (which asks contrast invariance). In addition, the contrast invariance means that the boundary of every level set of the image is analysed as a shape and we get a common multiscale analysis for shapes and images.

A multiscale formalization of *image segmentation* can be developed with analogous principles and leads to multiscale segmentation algorithms. A multiscale segmentation associates with an initial image I_0 a sequence

$I(t, x)$, $K(t, x)$, where $I(t, x)$ is the image simplified at scale t and $K(t, x)$ the set of boundaries of the homogeneous regions of I_0 at scale t. Thanks to the formalization, many segmentation algorithms can be reduced to one. We finally shall devote some pages to the above mentioned Affine Morphological Galilean Scale Sapce of movies. The underlying equation,

$$\frac{\partial I}{\partial t} = (|\nabla I| \; curv^{\frac{1}{3}}(I))^{1-q} \; ((|\nabla I|\mathrm{sgn}(curv(I))accel(I))^{+})^{q}, \quad (10.2)$$

also is a parabolic equation, where $curv$ denotes the nonlinear differential operator computing the curvature of the level lines, and $accel$ represents the "apparent acceleration" observed at a given space-time point of the movie.

2. Image multiscale analysis

2.1. A short story of the subject

Computer Vision is dealing with a philosophical, psychological, physiological and technical question which can be stated in a few words : how the local brightness information arriving to the retina of some individual (or to any optical sensor) can be transformed into a global percept of the objects surrounding him, with their distance, colour and shape ? In the sixties, this question has been translated into a very practical framework with the new possibilities of experimentation on digital pictures with computers. The new technology has allowed accurate measurements of the human visual performance on digital pictures and the first experiments in "computer vision". The joint developments of pschophysics and computer vision have led to a new doctrine: the existence of *low level vision*. The story of the doctrine is well explained (e.g.) in David Marr's book Vision [247] and we shall just give a few hints of how this doctrine developed.

On the one side, several psychophysical experiments due to Bela Julesz and his school [170] proved that the reconstruction of the spatial environment from binocular information was some automated, reflex process, not depending on any learning. Julesz also studied the "preattentive" perception of textures and proved the existence of a process for discriminating textures independently of any a priori knowledge. The discrimination process is fast, parallel and Julesz and his school discussed it in *mathematical terms* coming from statistics and geometry. These experiments, as well as the neurobiological experiments of Hubel and Wiesel [160, 161, 159], gave a proof of the existence, in the first milliseconds of the perception process, of a series of parallel, fast and irreversible operations applied to the retina information and already yielding a very rich and useful information to further understanding of the "image".

2.2. The visual pyramid as an algorithm

This series of operations, we shall call "visual pyramid". In mathematical terms, it may be thought of as an algorithm, but not in the Turing sense; in a more general sense where we define an algorithm as a black box transforming its input into a output in a deterministic way by a physical process: in any case, this machine is assumed to be physically implemented in the brain. We must distinguish the problem of how this machine works in the case of the brain, and what it really does as an information process (What Chomsky called performance versus competence). Indeed, the second question is simply a mathematical question, while the first one has a large relevance in neurobiology. We shall now focus on the mathematical question and treat it in a rather rough way by answering the three questions
a) What is the input of the visual pyramid ?
b) What is its output ?
c) What basic principles must obey the visual pyramid, if thought of as a physical system ?

2.3. What is the *input* of the visual pyramid ?

A simple model to discuss image processing is to define an "image" as a "brightness" function $I_0(x)$ at each point x of a domain of the plane. This domain, which may be the plane itself, is a model of the retina or of any other photosensitive surface. In what follows, we shall take the plane for simplicity. For commodity, in the discussion, we shall always assume that $I_0(x)$ is in the space $\mathcal{F}$ of all continuous real functions $I(x)$ on $\mathbb{R}^N$ such that $\|(1 + |x|)^{-N} I(x)\| \leq C$ for some N and C. Of course, the datum of $I_0(x)$ is not absolute in perception theory, but can be considered as the element of an equivalence class. If y is a vector of the plane, the shifted datum $I_0(x - y)$, which is the image shifted by y, is an equivalent datum. In the same way, the change of $I_0(x)$ into $I_0(Rx)$ where R is an isometry of the plane should not change the visual analysis. Finally, we can think of I_0 as belonging to a projective class, that is, as a representative of the class $I_0(Ax)$ where A is any projective map of the plane. Indeed, a plane image can be viewed by an observer from any distance and orientation in space (think of a painting in a gallery). Therefore, the input $I_0(x)$ is assumed to be equivalent to any of its anamorphoses $I_0(Ax)$. We shall assume in the following A to only be any affine map, which makes sense when the ratio between the size of the observed objects and the distance to the sensor is small. Last but not least, the observation of $I_0(x)$ does not generally give any reliable information on the number of photons sent by any visible place to the optical sensor. Therefore, the equivalence class in consideration will be $g(I_0(x))$, where g stands for any (unknown) contrast function depending on the sensor. This last assumption, that only isophotes matter (see also [108]), is associated with the "mathematical morphology" school. So we shall call it the "morphological" assumption.

To summarize, an image is an equivalence class of functions $I_0(Ax)$ where A is a translation or an isometry in most classical geometrical models, A is any affine map in the simplified projective model and it is a class of functions $g(I_0(x))$ where g is any continuous nondecreasing function in the morphological model. We can combine these models and consider a morphological projective model, that is, an equivalence class under the action of all g's and A's: $g(I_0(Ax))$.

2.4. What is the *output* of the visual pyramid ?

Starting from the local brightness information, each layer of the visual pyramid is assumed to yield more and more global "low level" information about the image. This information is assumed to be usable for the geometric reconstruction (stereovision) as well as for the "high level vision", that is, the interpretation of the scene. Whatever its use might be, most models define the basic output as either

- a smoothed image (from which reliable "features" can be extracted by local and therefore differential operators)
- a segmentation, that is, either a decomposition of the image domain into homogeneous regions ("strong segmentation"), with boundaries or a set of boundary points or "edge map".

In both cases, the output depends on two variables: a variable x which denotes the center of a spatial neighborhood and a variable t which can be identified or correlated with

- the "height" in the visual pyramid (or distance from the first layer: the retina). This distance corresponds to the biological time between the "arrival" at the retina and the first arrival at a given layer.
- the degree of globality of the local information in the considered layer, that is, the size of the neighborhood in the retina which influences what happens at x.

To summarize, the output of the vision pyramid is

- either a multiscale image $I(t, x)$
- or a multiscale "edge map" $K(t, x)$, where t is a parameter which can be identified with a time of analysis or with a measure of the spatial globality of the information provided by $I(t, x)$. The bottom of the output is the original image $I(0, x) = I_0(x)$.

2.5. What basic principles must obey the visual pyramid ?

The causality.

These principles come first from the "preattentive" assumption, that no feedback is allowed, the visual information being processed in parallel through a sequence of filters. So what happens in higher scales cannot influence what happens at lower scales: the pyramid acts "from fine to coarse". Furthermore, there is no time for taking into account at scale t what happens at a significantly smaller scale s. So we assume that the output at scale t can be computed from the output at a scale $t - h$ for very small h. (Take into account that the visual pyramid is a series of filters through which new visual information is constantly being processed. So, to look at what happens at a smaller scale means to "look in the future", at new arriving perceptual images.) To formalize this relation, we call $T_t : \mathcal{F} \to \mathcal{F}$ the map which associates with an image I_0 its "smoothed image" at the scale t, $T_t I_0$. (In the same way, we denote by T_t the map associating with a set K of boundaries a set of boundaries $T_t K$ simplified at scale t). This mapping is obtained by composition of "transition filters" which we call $T_{t+h,t} : \mathcal{F} \to \mathcal{F}$ and we have the

Pyramidal Structure (Causality 1) $T_{t+h} = T_{t+h,t} T_t, \quad T_0 = Id.$

Furthermore, the operator $T_{t+h,t}$ will always be assumed to act "locally", that is, to look at a small part of the processed image. In other terms, $(T_{t+h,t} I_0)(x)$ must essentially depend upon the values of $I_0(y)$ when y lies in a small neighborhood of x.

We shall give two formal versions of this "locality assumption". Let us now just give its "physical" interpretation: if the basic elements of the pyramid are assumed to be "neurons", this only means that a neuron is primarily influenced by its neighbours. A clear argument for that is time: only neurons which are close can influence without transmission delay. Let us finish with an intuitive requirement which is called in image processing "causality". Since the visual pyramid is assumed to yield more and more global information about the image and its features, it is clear that when the scale increases, no new feature should be created by the multiscale analysis: the image and the boundaries at scale $t' > t$ must be simpler than the boundaries at scale t. The causality assumption must of course be formalized. Its formalization has been discussed by Hummel [163], Koenderink [187, 190, 189], Yuille [394], Witkin [386], Perona and Malik [291], in the framework of image processing, by Kimia, Tannenbaum and Zucker [181] in the framework of shape analysis and by Muerle and Allen (1968), Brice and Fennema (1970), Horowitz, Pavlidis (1972) in early works on image segmentation.

The result of the discussion in the case of image processing is that causality must be formalized as *pyramidality* plus a *local comparison principle*: if an image I is locally brighter than another v, then this order must be conserved some time by the analysis (prevalence of local behaviour on global behaviour). In formal terms,

Local Comparison Principle (Causality 2)
If $I(y) > v(y)$ for y in a neighborhood of x and $y \neq x$, then for h small enough,

$$(T_{t+h,t}I)(x) \geq (T_{t+h,t}v)(x)$$

In the case of edge detection, there are several formalizations, but the simplest states that no new boundary is created when the scale increases, that is, $T_{t'}K$ is contained in $T_t K$ if $t' > t$.

We finally need some assumption stating that a very smooth image must evolve in a smooth way with the multiscale analysis. Somehow, this belongs to the "causality" galaxy, but we prefer to call it regularity and it clearly corresponds to the assumption of the existence of an infinitesimal generator for the multiscale analysis.

Regularity Let $I(y) = \frac{1}{2}(A(x - y), x - y) + (p, x - y) + c$ be a quadratic form of $\mathbb{R}^N$. There exists a function $F(A, p, x, c, t)$, continuous with respect to A, such that

$$\frac{(T_{t+h,t}I - I)(x)}{h} \to F(A, p, x, c, t) \quad \text{when} \quad h \to 0.$$

Morphological and affine invariance.
In addition to the causality requirement, we must keep in mind that the pyramid acts on equivalence classes of images of the form $g(I_0(Ax))$, where g is any nondecreasing continuous function and A any isometry (or any affine map) of the plane. Therefore, the output should not depend upon I_0 but on the equivalence class. So the transition operators $T_{t+h,t}$ must somehow commute with the perturbations g and A. In the case of a change of contrast g, this is easily translated into the

Morphological invariance $gT_{t+h,t} = T_{t+h,t}g$,

which means that change of contrast and multiscale analysis can be applied in any order. If A is an isometry, the same kind of relation must be true.

Denote by Au the function $AI(x) = I(Ax)$. Then we state the

Euclidean invariance $AT_{t+h,t} = T_{t+h,t}A$

Let us now examine the case of an arbitrary linear map A. The commutation relation cannot be so simple because A can reduce or enlarge the image. (Think of the case where A is a zoom defined by $AI(x) = I(\lambda x)$ for some positive constant λ). Since the zoom has changed the scale of the image, we can just impose a weak commutation property

Affine invariance For any A and $t \geq 0$, there exists a C^1 function $t'(t, A) \geq 0$ such that $AT_{t'(t,A),t'(s,A)} = T_{t,s}A$. Moreover, the function $\phi(t) = \frac{\partial t'}{\partial \lambda}(t, \lambda Id)$ is positive for $t > 0$.

This relation means that the result of the multiscale analysis T_t is independent of the size and position in space of the analyzed features: An affine map corresponds to the anamorphosis of a plane image when it is presented to the eye at any distance large enough with respect to its size and with an arbitrary orientation in space. (The general visual invariance should be projective, but for small objects at some distance, we shall be contented with the affine invariance). (See Forsyth and al. [117], Lambdan and al. (1988)) The assumption on t', $\phi(t) = \frac{\partial t'}{\partial \lambda}(t, \lambda Id) > 0$, can be interpreted by looking at the relation $\lambda(Id)T_{t'} = T_t(\lambda Id)$ when λ increases, i.e when the image is shrunk before analysis by T_t. Then, the corresponding analysis time before shrinking is increased. In more informal terms we can say that the analysis scale increases with the size of the picture.

Let us point out the fact that the affine invariance must be stated in such a general framework because we have until now made no attempt to fix the relation between the abstract "scale" parameter and the concrete scale understood as having some relation with the size of objects. As for all results to come, they will be true whatever the change of abstract scale $T_t \rightarrow T_{\sigma(t)}$ provided σ is a smooth increasing function: $\mathbb{R}^+ \rightarrow \mathbb{R}^+$. Now, the next lemma will permit a full normalization of the scale, thank to the affine invariance.

Lemma 1 *(Normalization of scale) Assume that $t \rightarrow T_t$ is a one to one family of operators satisfying pyramidality and affine invariance. Then the function $t'(t, B)$ only depends on t and $|detB|$: $t'(t, B) = t'(t, |detB|^{1/2})$ and is increasing with respect to t. Moreover, there exists an increasing differentiable rescaling function $\sigma : [0, \infty] \rightarrow [0, \infty]$, such that $t'(t, B) = \sigma^{-1}(\sigma(t)|detB|^{1/2})$ and if we set $S_t = T_{\sigma^{-1}(t)}$ we have $t'(t, B) = t|detB|^{1/2}$ for the rescaled analysis.*

We shall give a proof of this Lemma in Appendix B, because it is of particular relevance in image processing.

To summarize, the multiscale analysis T_t must (or may) satisfy:

- **Causality:** $T_{t+h} = T_{t+h,t}T_t$, $T_{t,t} = T_0 = Id$, and $(T_{t+h,t}I)(x) > (T_{t+h,t}v)(x)$ if $I(y) > v(y)$ for y in a neighborhood of x and $y \neq x$. In the case of boundary multiscale analysis, this last assumption is replaced by $T_t K \subset T_s K$ if $t > s$.
- **Regularity:** Let $I(y) = \frac{1}{2}(A(x-y), x-y) + (p, x-y) + c$ be a quadratic form of $\mathbb{R}^N$. There exists a function $F(A, p, x, c, t)$, continuous with respect to A, such that

$$\frac{(T_{t+h,t}I - I)(x)}{h} \to F(A, p, x, c, t) \quad \text{when} \quad h \to 0.$$

- **Morphological invariance:** $gT_{t+h,t} = T_{t+h,t}g$ for any change of contrast g.
- **Euclidian invariance:** $AT_{t+h,t} = T_{t+h,t}A$ for any isometry A of $\mathbb{R}^N$.
- **Affine invariance:** $AT_{t'+h',t'} = T_{t+h,t}A$ with $t' = |det(A)|^{1/2}t$ and $h' = |det(A)|^{1/2}h$. Notice that the affine invariance implies the Euclidean invariance.
- **(Optional) linearity:** $T_{t+h,t}(aI + bv) = aT_{t+h,t}(I) + bT_{t+h,t}(v)$. We add this property because it has been very in use in computer vision.

There are therefore five main axioms and we shall see that they allow to completely classify and characterize the theories of multiscale image and shape processing, to unify and to improve several of them.

3. Axiomatization of image multiscale analysis and classification of the main models

Theorem 1 *(Koenderink [190], Hummel [163], Yuille-Poggio [396]). If a multiscale analysis is causal, Euclidean invariant and linear, then it obeys (up to a rescaling $t \to \sigma(t)$) the heat equation*

$$\frac{\partial I}{\partial t} = \Delta u$$

(That is, $I(t, x) = (T_t I)(x)$ obeys the heat equation).

This model of multiscale analysis (the *"raw primal sketch"*) is due (among others) to Hildreth et Marr (see Marr [247]) and Witkin (the *scale-space* [386]). See also Koenderink [187] who is first to explicitly state the heat equation, more recently Lindeberg [213] who studied the associated discrete scale-space, and Florack and al. [110] who show how to use heat equation to find corners [36], T-junctions [359], etc... by simple differential operators.

What happens if we remove the linearity axiom ? As noticed in Perona and Malik (1988) in their nonlinear theory of scale-space, "anisotropic diffusion", we can get nonlinear heat equations

$$\frac{\partial I}{\partial t} = F(\nabla^2 I, \nabla I, u, x, t) \tag{10.3}$$

The converse implication and a complete study of the nonlinear models is done in Alvarez and al. [6, 9]:

Theorem 2 *(Fundamental theorem) If an image multiscale analysis T_t is causal and regular then $I(t, x) = (T_t I)(x)$ is a viscosity solution of (10.3), where the function F, defined in the regularity axiom, is nondecreasing with respect to its first argument $\nabla^2 I$. Conversely, if I_0 is a bounded uniformly continuous image, then the equation (10.3) has a unique viscosity solution.*

(For a quick proof of this theorem, see Appendix A).
The particular case of the heat equation corresponds to $F(A, p, c, x, t) = trace(A)$. As a consequence of this theorem, all multiscale models can be classified and new, more invariant models can be proposed:

Theorem 3 *Let $N = 2$. If a multiscale analysis is causal, regular, Euclidean invariant and morphological, then it obeys an equation of the form*

$$\frac{\partial I}{\partial t} = |\nabla I| F(div(\nabla I/|\nabla I|), t) \tag{10.4}$$

where $div(\frac{\nabla I}{|\nabla I|})(x)$ can be interpreted as the curvature of the level line of the image $I(t, x)$ passing by x and $F(s, t)$ is nondecreasing with respect to the real variable s.
An important particular case is when F is a constant function: if $F = +1$ or $F = -1$, then the equation becomes $\frac{\partial I}{\partial t} = |\nabla I|$ (resp. $\frac{\partial I}{\partial t} = -|\nabla I|$), which corresponds to the so called morphological erosion when the sign is "-" and to a morphological dilation when the sign is "+". (See Brockett and Maragos [44])). *Dilation* and *erosion* are the basic operators of the Morphologie Mathématique, founded by Matheron (1975) and his "Fontainebleau School". The dilation at scale t is defined by

$$D_t I_0(x) = \sup_{y \in B(x,t)} I_0(y),$$

where $B(x, t)$ is a set centered at x: **the "structuring element"** which generally is a ball with radius t. Assume for instance that I_0 is the characteristic function of a set X, then $D_t I_0$ is the characteristic function of the t-neighborhood of X. For the erosion, one simply replaces "sup" by "inf".

As has let us notice Michel Rascle, another relevant example satisfying the multiscale morphological axioms and therefore a parabolic P.D.E. is the family of zooms with ratio t, $(T_t I_0)(x) = I(t, x) = I_0(tx)$. Indeed, the preceding theorem applies and it is easily seen that the underlying equation is

$$\frac{\partial I}{\partial t} = \frac{1}{t}(\nabla I.x).$$

Notice that this formulation may be useful, the zooming operators on a digital picture being in no way easy to implement. Now, the preceding examples have been very particular instances of the equation and there are many other possibilities for morphological multiscale filtering ! If we set $F(s, t) = s$, we obtain the "mean curvature equation" (M.C.M.).

$$\frac{\partial I}{\partial t} = t.|\nabla I| div(\nabla I/|\nabla I|) \qquad (10.5)$$

This equation comes from a reformulation by Osher and Sethian [285] of a differential geometry model studied by Grayson [139, 140] and Gage-Hamilton [125]. It is also very close to the "anisotropic diffusion" of Perona and Malik [291] and to an image restoration equation due to Rudin, Osher and Fatemi [313]:

$$\frac{\partial I}{\partial t} = div(\nabla I/|\nabla I|)$$

Now, the most invariant model is new: it is proved in Alvarez, Guichard, Lions, Morel [6] that

Theorem 4 *(A.M.S.S. Model). Let* $\mathbb{N} = 2$*. There is a single causal, regular, morphological and affine invariant multiscale analysis. Its equation is*

$$\frac{\partial I}{\partial t} = |\nabla I|(t.div(\nabla I/|\nabla I|))^{1/3} \qquad (10.6)$$

We shall better understand this equation in the framework of Shape analysis: Sapiro and Tannenbaum [322, 328], who independently discovered the model as a *shape scale-space* have given it a remarkable geometric interpre- tation in this framework. Of course, the Fundamental Theorem 2 can be applied in any dimension. Let us just state a last example of scale-space in dimension 3 (of particular relevance for medical solid images). See again Alvarez, Guichard, Lions and Morel [6] and Caselles and Sbert [56].

Theorem 5 *Let* $N = 3$*. There is (up to a rescaling) a single causal, regular, affine invariant and morphological multiscale analysis, associated with the* *equation*

$$\frac{\partial I}{\partial t} = |\nabla I| sign(M(I))(G(I)^+)^{\frac{1}{4}} \tag{10.7}$$

By $G(I)$ we denote the Gaussian curvature, that is the determinant of $\nabla^2(I)$ restricted to $\nabla I^\perp$, and by $M(I)$ we denote the Mean curvature. We shall see more on this subject when looking for movie analysis equations.

4. Shape multiscale analyses

We could deduce the shape analysis statements from image analysis statements. However, since in this case the axiomatics is particularly simple and intuitive, we shall list well-adapted principles, equivalent however to the general image analysis principles. For more details, see Kimia, Tannenbaum and Zucker [181], Lopez and Morel [234], Mackworth and Mockhtarian [256]. We define a shape or ("silhouette") as a closed set X whose boundary is a Jordan curve of $\mathbb{R}^2$. We denote by $T_t(X)$ the *shape analysed at scale t*. X is identified with its characteristic function $X(x) = 1$ if $x \in X$ and 0 else. We call multiscale analysis any family of operators $(T_t)_{t\geq 0}$ acting on shapes and we set $X(t) = T_t(X)$. As before, we shall state *causality* principles, the first of which remains unchanged:

Pyramidal Structure (Causality 1) $T_{t+h} = T_{t+h,t}T_t$, $\quad T_0 = Id$.

Instead of the Local comparison principle, we shall give a very intuitive statement: the shape local inclusion principle.
Assume that X and Y are two silhouettes and that for some $x \in \partial Y$ and some $r > 0$, one has $X \cap B(x,2r) \subset Y \cap B(x,2r)$. Assume further that the inclusion is strict in the sense that ∂X and ∂Y only meet possibly at x. Then we shall say that *the shape X is included in the shape Y around x*.

[Shape local inclusion] If X is included in Y around x, then for h small enough, $T_{t+h,t}(X) \cap B(x,r) \subset T_{t+h,t}(Y) \cap B(x,r)$.
This last axiom implies that the value of $T_{t+h,t}(X)$ for h small, at any point x, is determined by the behaviour of X near x. We are allowed to take r infinite. Therefore the shape local inclusion principle also implies that if a shape is globally contained in another, this order is preserved for every scale (Mackworth and Mokhtarian [256]).
Both preceding principles allow, as we shall see, to localize the shape analysis in space and time and we therefore only need to state what the multiscale analysis makes of very simple shapes in order to specify it. So we add a "basic principle" which will state what happens to disks. As we shall see,

disks are somehow a "basis" for shape analysis because whenever we know how they are analyzed, we know what will happen to every other shape.

[**Basic principle**] Let $D = D(x, 1/r)$ be a disk with curvature $1/r$ and center x. Then $T_{t+h,t}(D)$ is a disk with radius $\rho(t, h, 1/r)$ and center x. Moreover, the function $h \to \rho(t, h, 1/r)$ is differentiable with respect to h at $h = 0$ and the differential is continuous with respect to $1/r$.

The "basic principle" implies that the multiscale analysis behaves in a smooth and isotropic way. In the following, we set

$$g(t, 1/r) = \frac{\partial \rho}{\partial h}(t, 0, 1/r) \tag{10.8}$$

Note that $g(t, s)$ is defined for $t \geq 0$ and $s \in \mathbb{R}$. Now in order to define $g(t, 0)$ we must assume that $lim\ g(t, 1/r)$ exists when r tends to $+\infty$ or $-\infty$. The radius r may be positive or negative, according to the orientation of the normal $\vec{n}(x)$. In the case where the curve is a Jordan curve enclosing a set X, we take as a *convention that $\vec{n}(x)$ is pointing outside X and the curvature is negative if X is convex at x, positive else.* It may seem natural to assume therefore that $g(t, \kappa)$ is odd with respect to κ. Indeed, this corresponds to the assumption that a black disk on grey background and a white disk behave in the same way.

4.1. The fundamental equation of Shape Analysis

When a point x belongs to an evolving curve, we denote by $\dot{x}$ the time derivative of x, which is a vector of $\mathbb{R}^2$. By $curv(x)$ we denote the curvature of a curve which is C^2 at x. Recall that the curvature is defined as the inverse of the radius of the osculatory circle to the curve at x. The curvature is zero if the radius is infinite.

Theorem 6 *(i) Under the three principles (pyramidal, local shape inclusion, "basic"), the multiscale analysis of shapes is governed by the curvature motion equation*

$$\dot{x} = g(t, curv(x))\vec{n}(x) \tag{10.9}$$

where g is defined by (10.8).
(ii) If the analysis is affine invariant, then the equation of the multiscale analysis is, up to rescaling:

$$\dot{x} = \gamma(t.curv(x))\vec{n}(x) \tag{10.10}$$

where γ is defined by $\gamma(x) = a.x^{1/3}$ if $x \geq 0$ and $\gamma(x) = b.x^{1/3}$ if $x \leq 0$ and a, b are two nonnegative values.
(iii) If we add that $T_t(X^c) = T_t(X)^c$ [Reverse contrast invariance] then the function g in (i) is odd and we get

$$\dot{x} = (t.curv(x))^{\frac{1}{3}} \vec{n}(x) \tag{10.11}$$

We prove this theorem in Appendix C. By the expression "governed by", we mean that the equation must be satisfied at any point (t, x) of $\partial X(t)$ where the boundary of the silhouette is smooth enough to give a classical sense to both terms of the equation.

Remark: It has been proved (Gage-Hamilton [125], Grayson [139], Angenent [18, 16, 19]) that the equation (10.9) with $g(t, s) = s$ has smooth solutions. This equation has been introduced in picture processing by Kimia-Tannen- baum-Zucker [180, 181], Mackworth-Mockhtarian [256] and Alvarez-Lions-Morel [9] in different contexts. An early version of an algorithm leading to equation (10.9) is Koenderink and Van Doorn [195]. See also Yuille [396]. The equation (10.11) has been introduced and axiomatically justified (with a more complicated axiomatic however) in Alvarez-Guichard-Lions-Morel [6]. It has also been proposed in Sapiro-Tannenbaum [328]. Existence and regularity of the solution are proved in Sapiro-Tannebaum [321]. The axiomatic presenta- tion adopted here follows Cohignac, Lopez and Morel [70].

5. Relation between image and shape multiscale analyses

In this section, we shall show how the AMSS model (10.1) can be deduced from the shape evolution (10.11). We give in appendix C proofs of the axiomatic deduction of this last equation, so that our exposition will be rather complete. In addition, the shape multiscale analysis equation has an easy geometric interpretation as an "intrinsic diffusion" which we shall explain at the end of this section. Since the multiscale analysis satisfies the obvious inclusion principle that

$$\text{If} \quad A \subset B \quad \text{then} \quad T_{t+h,t}(A) \subset T_{t+h,t}(B),$$

we can, as well known in the "mathematical morphology school" (Matheron 1975, Maragos 1987), associate with a picture I the set of its level sets

$$X_a I = \{(x, y), I(x, y) \geq a\}$$

Then, assuming that each level set is a union of silhouettes, we can simply define $T_t(I)$ from the multiscale analysis $T_t(X)$ of silhouettes by

Morphological principle For any a, t, h and I: $X_a T_{t+h,t}(I) = T_{t+h,t}(X_a I)$

Theorem 7 *Assume that a multiscale image analysis T_t satisfies the morphological principle and that each of the level sets is governed by equation (10.9). Then $T_t I$ satisfies*

$$\frac{\partial I}{\partial t} = g(t, curv(I))|\nabla I| \qquad (10.12)$$

Proof Let us prove (10.4) at any (x, a) such that $I(t, x) = a$ and $I(t, y) = a$ implies $\nabla I(t, y) \neq 0$ and I is C^2 at (t, y). The first condition implies by the implicit function theorem that x belongs to a Jordan curve Γ enclosing two regions Γ^+ and Γ^-, with $I(t, x) < a$ on Γ^-, $I(t, x) > a$ on Γ^+ and $I(t, x) = a$ on Γ. Set $I(t) = I(t, .)$. The relation $X_a T_{t+h,t} I(t) = T_{t+h,t} X_a I(t)$ implies that $T_{t+h,t} \Gamma^+$ is equal to a connected component of $X_a I(t+h)$. Therefore, the point $x(t+h)$ defined by equation (10.9) with initial value $x(t) = x$ belongs to the boundary of $X_a I(t+h)$. So we obtain $I(t+h, x(t+h)) - I(t, x(t)) = 0$. Dividing by h and passing to the limit yields

$$\frac{\partial I}{\partial t}(t, x) + \nabla I(t, x)\dot{x}(t) = 0$$

and using equation (10.9) and $\vec{n}(x) = -\frac{\nabla I}{|\nabla I|}$ we obtain equation (10.4).

The Affine Scale-Space model (10.11) of Sapiro and Tannenbaum yields a simple geometric interpretation of the AMSS model (10.1). Let us consider two ways of parametrizing a smooth Jordan curve $x(s)$:

– Either in an Euclidean-invariant way by imposing $|x_s| = 1$.
– Or in an affine-invariant way by setting $|det(x_s, x_{ss})| = 1$.

In the second case, we say that the curve has been parametrized by its "affine length parameter" s. This parametrization is affine covariant because $|det(Ax_s, Ax_{ss})| = |\det A||det(x_s, x_{ss})|$ for any affine map A. Sapiro-Tannenbaum [323] proved (it is an easy computation) that the A.S.S model is equivalent to the following "intrinsic heat equation":

$$\frac{\partial x(s,t)}{\partial t} = t^{1/3} \frac{\partial^2 x(s,t)}{\partial s^2}, \quad x(0, s) = x_0(s)$$

So the application of the AMSS model to an image I can be interpreted as the affine invariant diffusion of all the level lines (isophotes) of I. Therefore we get a nonlinear generalization of the linear classical scale-spaces.

6. Multiscale segmentation

Segmentation is acknowledged as a main tool in image interpretation. As we shall see, the segmentation problem is quite well understood in the framework of multiscale analysis. To define this problem in two lines, let us say that the objective of segmentation is to find the homogeneous regions of an image as well as their boundaries. However, the term "homogeneous" is extremely vague and in order to state what homogeneity is, one has to rely either on perceptual experiments or on axiomatic definitions. In any case, homogeneity may concern clues as different as brightness, colour and texture. In this section, we shall not discuss what these clues are (this will be axiomatically introduced in the next section for textures). We shall assume that the image datum is composed of a finite set of k "channels", each channel being itself a real image. In the simplest case, there is a single channel, the brightness. In the colour image case, the actual technology (partly based on perceptual criteria) yields three channels, i.e. three images of the same size (Red, Green, Blue). In the case of other clues, like texture elements, the number of channels is unlimited and experiments showed in the next section involve up to fourty channels computed from an initial grey level image and with the same size. The segmentation problem assumes that an initial multichannel image I_0,   with $I_0(x) \in \mathbb{R}^k$, and an initial boundary map K_0, where K_0 is a subset of the image domain with finite Hausdorff length, are given. The initial boundaries can simply be the boundaries of all pixels. The segmentation process computes a multiscale sequence $(K(t))$ of segmentations, as well as a multiscale sequence of piecewise homogeneous images $I(t)$. $K(t)$ is assumed to be the set of the boundaries of the homogeneous regions of $I(t)$.

We shall give simple multiscale principles which closely determine what the segmentation process is. Setting as now usual $T_t K_0 = K(t)$, we impose

-Causality: $K(t') \subset K(t)$ if $t < t'$ and $T_{t,s} T_s = T_t$, $T_0 = Id$.

-Euclidean Invariance: $T_{t+h,t} R = R T_{t+h,t}$ for every isometry R.

-Invariance by Zooming: Let Z_λ denote the zoom with ratio λ. Then

$$T_t Z_\lambda = Z_\lambda T_{\lambda^{1/2} t}$$

We finally need an axiom fixing the relations between the boundaries $K(t)$ at scale t and the image $I(t)$. Our choice (see Morel and Solimini [259]) is to take the simplest principle, which Mumford and Shah [264] called the "cartoon principle".

-Cartoon principle: $I(t)$ is locally constant in $\mathbb{R}^2 \setminus K(t)$ and equal to the mean value of I_0 on each connected component of $\mathbb{R}^2 \setminus K(t)$.

The causality and cartoon principles nearly fix the kind of segmentation algorithm to be used: it is a *region growing* algorithm. Let us call *region* every connected component of $\mathbb{R}^2 \setminus K(t)$. Since, for $t' > t$, $K(t') \subset K(t)$, we deduce that the regions at scale t' are unions of regions at scale t (which is another way of stating the causality axiom). Thus, in order to completely fix the multiscale analysis, we only need a criterion for region "merging". Mumford and Shah [264] proposed the following Euclidean and zoom-invariant criterion: with every segmentation is associated an energy (which somehow measures its complexity),

$$E(I(t), K(t)) = \int_{\mathbb{R}^2 \setminus K(t)} (I(t, x) - I_0(x))^2 dx + t.length(K(t))$$

Then two regions of the segmentation will be merged at scale t if and only if the energy of the resulting segmentation decreases. The associated "recursive merging" algorithm is extremely simple. We start with an initial trivial segmentation of the image at scale $t = 0$. In this case, the image simply is (e.g.) divided into "pixels" (small squares in the actual technology).

Region growing variational algorithm

- Fix I_0, K_0 as the initial trivial segmentation, where K_0 is the union of boundaries of all pixels.
- A scale t being fixed: For every pair of regions, check whether their merging decreases the Mumford-Shah energy and, the case being, merge them. Then the new K is obtained by removing the common boundary of both regions and I takes as new value the mean value of I_0 on the union of both regions.
- Increment the scale t and go back to the preceding step.

Mumford and Shah [264] proved that a segmentation which is minimal for their energy has a finite number of regions with smooth boundaries. It is however impossible to find a minimal segmentation, because the energy is highly not convex. Therefore, it is useful to get information about the segmentations obtained by a concrete algorithm computing local minima. Here is such a theorem, which justifies the use of region growing associated with the Mumford-Shah energy.

Theorem 8 *(Koepfler et al. [202]) Let us say that a segmentation of a bounded vectorial image is 2- normal at scale t if no pair of regions can be merged without increasing the Mumford-Shah energy. Then the set of 2-normal segmentations is compact for the Hausdorff distance and there is a bound, only depending on t, for the number of regions of a 2-normal segmentation.*

This section has been short, unless more than thousand papers have been written on segmentation algorithms. Now, Morel and Solimini [237], following and updating the terminology of Zucker (1976) discussed more than ten classes of algorithms in image processing and showed that they merely are variants of the Mumford-Shah energy minimizing process. Among the many theories which lead to the Mumford-Shah formalization (or variants), let us mention the "snakes" (Kass and al. [174]), the survey by Haralick and Shapiro [148], stochastic segmentation of Geman-Geman (1984), the Blake-Zisserman "weak membrane" model (1987), and the region growing algorithms of Brice-Fennema (1970), Muerle-Allen (1968), Pavlidis (1972), Horowitz and Pavlidis (1974). An affine invariant version of the Mumford-Shah theory has been proposed by Gonzalez and Ballester [25].

7. Movies multiscale analysis

In this section, we return to scale-space theory and look for its adaptation to movies. It is rather easy, since we still stay in the framework of the Fundamental Theorem (2). Indeed, a movie can be modelized as a 3D-image, $I(x, \theta)$, where $x \in \mathbb{R}^2$ and $\theta \in \mathbb{R}$ is the time parameter. The statement and justification of the causality axioms remain unchanged, as well as the morphological invariance. Of course, the Euclidean and affine invariance make less sense in $3D$, and we shall keep their 2D-versions, the $2D$-Euclidean invariance and the $2D$-affine invariance, where isometries (resp. affine maps) are restricted to the image plane. We however add two specific axioms related to motion.

- **Time affine invariance** For any affine time rescaling $A_{a,b}$: $I(x, \theta) \rightarrow I(x, a\theta + b)$, there exist $t'(a, t)$ such that $T_t A_{a,b} = A_{a,b} T_{t'}$.
- **Galilean invariance** Denote by B_v any Galilean motion operator defined by $B_v I(x, \theta) = I(x - v\theta, \theta)$. Then $B_v(T_t I) = T_t(B_v I)$.

The Galilean invariance means that the analysis is invariant under "traveling", that is a motion of the whole picture at constant velocity v does not alter the analysis. In the following, we distinguish the "spatial gradient" $\nabla I = (I_x, I_y)$, and the space-time gradient, $\nabla I = (I_x, I_y, I_\theta)$.

Theorem 9 *If a multiscale analysis of movies is causal, regular, time-translation invariant, space-Euclidean invariant and morphological, then it is governed by an equation*

$$\frac{\partial I}{\partial t} = |\nabla I| \, F(curv(I), accel(I), t) \qquad (10.13)$$

If in addition we assume the analysis to be time-affine and space- affine invariant, the equation is (up to a rescaling)

$$\frac{\partial I}{\partial t} = (|\nabla I|\ curv^{\frac{1}{3}}(I))^{1-q}\ ((|\nabla I|sgn(curv(I))accel(I))^+)^q, \quad (10.14)$$

For some $q \in [0,1]$.

(See Alvarez-Guichard-Lions-Morel 1992). In the above formulae, we use the convention that the power preserves the sign, that is $a^q = |a|^q sgn(a)$. Hence, when $q = \frac{1}{4}$, we obtain the equation

$$\frac{\partial I}{\partial t} = |\nabla I|sign(M(I))(G(I)^+)^{\frac{1}{4}} \quad (10.15)$$

This equation was mentioned in section 2 as the only affine invariant morpho- logical scale-space in $\mathbb{R}^3$. Of course, this full affine invariance doesn't have sense for classical movies : What is the meaning of a rotation involving spatial and time variables ? Now, in the field of relativity theory, such invariance makes sense because the Lorentz transform is nothing but a spatial-time rotation. In other words, when $q = \frac{1}{4}$, we have an equation which is both Galilean invariant and relativist invariant. We did not explain what the differential operator $accel(I)$ is. (As for $curv(I)$, it simply is the curvature of the spatial level curves, as in section 2). We could give the explicit formula of $accel(I)$ in terms of the first and second derivatives of I, but this would prove disastrous, since the formula takes several lines. We shall use two ways to characterize $accel$ and justify its name of "apparent acceleration". Let us first explain a classical notion in motion analysis: the apparent velocity. Assuming that a movie displays moving objects, let us call $x(\theta)$ the trajectory of a point of one of these objects. If we make the assumption that the object is *lambertian*, which means that the light that it is sending to the camera is constant, then we can ensure that $I(x(\theta), \theta) = C$ for some constant C. Differentiating with respect to θ yields $< \nabla I, \dot{x}(\theta) > + I_\theta = 0$. So we see that the component of the velocity in the direction of the spatial gradient can be recovered from the partial derivatives of I:

Definition 1 *We call apparent velocity of a movie $I(x,\theta)$ at point (x,θ) the scalar*

$$v_1 = -\frac{I_\theta}{|\nabla I|}$$

Now we are in a position to justify (and define) $accel(I)$:

Lemma 2 (Interpretation of *accel***)** *Consider a picture in translation motion* $I(x,\theta) = w(x - \int_0^\theta \vec{v}(\tau)d\tau)$, *where* $\vec{v}(\theta)$ *is the instantaneous real velocity vector. Then the apparent velocity* v_1 *is equal to the true velocity in the direction of the gradient,* $< \vec{v}, \frac{\nabla u}{|\nabla I|} >$. *Let* $V = (\vec{v}, 1)$ *be the real space-time velocity. Then*

$$accel(I) = - < Dv_1, V > \qquad (10.16)$$

The first result is easy to check. The second formula can be taken as a definition of accel and shows that in the case of objects in translation motion, accel(I) is the derivative of the apparent velocity in the direction of $-\nabla I$. *This is why we call it "apparent acceleration". For a proof, see Alvarez, Guichard, Lions, Morel [6] or Guichard [144].*

The second way of explaining what *accel* is consists in using Guichard's (1993) numerical approximate of it, which has proved essential when dealing with concrete movies. Normal movies display series of frames where objects may jump more than 3 (and up to 50...) pixels from one frame to the next. Indeed, quick motions may let an object jump from one side of the screen to the other side in a very few frames. So, real movies are dramatically undersampled in time, which does seem to affect the trained zapper, but makes classical (local) numerical schemes impossible. So we define a nonlocal search space for the "possible velocity vectors",

$$\mathcal{W} = \{V = (\alpha, \beta) \text{ for all } \alpha \text{ and } \beta \text{ in } \mathbb{R}\}$$

Theorem 10

$$|\nabla I|(sgn(curv(I))\, accel(I))^+ =$$

$$min_{w_1,w_2 \in \mathcal{W}} \frac{1}{\Delta\theta^2}(|I(x - w_1, \theta - \Delta\theta) - I(x,\theta)| + \qquad (10.17)$$

$$|I(x + w_2, \theta + \Delta\theta) - I(x,\theta)| + | < \nabla I, w_1 - w_2 > |) + o(1)$$

Interpretation. Of course, for numerical experiments, we shall not compute the minimum for all vectors in $\mathcal{W}$, but only for the vectors on the grid. We have two differents parts in the second term : The first part is the variation of the grey level value of the point x, for candidate velocity vectors : w_1 between $\theta - \Delta\theta$ and θ, and w_2 between θ and $\theta + \Delta\theta$. This variation must be as small as possible, because a point is not supposed to change its grey level value during its motion. The second part is nothing but an "apparent acceleration", that is, the difference between w_1 and w_2 in the direction of the spatial gradient $|\nabla I|$.

Since the movie analysis theory is very recent and computationally heavy (but not desperate), it has not received yet technological applications. The most promising applications at hand are detection of hidden trajectories and denoising of movies. Both are done in the same way, by simply applying the equation to a movie at some small scale. As can be deduced from the equation itself, a trajectory will be easily eliminated if either its acceleration is high or the moving object is small (and therefore with high curvature). The multiscale analysis acts like a sieve, by removing first all most erratic trajectories, and leaving as $t \to +\infty$ only the Galilean trajectories. (Which, by the Galilean invariance principle, remain unaltered).

Appendix

1. Appendix A. The "fundamental theorem" of image analysis

Theorem 2. If an image multiscale analysis T_t is causal and regular then $I(t,x) = (T_t I)(x)$ is a viscosity solution of (10.3), where the function F, defined in the regularity axiom, is nondecreasing with respect to its first argument $\nabla^2 I$. Conversely, if I_0 is a bounded uniformly continuous image, then the equation 10.3 has a unique viscosity solution.

Proof Assume for simplicity that $I(t,x)$ is C^2 in a neighbourhood of (t,x). Then, we have

$$I(t,y) = I(t,x) + (\nabla I, y - x) + \frac{1}{2} D^2 I(y-x,y-x) + o(|y-x|^2)$$

Let $\epsilon > 0$ and Q_ϵ a quadratic form given by

$$Q_\epsilon(y) = I(t,x) + (\nabla I, y - x) + \frac{1}{2} D^2 I(y-x,y-x) + \epsilon.|y-x|^2$$

Then, in a neighborhood of (t,x)

$$Q_\epsilon(y) < I(t,y) < Q_\epsilon(y) \qquad \text{for } y \neq x,$$

and by using the causality principle we obtain

$$(T_{t+h,t}Q_{-\epsilon})(x) \leq (T_{t+h,t}I(t))(x) \leq (T_{t+h,t}Q_\epsilon)(x).$$

On the other hand, we also have

$$Q_{-\epsilon}(x) = Q_\epsilon(x) = I(t,x) = (T_{t,t}Q_{-\epsilon})(x) = (T_{t,t}Q_\epsilon)(x)$$

Therefore we deduce from the above relations

$$\frac{\partial(T_{t+h,t}Q_{-\epsilon})}{\partial h}(x) \leq \liminf \frac{T_{t+h,t}I(,x) - T_{t,t}I(x)}{h} \leq$$

$$\limsup \frac{T_{t+h,t}I(,x) - T_{t,t}I(x)}{h} \leq \frac{\partial(T_{t+h,t}Q_\epsilon)}{\partial h}(x).$$

By using the regularity principle and the continuity of the function F, and taking $\epsilon \to 0$ we obtain that $I(t,x)$ satisfies equation (10.3). Finally, in order to obtain that $F(A,p,c,x,t)$ is nondecreasing with respect to A, we notice that if $A \leq B$ then the quadratic forms

$$Q_A(y) = \frac{1}{2}(A(y-x),(y-x)) + (p,y-x) + c$$

$$Q_B(y) = \frac{1}{2}(B(y-x),(y-x)) + (p, y-x) + c$$

satisfy $Q_A(y) \le Q_B(y)$ and by using an obvious adaptation of the above proof, we obtain $F(A, p, c, x, t) \le F(B, p, c, x, t)$ if $A \ge B$.

To simplify the exposition, we have showed that equation (10.3) is true in the case where I is a C^2 function. By using the same ideas in the framework of viscosity solutions (see Crandall, Ishii and Lions [72]), it is possible to show that equation (10.3) is true in the sense of viscosity solutions for any $I(t, x)$ uniformly continuous satisfying the causality and regularity principles. The fact that if $I_0(x)$ is a bounded uniformly continuous function, the equation (10.3) has a unique viscosity solution is proved in Chen, Giga and Goto [60], Crandall, Ishii and Lions [72] and Evans and Spruck [97].

2. Appendix B. Proof of the scale normalization lemma

Normalization Lemma *(Normalization of scale .)* Assume that $t \to T_t$ is a one to one family of operators satisfying [affine invariance]. Then the function $t'(t, B)$ only depends on t and $|det B|$: $t'(t, B) = t'(t, |det B|^{1/2})$ and is increasing with respect to t. Moreover, there exists an increasing differentiable rescaling function $\sigma : [0, \infty] \to [0, \infty]$, such that $t'(t, B) = \sigma^{-1}(\sigma(t)|det B|^{1/2})$ and if we set $S_t = T_{\sigma^{-1}(t)}$ we have $t'(t, B) = t|det B|^{1/2}$ for the rescaled analysis.

Proof

First we notice that for any linear transforms B and C and any t one has the semigroup property

$$(i) \qquad t'(t, BC) = t'(t'(t, B), C).$$

Indeed, we have $BCT_{t'(t,BC)} = T_t BC = BT_{t'(t,B)}C = BCT_{t'(t'(t,B),C)}$. The map which associates T_t with t being one to one, this implies the stated relation.

Next, we show that

$$(ii) \qquad t'(t, A) \text{ is increasing with respect to } t.$$

Let us prove that $t'(t, A)$ is one to one with respect to t for any A. Indeed, if not, there would be some A and some (s, t) such that $t'(t, A) = t'(s, A)$. Thus $T_t A = AT_{t'(t,A)} = AT_{t'(s,A)} = T_s A$ and therefore $t = s$ because T_t is one to one. Notice that this implies, in particular, that $t'(0, A) = 0$. Therefore, since $t'(t, A)$ is non negative (by definition), one to one, and continuous with respect to t, we can deduce that it is increasing with respect to t.

Moreover $t'(t, A)$ satisfies

(iii) $t'(t, R) = t$ for any orthogonal transform R

Indeed, let R be an orthogonal transform. Then iterating the formula of (i) we have

$$t'(t'(t'(t'(...t'(t, R)..., R), R), R) = t'(t, R^n)$$

Remark that there is a subsequence of R^n tending to Id. (Indeed, there is a subsequence R^{n_k} which converges to some H, orthogonal, because the orthogonal group is compact. Therefore, the subsequence $R^{n_{k+1}-n_k}$ converges to Id.) Since there exists a subsequence of R^n tending to Id and since t' is continuous we have for this subsequence $lim\ t'(t, R^n) = t'(t, Id) = t$. Assume by contradiction that $t'(t, R) = t$" with t" $< t$ then $t'(t'(t, R), R) = t'(t$", $R) \leq t'(t, R) = t$" and by recursion,

$$t'(t, R^n) = t'(t'(t'(t'(...t'(t, R)..., R), R), R) \leq t$" $< t.$$

This is a contradiction. Thus $t'(t, R) \geq t$. We prove the converse inequality in the same way and we obtain $t'(t, R) = t$.

We note that any linear transform B of $\mathbb{R}^2$ can be obtained as a product of orthogonal transforms and of linear transforms of the kind $A(\lambda)$: $(x, y) \to (\lambda x, y)$ where λ is non negative. We only need to make a singular value decomposition of B: $B = R_1 D R_2$, where R_1 and R_2 both are orthogonal transforms and D is a transform of the kind $(x, y) \to (\lambda_1 x, \lambda_2 y)$ where λ_i are non negative. Now, it is clear that D can be decomposed as $D = A(\lambda_1) R A(\lambda_2) R^{-1}$ where R is the orthogonal transform: $(x, y) \to (-y, x)$. Using (i), $t'(t, R_i) = t$, the singular value decomposition and $A(\lambda_1) A(\lambda_2) = A(\lambda_1 \lambda_2)$, we obtain

$$t'(t, B) = t'(t, \lambda_1 \lambda_2) = t'(t, |det B|^{1/2})$$

Using (i) and (ii), we have

(iv) $t'(t, \lambda\mu) = t'(t'(t, \mu)\lambda)$

for any positive λ and μ. Differentiating this relation with respect to μ at $\mu = 1$ yields

$$\lambda \frac{\partial t'}{\partial \lambda}(t, \lambda) = \frac{\partial t'}{\partial \lambda}(t, 1) \frac{\partial t'}{\partial t}(t, \lambda) \qquad (10.18)$$

Choose σ such that $\phi \sigma' = \sigma$ and set $t'(t, \lambda) = G(t, \sigma(t)\lambda)$, where

$$\sigma(t) = exp(\int_1^t ds/\phi(s))$$

Then the preceding relation (10.18) yields $\frac{\partial G}{\partial x}(x,y) = 0$. Thus $G(x,y) = \beta(y)$ for some differentiable nondecreasing function β. We obtain that $t'(t,\lambda) = \beta(\sigma(t)\lambda)$. Returning to the definition of $\phi(t)$, we have $\phi(t) = \frac{\partial t'}{\partial \lambda}(t,1) = \frac{\partial \beta(\sigma(t)\lambda)}{\partial \lambda}(t,1)$ and $\phi(t) = \sigma(t)\beta'(\lambda\sigma(t)) = \phi(t)\sigma'(t)\beta'(\sigma(t)\lambda)$. Thus the derivative of $\beta(\sigma(t))$ is 1 and integrating this last relation between 0 and t yields $\beta(\sigma(t)) = t + \beta(\sigma(0))$. Using the fact that $t'(0,\lambda) = 0$ (which derives from the injectivity of the T_t), we obtain $\beta(\sigma(0)) = 0$ and therefore $t'(t,\lambda) = \sigma^{-1}(\lambda\sigma(t))$. To finish the proof, we set $S_t = T_{\sigma^{-1}(t)}$ and we prove that the affine invariance is true for S_t with $t'(t,\lambda) = \lambda t$. $S_t B = T_{\sigma^{-1}(t)} B = BT_{t'(\sigma^{-1}(t),\lambda)} = BT_{\sigma^{-1}(\lambda\sigma(\sigma^{-1}(t)))} = BT_{\sigma^{-1}(\lambda t)} = BS_{\lambda t}$.

3. Appendix C. Classification of shape multiscale analyses

Theorem 11 *(i)* *Under the three principles (pyramidal, local shape inclusion, "basic"), the multiscale analysis of shapes is governed by the curvature motion equation*

$$\dot{x} = g(t, curv(x))\vec{n}(x) \tag{10.19}$$

where g is defined by (10.8).
(ii) If the analysis is affine invariant, then the equation of the multiscale analysis is, up to rescaling:

$$\dot{x} = \gamma(t, curv(x))\vec{n}(x) \tag{10.20}$$

where $\gamma(x) = a.x^{1/3}$ if $x \geq 0$ and $\gamma(x) = b.x^{1/3}$ if $x \leq 0$ and a, b are two nonnegative values.
(iii) If we add that $T_t(X^c) = T_t(X)^c$ [Reverse contrast invariance] then the function g in (i) is odd and we get

$$\dot{x} = (t, curv(x))^{\frac{1}{3}}\vec{n}(x) \tag{10.21}$$

Proof

(i) Let X be a silhouette and assume that $T_t(X)$ has a boundary which is a C^2 manifold in a neighbourhood of a point x of ∂X. Then it has a curvature κ at point x and we consider a subosculatory and a surosculatory disk, that is, a disk D with curvature $\kappa - \epsilon$ and a disk D' with curvature $\kappa + \epsilon$, both tangent to the silhouette at x. Applying the same two principles as in the lemma, we see that

$$T_{t+h,t}(D) \cap B(x,r) \subset T_{t+h,t}(X) \cap B(x,r) \subset T_{t+h,t}(D') \cap B(x,r)$$

Thus, denoting by $x(t+h)$ the point of $\partial T_{t+h}(X)$ such that $x(t+h) - x(t)$ is parallel to $\vec{n}(x)$, we obtain

$$\rho(t,h,\kappa-\epsilon) - \rho(t,0,\kappa-\epsilon) \leq (x(t+h)-x(t)).\vec{n}(x) \leq \rho(t,h,\kappa+\epsilon) - \rho(t,0,\kappa+\epsilon)$$

Dividing by h and passing to the limit when h tends to 0 yields

$$\frac{\partial \rho}{\partial h}(t, 0, \kappa - \epsilon) \leq \liminf \frac{x(t+h) - x(t)}{h}.\vec{n}(x)$$

$$\limsup \frac{x(t+h) - x(t)}{h}.\vec{n}(x) \leq \frac{\partial \rho}{\partial h}(t, 0, \kappa + \epsilon)$$

We obtain equation (10.19) by passing to the limit when ϵ tends to 0 and using the fact that $\kappa \to \frac{\partial \rho}{\partial h}(t, 0, \kappa)$ is continuous.

(ii) After renormalization, we can use the identity

$$T_{t+h,t}D_\lambda = D_\lambda T_{(t+h)\lambda,t\lambda}$$

(where $D_\lambda = \lambda Id$) and so, we can deduce that the function ρ of the basic principle must satisfy $\rho(t, h, \frac{\lambda}{r}) = \frac{1}{\lambda}\rho(\lambda t, \lambda h, 1/r)$ Therefore, we obtain the relation (after differentiation with respect to h at 0)

$$g(t, \lambda s) = g(\lambda t, s)$$

for any $t > 0$, $\lambda > 0$ and $s \in \mathbb{R}$. Changing t in t/λ and taking $\lambda = 1/t$ we get $g(t, s) = g(1, ts) = \beta(ts)$ for the function β defined as $\beta(x) = g(1, x)$. On the other hand, we can use the identity $T_{t+h,t}A = AT_{t+h,t}$, where A is the linear transform whose determinant is one,

$$(x, y) \to (\lambda x, (1/\lambda)y) \qquad \lambda > 0.$$

Let us apply this identity to the unit disk Δ. Look at the point $x_0 = (1, 0)$ on the boundary of Δ. Then the velocity of x_0 is $\beta(-t)$, and this velocity is transformed into $\lambda\beta(-t)$. Now, look at $A\Delta$. Since $A\Delta$ is an ellipse with curvature $-\lambda^3$ at point Ax_0, the velocity of Ax_0 is $\beta(-t.\lambda^3)$. Using the first identity, we obtain $\beta(-t.\lambda^3) = \lambda\beta(-t)$. Taking $t = 1$, we get $\beta(x) = b.x^{1/3}$ for $x < 0$ ($b = \beta(-1)$). Now, apply the same technique to Δ^c and we get the result $\beta(x) = a.x^{1/3}$ for $x > 0$ ($a = \beta(1)$).

(iii) With the same technique as above we obtain that the function β is odd.

DIFFERENTIAL INVARIANT SIGNATURES AND FLOWS IN COMPUTER VISION: A SYMMETRY GROUP APPROACH

Peter Olver

University of Minnesota
Department of Mathematics
Minneapolis, MN 55455, USA

and

Guillermo Sapiro

Massachusetts Institute of Technology
Center for Intelligent Control Systems
Laboratory for Information and Decision Systems
Cambridge, MA 02139, USA

and

Allen Tannenbaum

University of Minnesota
Department of Electrical Engineering
Minneapolis, MN 55455, USA

Abstract. [1] Computer vision deals with image understanding at various levels. At the low level, it addresses issues such us planar shape recognition and analysis. Some classical results on differential invariants associated to planar curves are relevant to planar object recognition under different views and partial occlusion, and recent results concerning the evolution of planar shapes under curvature controlled diffusion have found applications in geo-

[1]This work was supported in part by grants from the National Science Foundation DMS-9116672, DMS-9204192, DMS-8811084, and ECS-9122106, by the Air Force Office of Scientific Research AFOSR-90-0024 and F49620-94-1-00S8DEF, by the Army Research Office DAAL03-92-G-0115, DAAL03-91-G-0019, DAAH04-93-G-0332, DAAH04-94-G-0054, by the Rothschild Foundation-Yad Hanadiv, and by Image Evolutions, Ltd.

metric shape decomposition, smoothing, and analysis, as well as in other image processing applications. In this work we first give a modern approach to the theory of differential invariants, describing concepts like Lie theory, jets, and prolongations. Based on this and the theory of symmetry groups, we present a high level way of defining invariant geometric flows for a given Lie group. We then analyze in detail different subgroups of the projective group, which are of special interest for computer vision. We classify the corresponding invariant flows and show that the geometric heat flow is the simplest possible one. This uniqueness result, together with previously reported results which we review in this chapter, confirms the importance of this class of flows. Results on invariant geometric flows of surfaces are presented at the end of the chapter as well.

1. Introduction

Invariant theory has recently become a major topic of study in computer vision (see [265] and references therein). Indeed, since the same object may be seen from a number of points of view, one is motivated to look for shape invariants under various transformations.

Indeed, the problem of recognizing and locating a partially visible planar object, whose shape underwent a geometric viewing transformation, often arises in machine vision tasks. Attempts to address such shape recognition problems raise the question of invariants under viewing transformations [46, 117, 265].

Work in model based shape analysis and recognition has already resulted in many useful products, such as optical character recognizers, handwriting recognizer interfaces to computers, printed-circuit board inspection systems and quality control devices. In spite of such successes many low-level problems remain to be addressed. Efficient ways for analyzing, recognizing and understanding planar shapes, when they do not come from a well-defined and documented catalogue of shapes, when they are distorted by a geometric viewing transformation, such as perspective projection or when they are partially occluded, must still be developed.

Another topic that has been receiving much attention from the image analysis community is the theory of scale-spaces or multiscale representation. This was introduced by Witkin in [386] and developed after that by several authors in different frameworks [9, 24, 58, 90, 109, 110, 163, 180, 187, 195, 196, 190, 226, 235, 257, 256, 293, 320, 323, 394, 396]. Initially, most of the work was devoted to linear scale-spaces derived via linear filtering. In the last years, a number of non-linear and geometric scale-spaces have been investigated as well.

The combination of invariant theory with geometric multiscale analysis

was first investigated in [328, 322, 323]. There, the authors introduced
an affine invariant geometric scale-space, and extended part of the work
to other groups as well in [325, 327, 324]. Related work was also carried
out in [7, 69]. As we will see in future sections, this kind of multiscale
analysis replaces for some applications the originally used linear ones. The
obtained representations allows for example to compute invariant signatures
at different scales and in a robust way. These flows are already being used
with satisfactory results in different applications [110, 320, 323, 326]

In this chapter, we would like first of all to set out the basic theory of
differential invariants for computer vision. It will therefore have a tutorial
nature. We will sketch enough of the language of differential geometry in
order to make precise the notion of infinitesimal invariance. While the theory
may be stated in classical terms (it was developed after all by Sophus Lie in
the previous century), we believe that there is a strong advantage to working
with the more modern machinery if only to avoid the classical proliferation
of multi-indices. Moreover, this will give the interested reader a guide to
the modern literature on the subject. Other approaches to the formulation
of differential invariants, as those based on Cartan moving frames, can be
found in [54, 99, 101, 141, 280], as well as in some of the papers in [265].

Our main application of this theory, will be to the new theory of geometric-
invariant scale-spaces, based on invariant geometric diffusion equations.
These give geometric multi-resolution representations of shape which are
invariant to a number of the typical viewing transformations in vision:
Euclidean, affine, similarity, projective. The theory of differential invariants
allows a unification of all these scale-spaces. Moreover, using this theory, we
classify the flows and show that the geometric heat flows are the simplest
possible amongst all invariant equations.

We now briefly summarize the contents of this chapter. In Section 2, we give
an outline of a modern treatment of the theory of differential invariants. In
particular, we discuss manifolds, vector fields, Lie groups and Lie algebras
and their actions on spaces, and the relevant notions of jets, prolongations,
and symmetry groups. These will allow us to give a rigorous definition of
"differential invariant." The reader familiar with this topic, or interested in
having a less formal presentation of the invariant geometric flows, can skip
this section in a first reading, since the following sections are almost self
contained. In Section 3, we give the theory of invariant diffusion equations
as applied to the affine, Euclidean, projective, and similarity groups. In
particular, using the invariant theory developed in Section 2, we prove that
for any of the preceding subgroups of the projective group $SL(\mathbb{R}, 3)$, say G,
the flows we have defined give the unique G-invariant evolution equation of
lowest order (up to constant factor). See Theorem 9 and Corollary 1 below.
Next in Section 4, we show how our flows may be made area or length

preserving which effectively solves the problem of shrinkage in computer vision. In Section 5 we present the extension of the theory to invariant surface flows. Finally, in Section 6, we give some concluding remarks.

2. Basic Invariant Theory

In this section, we review the classical theory of differential invariants. In order to do this in a rigorous manner, we first sketch some relevant facts from differential geometry and the theory of Lie groups. The material here is based on the books by Olver [280, 279] to which we refer the interested reader for all the details. See also the classical work of Sophus Lie on the theory of differential invariants [210] as well as [54, 101, 280, 279, 349].

We will assume a certain mathematical background, i.e., the reader should be familiar with the basic definitions of "manifold" and "smooth function." Accordingly, all the manifolds and mappings we consider below are C^∞. (This type of foundational material may be found in [143, 349].) We should add that in this section no proofs will be given, and the results will just be stated.

2.1. Vector Fields and One-Forms

Since we will be considering the theory of differential invariants, we will first review the infinitesimal (differential) structure of manifolds. Accordingly, a *tangent vector* to a manifold M at a point $x \in M$ is geometrically given as the tangent to a (smooth) curve passing through x. The collection of all such tangent vectors gives the tangent space $TM|_x$ to M at x, which is a vector space of the same dimension m as M. In local coordinates, a curve is parametrized by $x = \phi(t) = (\phi^1(t), \ldots \phi^m(t))$, and has tangent vector $\mathbf{v} = \xi^1 \frac{\partial}{\partial x^1} + \cdots + \xi^m \frac{\partial}{\partial x^m}$ at $x = \phi(t)$, with components $\xi^i = \frac{d\phi^i}{dt}$ given by the components of the derivative $\phi'(t)$. Here, the tangent vectors to the coordinate axes are denoted by $\frac{\partial}{\partial x^i} = \partial_{x^i}$, and form a basis for the tangent space $TM|_x$. If $f : M \to \mathbb{R}$ is any smooth function, then its directional derivative along the curve is

$$\frac{d}{dt} f[\phi(t)] = \mathbf{v}(f)(\phi(t)) = \sum_{i=1}^{m} \xi^i(\phi(t)) \frac{\partial f}{\partial x^i}(\phi(t)),$$

which provides one motivation for using a derivational notation for tangent vectors. The tangent spaces are patched together to form the tangent bundle $TM = \bigcup_{x \in M} TM|_x$ of the manifold, which is an m-dimensional vector bundle over the m-dimensional manifold M. A *vector field* $\mathbf{v}$ is a smoothly (or analytically) varying assignment of tangent vector $\mathbf{v}|_x \in TM|_x$. In local coordinates, a vector field has the form

$$\mathbf{v} = \sum_{i=1}^{m} \xi^i(x) \frac{\partial}{\partial x^i},$$

where the coefficients $\xi^i(x)$ are smooth (analytic) functions.

A parametrized curve $\phi : \mathbf{R} \to M$ is called an *integral curve* of the vector field $\mathbf{v}$ if its tangent vector agrees with the vector field $\mathbf{v}$ at each point; this requires that $x = \phi(t)$ satisfy the first order system of ordinary differential equations

$$\frac{dx^i}{dt} = \xi^i(t), \qquad 1 \le i \le m.$$

Standard existence and uniqueness theorems for systems of ordinary differential equations imply that through each $x \in M$ there passes a unique, maximal integral curve. We use the notation $\exp(t\mathbf{v})x$ to denote the maximal integral curve passing through $x = \exp(0\mathbf{v})x$ at $t = 0$, which may or may not be defined for all t. The family of (locally defined) maps $\exp(t\mathbf{v})$ is called the *flow* generated by the vector field $\mathbf{v}$, and obeys the usual exponential rules:

$$\begin{aligned}
\exp(t\mathbf{v})\exp(s\mathbf{v})x &= \exp((t+s)\mathbf{v})x, \quad t, s \in \mathbf{R}, \\
\exp(0\mathbf{v})x &= x, \\
\exp(-t\mathbf{v})x &= \exp(t\mathbf{v})^{-1}x,
\end{aligned}$$

the equations holding where defined. Conversely, given a flow obeying the latter equalities, we can reconstruct a generating vector field by differentiation:

$$\mathbf{v}|_x = \frac{d}{dt}\exp(t\mathbf{v})|_{t=0}\, x, \qquad x \in M.$$

Applying the vector field $\mathbf{v}$ to a function $f : M \to \mathbb{R}$ determines the infinitesimal change in f under the flow induced by $\mathbf{v}$:

$$\mathbf{v}(f) = \sum_{i=1}^{n} \xi^i(x) \frac{\partial f}{\partial x^i} = \frac{d}{dt} f(\exp(t\mathbf{v})x)\Big|_{t=0},$$

so that

$$f(\exp(t\mathbf{v})x) = f(x) + t\mathbf{v}(f)(x) + \frac{1}{2}t^2\mathbf{v}(\mathbf{v}(f)) + \dots.$$

Next given a (smooth) mapping $F : M \to N$, we define the *differential* $dF : TM|_x \to TN|_{F(x)}$ by

$$[(dF)(\mathbf{v})](f)(F(x)) := \mathbf{v}(f \circ F)(x),$$

where $f : N \to \mathbf{R}$ is a smooth function, $\mathbf{v}$ is a vector field.
In general, given a point $x \in M$, a *one-form at x* is a real-valued linear map
on the tangent space

$$\omega : TM|_x \to \mathbb{R}.$$

In local coordinates $x = (x^1, \ldots, x^m)$, the differentials dx^i are characterized
by $dx^i(\partial_{x^j}) = \delta_{ij}$ (the Kronecker delta), where $\partial_{x^1}, \ldots, \partial_{x^m}$ denotes the
standard basis of $TM|_x$. Then locally,

$$\omega = \sum_{i=1}^{m} h_i(x) dx^i.$$

In particular, for $f : M \to \mathbb{R}$, we get the one-form df given by its differential,
so

$$df(\mathbf{v}) := \mathbf{v}(f).$$

The (vector space) of one-forms at x is denoted by $T^*M|_x$ and is called the
cotangent space. (It can be regarded as the dual space of $TM|_x$.) As for the
tangent bundle, the cotangent spaces can be patched together to form the
cotangent bundle T^*M over M.

2.2. Lie Groups

In this section, we collect together the basic necessary facts from the theory
of Lie groups which will be used below. Recall first that a *group* is a set,
together with an associative multiplication operation. The group must also
contain an identity element, denoted e and each group element g has an
inverse g^{-1} satisfying $g \cdot g^{-1} = g^{-1} \cdot g = e$. Historically, it was Galois who
made the fundamental observation that the set of symmetries of an object
forms a group (this was in his work on the roots of polynomials). However,
the groups of Galois were discrete; in this chapter we study the continuous
groups first investigated by Sophus Lie.

Definition. A *Lie group* is a group G which also carries the structure of a
smooth manifold so that the operations of group multiplication $(g, h) \mapsto g \cdot h$
and inversion $g \mapsto g^{-1}$ are smooth maps.

Example. The basic example of a real Lie group is the general linear
group $\mathrm{GL}(\mathbb{R}, n)$ consisting of all real invertible $n \times n$ matrices with matrix
multiplication as the group operation; it is an n^2-dimensional manifold, the
structure arising simply because it is an open subset (namely, where the
determinant is nonzero) of the space of all $n \times n$ matrices which is itself
isomorphic to $\mathbb{R}^{n^2}$.

A subset $H \subset G$ of a group is a subgroup if and only if it is closed under multiplication and inversion; if G is a Lie group, then a subgroup H is a *Lie subgroup* if it is also a submanifold. Most Lie groups can be realized as Lie subgroups of $\text{GL}(\mathbb{R}, n)$; these are the so-called "matrix Lie groups", and, in this chapter, we will assume that all Lie groups are of this type. One can also define a notion of *local Lie group* in the obvious way (see e.g., [143]).

Example. We list here some of the key "classical groups". The *special linear group* $\text{SL}(\mathbb{R}, n) = \{A \in \text{GL}(\mathbb{R}, n) : \det A = 1\}$ is the group of volume-preserving linear transformations. The group is connected and has dimension $n^2 - 1$. The *orthogonal group* $\text{O}(n) = \{A \in \text{GL}(\mathbb{R}, n) : A^T A = I\}$ is the group of norm-preserving linear transformations — rotations and reflections — and has two connected components. The *special orthogonal group* $\text{SO}(n) = \text{O}(n) \cap \text{SL}(\mathbb{R}, n)$ consisting of just the rotations is the component containing the identity. This is also called the *rotation group*.

2.2.1. Transformation Groups

In many cases in vision (and physical) problems, groups are presented to us as a family of transformations acting on a space. In the case of Lie groups, the most natural setting is as groups of transformations acting smoothly on a manifold. More precisely, we have the following:

Definition. Let M be a smooth manifold. A *group of transformations* acting on M is given by a Lie group G and smooth map $\Phi : G \times M \to M$, denoted by $\Phi(g, x) = g \cdot x$, which satisfies

$$e \cdot x = x, \quad g \cdot (h \cdot x) = (g \cdot h) \cdot x, \text{ for all } x \in M, \ g \in G.$$

One can also define in the obvious way the notion of a *local Lie group action*.

Example. The key example is the usual linear action of the group $\text{GL}(\mathbb{R}, n)$ of $n \times n$ matrices acting by matrix multiplication on column vectors $x \in \mathbb{R}^n$. This action includes linear actions (representations) of the subgroups of $\text{GL}(\mathbb{R}, n)$ on $\mathbb{R}^n$. Since linear transformations map lines to lines, there is an induced action of $\text{GL}(\mathbb{R}, n)$ on the projective space $\mathbf{RP}^{n-1}$.

The diagonal matrices λI (I denotes the identity matrix) act trivially, so the action reduces effectively to one of the projective linear group $\text{PSL}(\mathbb{R}, n) = \text{GL}(\mathbb{R}, n)/\{\lambda I\}$. If n is odd, $\text{PSL}(\mathbb{R}, n) = \text{SL}(\mathbb{R}, n)$ can be identified with the special linear group, while for n even, since $-I \in \text{SL}(\mathbb{R}, n)$ has the same effect as the identity, the projective group is a quotient $\text{PSL}(\mathbb{R}, n) = \text{SL}(\mathbb{R}, n)/\{\pm I\}$.

In vision, of particular importance is the case of $GL(\mathbb{R}, 2)$, so we discuss this in some detail. The linear action of $GL(\mathbb{R}, 2)$ on $\mathbb{R}^2$ is given by

$$(x, y) \longmapsto (\alpha x + \beta y, \gamma x + \delta y), \qquad A = \begin{bmatrix} \alpha & \beta \\ \gamma & \delta \end{bmatrix} \in GL(\mathbb{R}, 2).$$

As above, we can identify the projective line $\mathbf{RP}^1$ with a circle S^1. If we use the projective coordinate $p = x/y$, the induced action is given by the *linear fractional* or Möbius transformations:

$$p \longmapsto \frac{\alpha p + \beta}{\gamma p + \delta}, \qquad A = \begin{bmatrix} \alpha & \beta \\ \gamma & \delta \end{bmatrix} \in GL(\mathbb{R}, 2).$$

In this coordinate chart, the x–axis $\{(x, 0)\}$ in $\mathbb{R}^2$ is identified with the point $p = \infty$ in $\mathbf{RP}^1$, and the linear fractional transformations have a well-defined to include the point at ∞.

Example. Let $\mathbf{v}$ be a vector field on the manifold M. Then the flow $\exp(t\mathbf{v})$ is a (local) action of the one-parameter group $\mathbb{R}$, parametrized by the "time" t, on the manifold M.

Example. In general, if G is a Lie group which acts as a group of transformations on another Lie group H, we define the *semi-direct product* $G \times_s H$ to be the Lie group which, as a manifold just looks like the Cartesian product $G \times H$, but whose multiplication is given by $(g, h) \cdot (\tilde{g}, \tilde{h}) = (g \cdot \tilde{g}, h \cdot (g \cdot \tilde{h}))$, and hence is different from the Cartesian product Lie group, which has multiplication $(g, h) \cdot (\tilde{g}, \tilde{h}) = (g \cdot \tilde{g}, h \cdot \tilde{h})$.

The *(full) affine group* $A(n)$ is defined as the group of affine transformations $x \mapsto Ax + a$ in $\mathbb{R}^n$, parametrized by a pair (A, a) consisting of an invertible matrix A and a vector $a \in \mathbb{R}^n$. The group multiplication law is given by $(A, a) \cdot (B, b) = (AB, a + Ab)$, and hence can be identified with the semi-direct product $GL(\mathbb{R}, n) \times_s \mathbb{R}^n$. The affine group can be realized as a subgroup of $GL(\mathbf{R}, n + 1)$ by identifying the pair (A, a) with the $(n+1) \times (n+1)$ matrix

$$\begin{bmatrix} A & a \\ 0 & 1 \end{bmatrix}.$$

Let $GL_+(\mathbb{R}, n)$ denote the subgroup of $GL(\mathbb{R}, n)$ with positive determinant. Then the *group of proper affine motions* of $\mathbb{R}^n$ is the semidirect product of $GL_+(\mathbb{R}, n)$ and the translations. Similarly, the *special affine group* is given by the semidirect product of $SL(\mathbb{R}, n)$ and $\mathbb{R}^n$.

We may also define the *Euclidean group* $E(n)$ as the semi-direct product of $O(n)$ and translations in $\mathbb{R}^n$, and the *group of Euclidean motions* as the semidirect product of the rotation group $SO(n)$ and $\mathbb{R}^n$. The *similarity group* in $\mathbb{R}^n$, $Sm(n)$, is generated by rotations, translations, and isotropic scalings. In the sequel, we will usually not differentiate between the real affine group and the group of proper affine motions, and the Euclidean group and the group of Euclidean motions.

Example. In what follows, we will consider all the above subgroups for $n = 2$, i.e., acting on the the plane $\mathbb{R}^2$. In this case, they are all subgroups of $SL(\mathbb{R}, 3)$, the so-called *group of projective transformations on* $\mathbb{R}^2$. More precisely, $SL(\mathbb{R}, 3)$ acts on $\mathbf{R}^2$ as follows: for $A \in SL(\mathbf{R}, 3)$

$$(\tilde{x}, \tilde{y}) = \left(\frac{a_{11}x + a_{21}y + a_{31}}{a_{13}x + a_{23}y + a_{33}}, \frac{a_{12}x + a_{22}y + a_{32}}{a_{13}x + a_{23}y + a_{33}} \right)$$

where

$$A = [a_{ij}]_{1 \leq i,j \leq 3}.$$

2.2.2. *Representations*

Linear actions of Lie groups, that is, "representations" of the group, play an essential role in applications. Formally, a *representation* of a group G is defined by a group homomorphism $\rho : G \to GL(V)$ from G to the space of invertible linear operators on a vector space V. This means that ρ satisfies the properties: $\rho(e) = I$, $\rho(g \cdot h) = \rho(g)\rho(h)$, $\rho(g^{-1}) = \rho(g)^{-1}$.

One important method to turn a nonlinear group action into a linear representation is to look at its induced action on functions on the manifold. Given any action of a Lie group G on a manifold M, there is a naturally induced representation of G on the space $\mathcal{F} = \mathcal{F}(M)$ of real-valued functions $F : M \to \mathbb{R}$, which maps the function F to $\bar{F} := g \cdot F$ defined by

$$\bar{F}(\tilde{x}) = F(g^{-1} \cdot \tilde{x}),$$

or equivalently,

$$(g \cdot F)(g \cdot x) = F(x).$$

The introduction of the inverse g^{-1} in this formula ensures that the action of G on $\mathcal{F}$ is a group homomorphism: $g \cdot (h \cdot F) = (g \cdot h) \cdot F$ for all $g, h \in G$, $F \in \mathcal{F}$.

The representation of G on the function space $\mathcal{F}$ will usually decompose into a wide variety of important subrepresentations, e.g., representations on spaces of polynomial functions, representations on spaces of smooth (C^∞) functions, or L^2 functions, etc. In general, representations of a group containing (nontrivial) subrepresentations are called reducible. An *irreducible representation*, then, is a representation $\rho : G \mapsto GL(V)$ which contains no (non-trivial) sub-representations, i.e., there are no subspaces

$W \subset V$ which are invariant under the representation, $\rho(g)W \subset W$ for all $g \in G$, other than $W = \{0\}$ and $W = V$. The classification of irreducible representations of Lie groups is a major subject of research in this century.

Example. Consider the action of the group $GL(\mathbb{R}, 2)$ on the space $\mathbb{R}^2$ acting via matrix multiplication. This induces a representation on the space of functions

$$\bar{F}(\bar{x}, \bar{y}) = \bar{F}(\alpha x + \beta y, \gamma x + \delta y) = F(x, y), \text{ where } A = \begin{bmatrix} \alpha & \beta \\ \gamma & \delta \end{bmatrix} \in GL(\mathbb{R}, 2).$$

Note that is F is a homogeneous polynomial of degree n, so is $\bar{F}$, so that this representation includes the finite-dimensional irreducible representations $\rho_{n,0}$ of $GL(\mathbb{R}, 2)$ on $\mathcal{P}^{(n)}$, the space of homogeneous polynomials of degree n. For example, on the space $\mathcal{P}^{(1)}$ of linear polynomials, the coefficients of general linear polynomial $F(x, y) = ax + by$ will transform according to

$$\begin{bmatrix} \alpha & \gamma \\ \beta & \delta \end{bmatrix} \begin{pmatrix} \bar{a} \\ \bar{b} \end{pmatrix} = \begin{pmatrix} a \\ b \end{pmatrix}, \quad A = \begin{bmatrix} \alpha & \beta \\ \gamma & \delta \end{bmatrix} \in GL(\mathbb{R}, 2),$$

so that the representation $\rho_{1,0}(A) = A^{-T}$ can be identified with the inverse transpose representation.

2.2.3. Orbits and Invariant Functions

Let G be a group of transformations acting on the manifold M. A subset $S \subset M$ is called G–invariant if it is unchanged by the group transformations, meaning $g \cdot x \in S$ whenever $g \in G$ and $x \in S$ (provided $g \cdot x$ is defined if the action is only local). An *orbit* of the transformation group is a minimal (nonempty) invariant subset. For a global action, the orbit through a point $x \in M$ is just the set of all images of x under arbitrary group transformations $\mathcal{C}_{0x} = \{g \cdot x : g \in G\}$.

Clearly, a subset $S \subset M$ is G-invariant if and only if it is the union of orbits. If G is connected. its orbits are connected. The action is called *transitive* if there is only one orbit, so (assuming the group acts globally), for every $x, y \in M$ there exists at least one $g \in G$ such that $g \cdot x = y$.

A group action is called *semi-regular* if its orbits are all submanifolds having the same dimension. The action is called *regular* if, in addition, for any $x \in M$, there exist arbitrarily small neighborhoods U of x with the property that each orbit intersects U in a pathwise connected subset. In particular, each orbit is a regular submanifold, but this condition is not sufficient to guarantee regularity; for instance, the one-parameter group $(r, \theta) \mapsto (e^t(r - 1) + 1, \theta + t)$, written in polar coordinates, is semi-regular on $\mathbb{R}^2 \setminus \{0\}$, and has regular orbits, but is not regular on the unit circle.

Example. Consider the two-dimensional torus $T = S^1 \times S^1$, with angular coordinates (θ, φ), $0 \le \theta, \varphi < 2\pi$. Let α be a nonzero real number, and consider the one-parameter group action $(\theta, \varphi) \mapsto (\theta + t, \varphi + \alpha t) \bmod 2\pi$, $t \in \mathbb{R}$. If α/π is a rational number, then the orbits of this action are closed curves, diffeomorphic to the circle S^1, and the action is regular. On the other hand, if α/π is an irrational number, then the orbits of this action never close, and, in fact, each orbit is a dense subset of T. Therefore, the action in the latter case is semi-regular, but not regular.

The *quotient space* M/G is defined as the space of orbits of the group action, endowed with a topology induced from that of M. As the irrational flow on the torus makes clear, the quotient space can be a very complicated topological space. However, regularity of the group action will ensure that the quotient space is a smooth manifold.

Given a group of transformations acting on a manifold M, by a *canonical form* of an element $x \in M$ we just mean a distinguished, simple representative x_0 of the orbit containing x. Of course, there is not necessarily a uniquely determined canonical form, and some choice, usually based on one's æsthetic sense of "simplicity", must be employed for such forms.

Now orbits and canonical forms of group actions are characterized by the invariants, which are defined as real-valued functions whose values are unaffected by the group transformations.

Definition An *invariant* for the transformation group G is a function I : $M \to \mathbb{R}$ which satisfies $I(g \cdot x) = I(x)$ for all $g \in G$, $x \in M$.

Proposition 1 *Let I : $M \to \mathbb{R}$ be a function. Then the following three conditions are equivalent:*

1. *I is a G-invariant function.*
2. *I is constant on the orbits of G.*
3. *The level sets $\{I(x) = c\}$ of I are G-invariant subsets of M.*

For example, in the case of the orthogonal group $O(n)$ acting on $\mathbb{R}^n$, the orbits are spheres $|x| = $ constant, and hence any orthogonal invariant is a function of the radius $I = F(r)$, $r = |x|$. Invariants are essentially classified by their "quotient representatives": every invariant of the group action induces a function $\tilde{I} : M/G \to \mathbb{R}$ on the quotient space, and conversely. The canonical form x_0 of any element $x \in M$ must have the same invariants: $I(x_0) = I(x)$; this condition is also sufficient if there are enough invariants to distinguish the orbits, i.e., x and y lie in the same orbit if and only if $I(x) = I(y)$ for every invariant I which according to the next theorem is the case for regular group actions.

An important problem is the determination of *all* the invariants of a group of transformations. Note that if $I_1(x), \ldots, I_k(x)$ are invariant functions, and $\Phi(y_1, \ldots, y_k)$ is any function, then $\hat{I} = \Phi(I_1(x), \ldots, I_k(x))$ is also invariant. Therefore, to classify invariants, we need only determine all different functionally independent invariants. Many times globally defined invariants are difficult to find, and so one must be satisfied with the description of locally defined invariants of a group action.

Theorem 1 *Let G be a Lie group acting regularly on the m-dimensional manifold M with r-dimensional orbits. Then, locally, near any $x \in M$ there exist exactly $m - r$ functionally independent invariants $I_1, \ldots, I_{m-r}$ with the property that any other invariant can be written as a function of the fundamental invariants: $I = \Phi(I_1, \ldots, I_{m-r})$. Moreover, two points x and y in the coordinate chart lie in the same orbit of G if and only if the invariants all have the same value, $I_\nu(x) = I_\nu(y)$, $\nu = 1, \ldots, m - r$.*

This theorem provides a complete answer to the question of local invariants of group actions. Global and irregular considerations are more delicate; for example, consider the one-parameter isotropy group $(x, y) \mapsto (\lambda x, \lambda y)$, $\lambda \in \mathbb{R}^+$. Locally, away from the origin, x/y or y/x or any function thereof (e.g., $\theta = \tan^{-1}(y/x)$) provides the only invariant. However, if we include the origin, then there are no non-constant invariants. On the other hand, the scaling group

$$(x, y) \mapsto (\lambda x, \lambda^{-1} y), \quad \lambda \neq 0,$$

has the global invariant xy. In general, if G acts transitively on the manifold M, then the only invariants are constants, which are completely trivial invariants. More generally, if G acts transitively on a dense subset $M_0 \subset M$, then the only *continuous* invariants are constants. For example, the only continuous invariants of the irrational flow on the torus are the constants, since every orbit is dense in this case. Similarly, the only continuous invariants of the standard action of $\mathrm{GL}(\mathbb{R}, n)$ on $\mathbb{R}^n$ are the constant functions, since the group acts transitively on $\mathbb{R}^n \setminus \{0\}$. (A discontinuous invariant is provided by the function which is 1 at the origin and 0 elsewhere.)

2.2.4. *Lie Algebras*
Besides invariant functions, there are other important invariant objects associated with a transformation group, including vector fields, differential forms, differential operators, etc. We begin by considering the case of an invariant vector field, which will, in the particular case of a group acting on itself by right (or left) multiplication, lead to the crucially important concept of a Lie algebra or "infinitesimal" Lie group. A basic feature of (connected) Lie groups is the ability to work infinitesimally, thereby effectively linearizing complicated invariance criteria.

Definition. Let G act on the manifold M. A vector field $\mathbf{v}$ on M is called *G–invariant* if it is unchanged by the action of any group element: $dg(\mathbf{v}|_x) = \mathbf{v}|_{g \cdot x}$ for all $g \in G$, $x \in M$.

In particular, if we consider the action of G on itself by right multiplication, the space of all invariant vector fields forms the Lie algebra of the group. Given $g \in G$, let $R_g : h \mapsto h \cdot g$ denote the associated right multiplication map. A vector field $\mathbf{v}$ on G is *right-invariant* if it satisfies $dR_g(\mathbf{v}) = \mathbf{v}$ for all $g \in G$.

Definition. The *Lie algebra* $\mathcal{G}$ of a Lie group G is the space of all right-invariant vector fields.

Every right invariant vector field $\mathbf{v}$ is uniquely determined by its value at the identity e, because $\mathbf{v}|_g = dR_g(\mathbf{v}|_e)$. Therefore, we can identify $\mathcal{G}$ with $TG|_e$, the tangent space to the manifold G at the identity, and hence $\mathcal{G}$ is a finite-dimensional vector space having the same dimension as G.

The Lie algebra associated with a Lie group comes equipped with a natural multiplication, defined by the *Lie bracket* of vector fields given by:

$$[\mathbf{v}, \mathbf{w}](f) := \mathbf{v}(\mathbf{w}(f)) - \mathbf{w}(\mathbf{v}(f)).$$

By the invariance of the Lie bracket under diffeomorphisms, if both $\mathbf{v}$ and $\mathbf{w}$ are right invariant, so is $[\mathbf{v}, \mathbf{w}]$. Note that the bracket satisfies the *Jacobi identity*

$$[\mathbf{u}, [\mathbf{v}, \mathbf{w}]] + [\mathbf{v}, [\mathbf{w}, \mathbf{u}]] + [\mathbf{w}, [\mathbf{u}, \mathbf{v}]] = 0.$$

The basic properties of the Lie bracket translate into the defining properties of an (abstract) Lie algebra.

Definition. A *Lie algebra* $\mathcal{G}$ is a vector space equipped with a bracket operation $[\,\cdot\,,\,\cdot\,] : \mathcal{G} \times \mathcal{G} \to \mathcal{G}$ which is bilinear, anti-symmetric, and satisfies the Jacobi identity.

Theorem 2 *Let G be a connected Lie group with Lie algebra $\mathcal{G}$. Every group element can be written as a product of exponentials:* $g = \exp(\mathbf{v}_1) \exp(\mathbf{v}_2) \cdots \exp(\mathbf{v}_k)$, *for* $\mathbf{v}_1, \ldots, \mathbf{v}_k \in \mathcal{G}$.

Example. The Lie algebra $\mathcal{GL}_n$ of $GL(\mathbb{R}, n)$ can be identified with the space of all $n \times n$ matrices. Coordinates on $GL(\mathbb{R}, n)$ are given by the matrix entries $X = (x_{ij})$. The right-invariant vector field associated with a matrix $A \in \mathcal{GL}_n$ is given by $\mathbf{v}_A = \sum_{i,j,k} a_{ij} x_{jk} \partial_{x^k}$. The exponential map is the usual matrix exponential $\exp(t\mathbf{v}_A) = e^{tA}$. The Lie bracket of two such vector fields is found to be $[\mathbf{v}_A, \mathbf{v}_B] = \mathbf{v}_C$, where $C = BA - AB$. Thus the Lie bracket on $\mathcal{GL}_n$ is identified with the *negative* of the matrix commutator $[A, B] = AB - BA$.

The formula $\det \exp(tA) = \exp(t \, \mathrm{tr} A)$ proves that the Lie algebra $\mathcal{SL}_n$ of the unimodular subgroup $\mathrm{SL}(\mathbb{R}, n)$ consists of all matrices with trace 0. The subgroups $\mathrm{O}(n)$ and $\mathrm{SO}(n)$ have the same Lie algebra, $\mathcal{SO}(n)$, consisting of all skew-symmetric $n \times n$ matrices.

Finally, we want to define the key concept of an *invariant one-form*. In order to do this, we will first have to define the *pullback of a one-form*. Let $F : M \to N$ be a smooth mapping of manifolds, and let η denote a one-form in $T^*N|_{y=F(x)}$. Then $F^*(\eta) \in T^*M|_x$ is the one-form given by

$$F^*(\eta)(\mathbf{v}) := \eta(dF(\mathbf{v})),$$

where $\mathbf{v} \in TM|_x$.

Definition. Let G act on the manifold M. A one-form ω on M is called *G–invariant* if it is unchanged by the pull-back action of any group element

$$g^*(\omega|_{g \cdot x}) = \omega|_x, \quad \forall g \in G, \ x \in M.$$

Dual to the right-invariant vector fields forming the Lie algebra of a Lie group are the right-invariant one-forms known as the *Maurer-Cartan forms*. See [279, 349] for details.

The following result follows from the definitions:

Lemma 2 *Let G be a transformation group acting on M. Then:*

1. *If I is an invariant function, then dI is an invariant one-form.*
2. *If I is an invariant function, and ω an invariant one-form, then $I\omega$ is an invariant one-form.*

2.2.5. *Infinitesimal Group Actions*

Just as a one-parameter group of transformations is generated as the flow of a vector field, so a general Lie group of transformations G acting on the manifold M will be generated by a set of vector fields on M, known as the *infinitesimal generators* of the group action, whose flows coincide with the action of the corresponding one-parameter subgroups of G. More precisely, if $\mathbf{v}$ generates the one-parameter subgroup $\{\exp(t\mathbf{v}) : t \in \mathbb{R}\} \subset G$, then we identify $\mathbf{v}$ with the infinitesimal generator $\hat{\mathbf{v}}$ of the one-parameter group of transformations (or flow) $x \mapsto \exp(t\mathbf{v}) \cdot x$. Note that the infinitesimal generators of the group action are found by differentiating the various one-parameter subgroups:

$$\hat{\mathbf{v}}|_x = \frac{d}{dt} \exp(t\mathbf{v})|_{t=0} \, x, \qquad x \in M, \quad \mathbf{v} \in \mathcal{G}. \tag{11.1}$$

If $\Phi_x : G \to M$ is given by $\Phi_x(g) = g \cdot x$ (where defined), so $\hat{\mathbf{v}}|_x = d\Phi_x(\mathbf{v}|_e)$, and hence $d\Phi_x(\mathbf{v}|_g) = \hat{\mathbf{v}}|_{g \cdot x}$. Therefore, resulting vector fields satisfy the same commutation relations as the Lie algebra of G, forming a finite-dimensional Lie algebra of vector fields on the manifold M isomorphic to the Lie algebra of G. Conversely, given a finite-dimensional Lie algebra of vector fields on a manifold M, we can reconstruct a (local) action of the corresponding Lie group via the exponentiation process.

Theorem 3 *If G is a Lie group acting on a manifold M, then its infinitesimal generators form a Lie algebra of vector fields on M isomorphic to the Lie algebra $\mathcal{G}$ of G. Conversely, any Lie algebra of vector fields on M which is isomorphic to $\mathcal{G}$ will generate a local action of the group G on M.*

Consequently, for a fixed group action, the associated infinitesimal generators will, somewhat imprecisely, be identified with the Lie algebra $\mathcal{G}$ itself, so that we will not distinguish between an element $\mathbf{v} \in \mathcal{G}$ and the associated infinitesimal generator of the action of G, which we also denote as $\mathbf{v}$ from now on.

Given a group action of a Lie group G, the infinitesimal generators also determine the tangent space to, and hence the dimension of the orbits.

Proposition 2 *Let G be a Lie group with Lie algebra $\mathcal{G}$ acting on a manifold M. Then, for each $x \in M$, the tangent space to the orbit through x is the subspace $\mathcal{G}|_x \subset TM|_x$ spanned by the infinitesimal generators $\mathbf{v}|_x$, $\mathbf{v} \in \mathcal{G}$. In particular, the dimension of the orbit equals the dimension of $\mathcal{G}|_x$.*

2.2.6. Infinitesimal Invariance

As alluded to above, the invariants of a connected Lie group of transformations can be effectively computed using purely infinitesimal techniques. Indeed, the practical applications of Lie groups ultimately rely on this basic observation.

Theorem 4 *Let G be a connected group of transformations acting on a manifold M. A function $F : M \to \mathbb{R}$ is invariant under G if and only if*

$$\mathbf{v}[F] = 0, \tag{11.2}$$

for all $x \in M$ and every infinitesimal generator $\mathbf{v} \in \mathcal{G}$ of G.

Thus, according to Theorem (4) the invariants of a one-parameter group with infinitesimal generator $\mathbf{v} = \sum_i \xi^i(x)\partial_{x^i}$ satisfy the first order, linear, homogeneous partial differential equation

$$\sum_{i=1}^{m} \xi^i(x) \frac{\partial F}{\partial x^i} = 0. \tag{11.3}$$

The solutions of (11.3) can be computed by the *method of characteristics*. We replace the partial differential equation by the characteristic system of ordinary differential equations

$$\frac{dx^1}{\xi^1(x)} = \frac{dx^2}{\xi^2(x)} = \cdots = \frac{dx^m}{\xi^m(x)}. \tag{11.4}$$

The general solution to (11.4) can be written in the form $I_1(x) = c_1, \ldots, I_{m-1}(x) = c_{m-1}$, where the c_i are constants of integration. It is not hard to prove that the resulting functions $I_1, \ldots, I_{m-1}$ form a complete set of functionally independent invariants of the one-parameter group generated by $\mathbf{v}$.

Example. We consider the (local) one-parameter group generated by the vector field

$$\mathbf{v} = -y\frac{\partial}{\partial x} + x\frac{\partial}{\partial y} + (1 + z^2)\frac{\partial}{\partial z}.$$

The group transformations are

$$(x, y, z) \longmapsto \left(x\cos\varepsilon - y\sin\varepsilon, x\sin\varepsilon + y\cos\varepsilon, \frac{\sin\varepsilon + z\cos\varepsilon}{\cos\varepsilon - z\sin\varepsilon} \right).$$

The characteristic system (11.4) for this vector field is

$$\frac{dx}{-y} = \frac{dy}{x} = \frac{dz}{1 + z^2}.$$

The first equation reduces to a simple separable ordinary differential equation $\frac{dy}{dx} = -x/y$, with general solution $x^2 + y^2 = c_1$, for c_1 a constant of integration; therefore the cylindrical radius $r = \sqrt{x^2 + y^2}$ is one invariant. To solve the second characteristic equation, we replace x by $\sqrt{r^2 - y^2}$, and treat r as constant; the solution is $\tan^{-1}z - \sin^{-1}(y/r) = \tan^{-1}z - \tan^{-1}(y/x) = c_2$, where c_2 is a second constant of integration. Therefore $\tan^{-1}z - \tan^{-1}(y/x)$ is a second invariant; a more convenient choice is found by taking the tangent of this invariant, and hence we deduce that $r = \sqrt{x^2 + y^2}$, $w = (xz - y)/(yz + x)$ form a complete system of functionally independent invariants, provided $yz + x \neq 0$.

2.2.7. *Invariant Equations*

In addition to the classification of invariant functions of group actions, it is also important to characterize invariant systems of equations. A group G is called a *symmetry group* of a system of equations

$$F_1(x) = \cdots = F_k(x) = 0, \tag{11.5}$$

defined on an m-dimensional manifold M, if it maps solutions to other solutions, i.e., if $x \in M$ satisfies the system and $g \in G$ is any group element such that $g \cdot x$ is defined, then we require that $g \cdot x$ is also a solution to the system. Knowledge of a symmetry group of a system of equations allows us to construct new solutions from old ones, a fact that is particularly useful when applying these methods to systems of differential equations; [280, 279]. Let $\mathcal{S}_{\mathcal{F}}$ denote the subvariety defined by the functions $\mathcal{F} = \{F_1, \ldots, F_k\}$, meaning the set of all solutions x to the system (11.5). (Note that G is a symmetry group of the system if and only if $\mathcal{S}_{\mathcal{F}}$ is a G–invariant subset.) Recall that the system is *regular* if the Jacobian matrix $\left(\frac{\partial F_i}{\partial x^k}\right)$ has constant rank n in a neighborhood of $\mathcal{S}_{\mathcal{F}}$, which implies (via the implicit function theorem), that the solution set $\mathcal{S}_{\mathcal{F}}$ is a submanifold of dimension $m - n$. In particular, if the rank is maximal, equaling k, on $\mathcal{S}_{\mathcal{F}}$, the system is regular.

Proposition 3 *Let* $F_1(x) = \cdots = F_k(x) = 0$ *be a regular system of equations. A connected Lie group G is a symmetry group of the system if and only if*

$$\mathbf{v}[F_\nu(x)] = 0, \qquad \text{whenever} \quad F_1(x) = \cdots = F_k(x) = 0, \qquad 1 \le \nu \le k,$$

for every infinitesimal generator $\mathbf{v} \in \mathcal{G}$ *of G.*

Example. The equation $x^2 + y^2 = 1$ defines a circle, which is rotationally invariant. To check the infinitesimal condition, we apply the generator $\mathbf{v} = -y\partial_x + x\partial_y$ to the defining function $F(x, y) = x^2 + y^2 - 1$. We find $\mathbf{v}(F) = 0$ everywhere (since F is an invariant). Since dF is nonzero on the circle, the solution set is rotationally invariant.

2.3. Prolongations

In this section, we review the theory of jets and prolongations, to formalize the notion of differential invariants.

2.3.1. *Point Transformations*

We have reviewed linear actions of Lie groups on functions. While of great importance, such actions are not the most general, and we will have to consider more general nonlinear group actions. Such transformation groups figure prominently in Lie's theory of symmetry groups of differential equations, and appear naturally in the geometrically invariant diffusion equations of computer vision that we consider below. The transformation groups will act on the basic space coordinatized by the independent and dependent variables relevant to the system of differential equations under consideration. Since we want to treat *differential equations*, we must be able to handle the derivatives of the dependent variables on the same footing as the independent and dependent variables themselves. In this section, we describe a suitable geometric space for this purpose — the so-called "jet space." We then discuss how group transformations are "prolonged" so that the derivative coordinates are appropriately acted upon, and, in the case of infinitesimal generators, state the fundamental prolongation formula that explicitly determines the prolonged action.

A general system of (partial) differential equations involves p independent variables $x = (x^1, \ldots, x^p)$, which we can view as local coordinates on the space $X \simeq \mathbf{R}^p$, and q dependent variables $u = (u^1, \ldots, u^q)$, coordinates on $U \simeq \mathbf{R}^q$. The total space will be an open subset $M \subset X \times U \simeq \mathbf{R}^{p+q}$.

A solution to the system of differential equations will be described by a smooth function $u = f(x)$. The graph of a function, $\Gamma_f = \{(x, f(x))\}$, is a p-dimensional submanifold of M which is *transverse*, meaning that it has no vertical tangent directions. A vector field is *vertical* if it is tangent to the vertical fiber $U_{x_0} \equiv \{x_0\} \times U$, so the transversality condition is $T\Gamma_f|_{z_0} \cap TU_{x_0}|_{z_0} = \{0\}$ for each $z_0 = (x_0, f(x_0))$ with x_0 in the domain of f. Conversely, the Implicit Function Theorem implies that any p-dimensional submanifold $\Gamma \subset M$ which is transverse at a point $z_0 = (x_0, u_0) \in \Gamma$ locally, coincides with the graph of a single-valued smooth function $u = f(x)$.

The most basic type of symmetry we will discuss is provided by a (locally defined) smooth, invertible map on the space of independent and dependent variables:

$$(\bar{x}, \bar{u}) = g \cdot (x, u) = (\varphi(x, u), \psi(x, u)). \qquad (11.6)$$

The general type of transformations defined by (11.6), are often referred to as *point transformations* since they act pointwise on the independent and dependent variables. Point transformations act on functions $u = f(x)$ by pointwise transforming their graphs; in other words if $\Gamma_f = \{(x, f(x))\}$ denotes the graph of f, then the transformed function $\bar{f} = g \cdot f$ will have graph

$$\Gamma_{\bar{f}} = \{(\bar{x}, \bar{f}(\bar{x}))\} = g \cdot \Gamma_f = \{g \cdot x(x, f(x))\}. \qquad (11.7)$$

In general, we can only assert that the transformed graph is another p-dimensional submanifold of M, and so the transformed function will not be well-defined unless $g \cdot \Gamma_f$ is (at least) transverse to the vertical space at each point. This will be guaranteed if the transformation g is sufficiently close to the identity transformation, and the domain of f is compact.

Example. Let

$$g_t \cdot (x, u) = (x \cos t - u \sin t, \ x \sin t + u \cos t),$$

be the one-parameter group of rotations acting on the space $M \simeq \mathbf{R}^2$ consisting of one independent and one dependent variable. Such a rotation transforms a function $u = f(x)$ by rotating its graph; therefore, the transformed graph $g_t \cdot \Gamma_f$ will be the graph of a well-defined function only if the rotation angle t is not too large. The equation for the transformed function $\bar{f} = g_t \cdot f$ is given in implicit form

$$\begin{aligned} \bar{x} &= x \cos t - f(x) \sin t, \\ \bar{u} &= x \sin t + f(x) \cos t, \end{aligned} \qquad (11.8)$$

and $\bar{u} = \bar{f}(\bar{x})$ is found by eliminating x from these two equations. For example, if $u = ax + b$ is affine, then the transformed function is also affine, and given explicitly by

$$\bar{u} = \frac{\sin t + a \cos t}{\cos t - a \sin t} \bar{x} + \frac{b}{\cos t - a \sin t},$$

which is defined provided $\cot t \neq a$, i.e., provided the graph of f has not been rotated to be vertical.

2.3.2. Jets and Prolongations

Since we are interested in symmetries of differential equations, we need to know not only how the group transformations act on the independent and dependent variables, but also how they act on the derivatives of the dependent variables. In the last century, this was done automatically, without worrying about the precise mathematical foundations of the method; in modern times, geometers have defined the "jet space" (or bundle) associated with the space of independent and dependent variables, whose coordinates will represent the derivatives of the dependent variables with respect to the independent variables. This gives a rigorous, cleaner, and more geometric interpretation of this theory.

Given a smooth, scalar-valued function $f(x_1, \ldots, x_p)$ of p independent variables, there are $p_k = \begin{pmatrix} p + k - 1 \\ k \end{pmatrix}$ different k-th order partial derivatives

$$\partial_J f(x) = \frac{\partial^k f}{\partial x^{j_1} \partial x^{j_2} \cdots \partial x^{j_k}},$$

indexed by all unordered (symmetric) multi-indices $J = (j_1, \ldots, j_k)$, $1 \leq j_k \leq p$, of order $k = \#J$. Therefore, if we have q dependent variables $(u^1, \ldots, u^q)$, we require $q_k = q p_k$ different coordinates u_J^α, $1 \leq \alpha \leq q$, $\#J = k$ to represent all the k-th order derivatives $u_J^\alpha = \partial_J f^\alpha(x)$ of a function $u = f(x)$. For the space $M = X \times U \simeq \mathbf{R}^p \times \mathbf{R}^q$, the n-th jet space $\mathrm{J}^n = \mathrm{J}^n M = X \times U^n$ is the Euclidean space of dimension $p + q \begin{pmatrix} p + n \\ n \end{pmatrix}$, whose coordinates consist of the p independent variables x^i, the q dependent variables u^α, and the derivative coordinates u_J^α, $1 \leq \alpha \leq q$, of orders $1 \leq \#J \leq n$. The points in the vertical space $U^{(n)}$ are denoted by $u^{(n)}$, and consist of all the dependent variables and their derivatives up to order n; thus a point in J^n has coordinates $(x, u^{(n)})$.

A smooth function $u = f(x)$ from X to U has n-th *prolongation* $u^{(n)} = \mathrm{pr}^{(n)} f(x)$ (also known as the *n-jet*), which is a function from X to $U^{(n)}$, given by evaluating all the partial derivatives of f up to order n; thus the individual coordinate functions of $\mathrm{pr}^{(n)} f$ are $u_J^\alpha = \partial_J f^\alpha(x)$. Note that the graph of the prolonged function $\mathrm{pr}^{(n)} f$, namely $\Gamma_f^{(n)} = \{(x, \mathrm{pr}^{(n)} f(x))\}$, will be a p-dimensional submanifold of J^n. At a point $x \in X$, two functions have the same n-th order prolongation, and so determine the same point of J^n, if and only if they have n-th order contact, meaning that they and their first n derivatives agree at the point. (This is the same as requiring that they have the same n-th order Taylor polynomial at the point x.) Thus, a more intrinsic way of defining the jet space J^n is to consider it as the set of equivalence classes of smooth functions using the equivalence relation of n-th order contact. If g is a (local) point transformation (11.6), then g acts on functions by transforming their graphs, and hence also acts on the derivatives of the functions in a natural manner. This allows us to naturally define an induced "prolonged transformation" $(\bar{x}, \bar{u}^{(n)}) = \mathrm{pr}^{(n)} g \cdot (x, u^{(n)})$ on the jet space J^n, given directly by the chain rule. More precisely, for any point $(x_0, u_0^{(n)}) = (x_0, \mathrm{pr}^{(n)} f(x_0)) \in J^n$, the transformed point $(\bar{x}_0, \bar{u}_0^{(n)}) = \mathrm{pr}^{(n)} g \cdot (x_0, u_0^{(n)}) = (\bar{x}_0, \mathrm{pr}^{(n)} \bar{f}(\bar{x}_0)$ is found by evaluating the derivatives of the transformed function $\bar{f} = g \cdot f$ at the image point $\bar{x}_0$, defined so that $(\bar{x}_0, \bar{u}_0) = (\bar{x}_0, \bar{f}(\bar{x}_0)) = g \cdot (x_0, f(x_0))$. This definition assumes that $\bar{f}$ is smooth at $\bar{x}_0$ — otherwise the prolonged transformation is not defined at $(x_0, u_0^{(n)})$. It is not hard to see that the construction does not depend on the particular function f used to represent the point of J^n; in particular, using the identification of the points in J^n with Taylor polynomials of order n, it suffices to determine how the point transformations act on polynomials of degree at most n in order to compute their prolongation.

Example. For the one-parameter rotation group considered above, the first prolongation $\mathrm{pr}^{(1)} g_t$ will act on the space coordinatized by (x, u, p) where p represents the derivative coordinate u_x. Given a point (x_0, u_0, p_0), we choose the linear polynomial $u = f(x) = p_0(x - x_0) + u_0$ to represent it, so $f(x_0) = u_0$, $f'(x_0) = p_0$. The transformed function is given by

$$\bar{f}(\bar{x}) = \frac{\sin t + p_0 \cos t}{\cos t - p_0 \sin t} \bar{x} + \frac{u_0 - p_0 x_0}{\cos t - p_0 \sin t}.$$

Then, by (11.8), $\bar{x}_0 = x_0 \cos t - u_0 \sin t$, so $\bar{f}(\bar{x}_0) = \bar{u}_0 = x_0 \sin t + u_0 \cos t$, and $\bar{p}_0 = \bar{f}'(\bar{x}_0) = (\sin t + p_0 \cos t)/(\cos t - p_0 \sin t)$, which is defined provided

$p_0 \neq \cot t$. Therefore, dropping the 0 subscripts, the prolonged group action is

$$\mathrm{pr}^{(1)} g_t \cdot (x, u, p) = \left(x \cos t - u \sin t, x \sin t + u \cos t, \frac{\sin t + p \cos t}{\cos t - p \sin t} \right), \quad (11.9)$$

defined for $p \neq \cot t$. Note that even though the original group action is globally defined, the prolonged group action is only locally given.

2.3.3. *Total Derivatives*

The chain rule computations used to compute prolongations are notationally simplified if we introduce the concept of a total derivative. The total derivative of a function of x, u and derivatives of u is found by differentiating the function treating the u's as functions of the x's.

Formally, let $F(x, u^{(n)})$ be a function on J^n. Then the *total derivative* $D_i F$ of F with respect to x^i is the function on $J^{(n+1)}$ defined by

$$D_i (x, \mathrm{pr}^{(n+1)} f(x)) = \frac{\partial F(x, \mathrm{pr}^{(n)} f(x))}{\partial x^i}.$$

For example, in the case of one independent variable x and one dependent variable u, the total derivative D_x with respect to x has the general formula

$$D_x = \frac{\partial}{\partial x} + u_x \frac{\partial}{\partial u} + u_{xx} \frac{\partial}{\partial u_x} + u_{xxx} \frac{\partial}{\partial u_{xx}} + \cdots.$$

In general, the total derivative with respect to the i-th independent variable is the first order differential operator

$$D_i = \frac{\partial}{\partial x^i} + \sum_{\alpha=1}^{q} \sum_{J} u_{J,i}^{\alpha} \frac{\partial}{\partial u_J^{\alpha}},$$

where $u_{J,i}^{\alpha} = D_i(u_J^{\alpha}) = u_{j_1 \cdots j_k i}^{\alpha}$. The latter sum is over all multi-indices J of arbitrary order. Even though D_i involves an infinite summation, when applying the total derivative to any function $F(x, u^{(n)})$ defined on the n-th jet space, only finitely many terms (namely, those for $\#J \leq n$) are needed. Higher order total derivatives are defined in the obvious manner, with $D_J = D_{j_1} \cdot \ldots \cdot D_{j_k}$ for any multi-index $J = (j_1, \ldots, j_k)$, $1 \leq j_\nu \leq p$.

2.3.4. *Prolongation of Vector Fields*

Given a vector field $\mathbf{v}$ generating a one-parameter group of transformations $\exp(t\mathbf{v})$ on $M \subset X \times U$, the associated n-th order prolonged vector field $\mathrm{pr}^{(n)} \mathbf{v}$ is the vector field on the jet space J^n which is the infinitesimal generator of the prolonged one-parameter group $\mathrm{pr}^{(n)} \exp(t\mathbf{v})$. Thus,

$$\mathrm{pr}^{(n)} \mathbf{v}|_{(x, u^{(n)})} = \frac{d}{dt} \mathrm{pr}^{(n)} [\exp(t\mathbf{v})]|_{t=0} \cdot (x, u^{(n)}). \quad (11.10)$$

The explicit formula for the prolonged vector field is given by the following very important "prolongation formula" (see [279], Theorem 2.36 for the proof):

Theorem 5 *The n–th prolongation of the vector field*

$$\mathbf{v} = \sum_{i=1}^{p} \xi^i(x, u) \frac{\partial}{\partial x^i} + \sum_{\alpha=1}^{q} \varphi^\alpha(x, u) \frac{\partial}{\partial u^\alpha}$$

is given explicitly by

$$pr^{(n)}\mathbf{v} = \sum_{i=1}^{p} \xi^i(x, u) \frac{\partial}{\partial x^i} + \sum_{\alpha=1}^{n} \sum_{j=\#J=0}^{n} \varphi_J^\alpha(x, u^{(j)}) \frac{\partial}{\partial u_J^\alpha}, \tag{11.11}$$

with coefficients

$$\varphi_J^\alpha = D_J Q^\alpha + \sum_{i=1}^{p} \xi^i u_{J,i}^\alpha, \tag{11.12}$$

where the "characteristics" of $\mathbf{v}$ *are given by*

$$Q^\alpha(x, u^{(1)}) := \varphi^\alpha(x, u) - \sum_{i=1}^{p} \xi^i(x, u) \frac{\partial u^\alpha}{\partial x^i}, \quad \alpha = 1, \ldots, q. \tag{11.13}$$

Remark. One can easily prove [280, 279] that a function $u = f(x)$ is invariant under the group generated by $\mathbf{v}$ if and only if it satisfies the characteristic equations

$$Q^\alpha(x, pr^{(1)} f(x)) = 0, \quad \alpha = 1, \ldots, q.$$

Example. Suppose we have just one independent and dependent variable. Consider a general vector field $\mathbf{v} = \xi(x, u)\partial_x + \varphi(x, u)\partial_u$ on $M = \mathbb{R}^2$. The characteristic (11.13) of $\mathbf{v}$ is the function

$$Q(x, u, u_x) = \varphi(x, u) - \xi(x, u)u_x.$$

From the above remark, we see that a function $u = f(x)$ is invariant under the one-parameter group generated by $\mathbf{v}$ if and only if it satisfies the ordinary differential equation $\xi(x, u)u_x = \varphi(x, u)$. The second prolongation $\mathbf{v}$ is a vector field

$$pr^{(2)}\mathbf{v} = \xi(x, u) \frac{\partial}{\partial x} + \varphi(x, u) \frac{\partial}{\partial u} + \varphi^x(x, u^{(1)}) \frac{\partial}{\partial u_x} + \varphi^{xx}(x, u^{(2)}) \frac{\partial}{\partial u_{xx}},$$

on J^2, whose coefficients φ^x, φ^{xx} are given by

$$
\begin{aligned}
\varphi^x &= D_x Q + \xi u_{xx} = \varphi_x + (\varphi_u - \xi_x) u_x - \xi_u u_x^2, \\
\varphi^{xx} &= D_x^2 Q + \xi u_{xxx} \\
&= \varphi_{xx} + (2\varphi_{xu} - \xi_{xx}) u_x + (\varphi_{uu} - 2\xi_{xu}) u_x^2 - \xi_{uu} u_x^3 \\
&\quad + (\varphi_u - 2\xi_x) u_{xx} - 3\xi_u u_x u_{xx}.
\end{aligned}
$$

For example, the second prolongation of the infinitesimal generator $\mathbf{v} = -u\partial_x + x\partial_u$ of the rotation group is given by

$$
\mathrm{pr}^{(2)}\mathbf{v} = -u\frac{\partial}{\partial x} + x\frac{\partial}{\partial u} + (1 + u_x^2)\frac{\partial}{\partial u_x} + 3u_x u_{xx}\frac{\partial}{\partial u_{xx}},
$$

where the coefficients are computed as

$$
\begin{aligned}
\varphi^x &= D_x Q + u_{xx}\xi = D_x(x + uu_x) - uu_{xx} = 1 + u_x^2, \quad \varphi^{xx} \\
&= D_x^2 Q + u_{xxx}\xi = D_x^2(x + uu_x) - uu_{xxx} = 3\,u_x u_{xx}.
\end{aligned}
$$

The group transformations can then be readily recovered by integrating the system of
ordinary differential equations

$$
\frac{dx}{dt} = -u, \quad \frac{du}{dt} = x, \quad \frac{dp}{dt} = 1 + p^2, \quad \frac{dq}{dt} = 3pq,
$$

where we have used p and q to stand for u_x and u_{xx} to avoid confusing derivatives with jet space coordinates. We find the second prolongation of the rotation group to be

$$
\left(x\cos t - u\sin t, x\sin t + u\cos t, \frac{\sin t + p\cos t}{\cos t - p\sin t}, \frac{q}{(\cos t - p\sin t)^3} \right),
$$

as could be computed directly.

2.4. Differential Invariants

At long last, we can precisely define the notion of "differential invariant." Indeed, recall that an invariant of a group G acting on a manifold M is just a function $I : M \to \mathbf{R}$ which is not affected by the group transformations. A *differential invariant* is an invariant in the standard sense for a prolonged group of transformations acting on the jet space J^n. Just as the ordinary invariants of a group action serve to characterize invariant equations, so differential invariants will completely characterize invariant systems of differential equations for the group, as well as invariant variational principles. As such they form the basic building block of many physical theories, where one often begins by postulating the invariance of the equations or the variational principle under an underlying symmetry group. In particular, they are essential in understanding the invariant heat-type flows presented below.

Suppose G is a local Lie group of point transformations acting on an open subset $M \subset X \times U$ of the space of independent and dependent variables, and let $\mathrm{pr}^{(n)}G$ be the n-th prolongation of the group action on the n-th jet space $J^n = J^n M$. A *differential invariant* is a real-valued function $I : J^n \to \mathbf{R}$ which satisfies $I(\mathrm{pr}^{(n)}g \cdot (x, u^{(n)})) = I(x, u^{(n)})$ for all $g \in G$ and all $(x, u^{(n)}) \in J^n$ where $\mathrm{pr}^{(n)}g \cdot (x, u^{(n)})$ is defined. Note that I may only be locally defined.

The following gives a characterization of differential invariants:

Proposition 4 *A function* $I : J^n \to \mathbf{R}$ *is a differential invariant for a connected group* G *if and only if*

$$pr^{(n)}\mathbf{v}(I) = 0,$$

for all $\mathbf{v} \in \mathcal{G}$ *where* $\mathcal{G}$ *denotes the Lie algebra of* G.

A basic problem is to classify the differential invariants of a given group action. Note first that if the prolonged group $\mathrm{pr}^{(n)}G$ acts regularly on (an open subset of) J^n with r_n- dimensional orbits, then, locally, there are $p + q^{(n)} - r_n = p + q \begin{pmatrix} p+n \\ n \end{pmatrix} - r_n$ functionally independent n-th order differential invariants. Furthermore, any lower order differential invariant $I(x, u^{(k)})$, $k < n$ is automatically an n –th differential invariant, and be included in the preceding count. (Here we are identifying $I : J^k \to \mathbf{R}$ and its composition $I \circ \pi_k^n$ with the standard projection $\pi_k^n : J^n \to J^k$.)

If $\mathcal{O}^{(n)} \subset J^n$ is an orbit of $\mathrm{pr}^{(n)}G$, then, for any $0 \le k < n$ its projection $\pi_k^n(\mathcal{O}) \subset J^n$ is an orbit of the k-th prolongation $\mathrm{pr}^{(k)}G$. Therefore, the generic orbit dimension r_n of $\mathrm{pr}^{(n)}G$ is a *nondecreasing* function of n, bounded by r, the dimension of G itself. This implies that the orbit dimension eventually stabilizes, $r_n = r^*$ for all $n \ge n_0$. We call r^* the *stable orbit dimension*, and the minimal order n_0 for which $r_{n_0} = r^*$, the *order of stabilization* of the group.

Now a transformation group G acts *effectively* on a space M if

$$g \cdot x = h \cdot x, \quad \forall x \in M,$$

if and only if $g = h$. Define the *global isotropy group*

$$G_M := \{g : g \cdot x = x \ \forall x \in M\}.$$

Then G acts effectively if and only if G_M is trivial. Moreover, G acts *locally effectively* if the global isotropy group G_M is a discrete subgroup of G in which case G/G_M has the same dimension and the same Lie algebra as G. We can now state the following remarkable result [287]:

Theorem 6 *The transformation group G acts locally effectively if and only if its dimension is the same as its stable orbit dimension, so that*

$$r_n = r^* = \dim G,$$

for all n sufficiently large.

There are a number of important results on the stabilization dimensions, maximal orbit dimensions, and their relationship to invariants; see [280, 281]. We will suffice with the following theorem which is very useful for counting the number of independent differential invariants of large order:

Theorem 7 *Suppose, for each $k \geq n$, the (generic) orbits of $pr^{(n)}G$ have the same dimension $r_k = r_n$. Then for every $k > n$ there are precisely $q_k = q \left(\begin{array}{c} p + k - 1 \\ k \end{array} \right)$ independent k-th order differential invariants which are not given by lower order differential invariants.*

Next we note that the basic method for constructing a complete system of differential invariants of a given transformation group is to use invariant differential operators [280, 279, 281]. A *differential operator* is said to be G-invariant if it maps differential invariants to higher order differential invariants, and thus, by iteration, produces hierarchies of differential invariants of arbitrarily large order. For sufficiently high orders, one can guarantee the existence of sufficiently many such invariant operators in order to completely generate all the higher order independent differential invariants of the group by successively differentiating lower order differential invariants. See [280, 279, 281] for details. Hence, a complete description of all the differential invariants is obtained by a set of low order *fundamental* differential invariants along with the requisite invariant differential operators.

In our case (one independent variable), the following theorem is fundamental:

Theorem 8 *Suppose that G is a group of point transformations acting on a space M having one independent variable and q dependent variables. Then there exist (locally) a G-invariant one-form $dr = g\,dx$ of lowest order, and q fundamental, independent differential invariants $J_1, \ldots, J_q$ such that* *every differential invariant can be written as a function of these differential invariants and their derivatives $\mathcal{D}^m J_\nu$, where*

$$\mathcal{D} := \frac{d}{dr} = \frac{1}{g}\frac{d}{dx},$$

is the invariant differential operator associated with dr. The parameter r gives an invariant parametrization of the curve and is called arc-length.

Remark. A version of Theorem 8 is true more generally. See [280, 281].

With this, we have completed our sketch of the theory of differential invariants. Once again, we refer the reader to the texts [280, 279, 281] for a full modern treatment of the subject, including methods for constructing and counting differential invariants.

With the above background, we are ready to turn to our main topic, namely invariant flows in vision.

3. Invariant Flows

In this section, a general approach for formulating invariant curve flows is described. In particular, we will consider the uniqueness of our models (see Theorem 9). Thus, given a certain transformation (Lie) group G, we show how to obtain the corresponding invariant geometric heat flow. We also show how to formulate this flow just in terms of Euclidean parameters such as the Euclidean curvature. This formulation permits us to employ already existing results and techniques for the analysis of such flows. This topic was first presented in [325, 327]. Here we emphasize a novel classification and uniqueness result.

3.1. Special Differential Invariants

In order to separate the geometric concept of a plane curve from its parametric description, it is useful to consider the *image* (or *trace*) of $\mathcal{C}(p)$, denoted by $\text{Img}[\mathcal{C}(p)]$. Therefore, if the curve $\mathcal{C}(p)$ is parametrized by a new parameter w such that $w = w(p)$, $\frac{\partial w}{\partial p} > 0$, we obtain

$$\text{Img}[\mathcal{C}(p)] = \text{Img}[\mathcal{C}(w)].$$

In general, the parametrization gives the "velocity" of the trajectory. Given a transformation group G, the curve can be parametrized using what is called the *group arc-length*, dr, which is a non-trivial G-invariant one-form of minimal order (see Theorem 7). This parametrization, which is an invariant of the group, is useful for defining differential invariant descriptors [143,

280, 279]. In order to perform this re-parametrization, the *group metric g* is defined by the equality

$$dr = g\, dp,$$

for any parametrization p. Then r is obtained via the relation (after fixing an arbitrary initial point)

$$r(p) := \int_0^p g(\xi)\, d\xi, \tag{11.14}$$

and the re-parametrization is given by $\mathcal{C} \circ r$. We have of course,

$$\mathrm{Img}[\mathcal{C}(p)] = \mathrm{Img}[\mathcal{C}(r)].$$

For example, in the Euclidean case we have

$$g_{euc} := \left\| \frac{\partial \mathcal{C}}{\partial p} \right\|, \tag{11.15}$$

and the Euclidean arc-length is given by

$$v := \int_0^p \left\| \frac{\partial \mathcal{C}}{\partial \xi} \right\| d\xi.$$

This parametrization is Euclidean invariant (since the norm is invariant), and implies that the curve $\mathcal{C}(s)$ is traversed with constant velocity ($\left\| \frac{\partial \mathcal{C}}{\partial v} \right\| \equiv 1$). For examples of other groups, see Sections 3.5, 3.6, 3.7 and [143, 280, 279, 384].

Based on the group metric and arc-length, the *group curvature* χ, is computed. (Note that g, r, and χ can be computed using Lie theory as well as Cartan moving frames method [54, 101, 143, 280].) The group curvature, as a function of the arc-length, is defined as the simplest non-trivial differential invariant of the group (see Theorem 7).

For example, in the Euclidean case, since

$$\left\| \frac{\partial \mathcal{C}}{\partial v} \right\| \equiv 1.$$

we have that $C_v \perp C_{vv}$, and the Euclidean curvature is defined as

$$\kappa := \| C_{vv} \|.$$

κ is also the rate of change of the angle between the tangent to the curve and a fixed direction. The corresponding invariants of the affine group will be presented below in Section 3.5.1.

3.2. Geometric Invariant Flow Formulation

We are now ready to describe the type of evolution that we want to deal with. First let $\mathcal{C}(p,t) : S^1 \times [0,\tau) \to \mathbf{R}^2$ be a family of smooth curves, where p parametrizes the curve and t the family. (In this case, we take p to be independent of t.) Assume that this family evolves according to the following evolution equation:

$$\frac{\partial \mathcal{C}(p,t)}{\partial t} = \frac{\partial^2 \mathcal{C}(p,t)}{\partial p^2},$$

$$\mathcal{C}(p,0) = \mathcal{C}_0(p),$$

(11.16)

which is the classical heat equation. If $\mathcal{C}(p,t) = [x(p,t), y(p,t)]^T$, then $[x(p,t), y(p,t)]$ satisfying (11.16) can also be obtained by convolution with a Gaussian filter of $[x_0(p), y_0(p)]$ whose variance depends on t. Equation (11.16) has been studied by the computer vision community, and is used for the definition of a linear scale-space for planar shapes [24, 90, 109, 187, 195, 196, 190, 226, 235, 386, 396].

The Gaussian kernel, being one of the most used in image analysis, has several undesirable properties, principally when applied to planar curves. One of these is that the filter is not intrinsic to the curve (see [324] for a detailed description of this problem). This can be remedied by replacing the linear heat equation by *geometric heat flows* [124, 139, 328, 322, 323, 325, 327, 324]. In particular, if the *Euclidean* geometric heat flow [122, 123, 124, 139, 140] is used, a scale-space invariant to rotations and translations is obtained. If the *affine* one is used [328, 322, 327], an affine invariant multi-scale representation is obtained [323]. This and other geometric heat flows are presented below.

Another problem with the Gaussian kernel is that the smoothed curve shrinks when the Gaussian variance (or the time) increases. Several approaches, briefly discussed in Section 4, have been proposed in order to partially solve this problem for Gaussian-type kernels (or linear filters). These approaches violate basic scale-space properties. In [324], the authors showed that this problem can be completely solved using a variation of the geometric heat flow methodology, which keeps the area enclosed by the curve constant. The flows obtained, precisely satisfy all the basic scale-space requirements. In the Euclidean case, the flow is local as well. The same approach can be used for deriving length preserving heat flows. In this case, the similarity flow exhibits locality. In short, *we can get geometric smoothing without shrinkage*. In order to give a complete picture of invariant geometric flows, basic results of this area/length preserving approach are given in Section 4 as well.

Assume that we want to formulate an intrinsic *geometric heat flow* for plane curves which is invariant under certain transformation group G. Let r denote the group arc-length (Theorem 7). Then, the *invariant geometric heat flow* is given by [325, 327, 324]

$$\frac{\partial \mathcal{C}(p,t)}{\partial t} = \frac{\partial^2 \mathcal{C}(p,t)}{\partial r^2},$$
$$\mathcal{C}(p,0) = \mathcal{C}_0(p).$$
(11.17)

If G acts linearly, it is easy to see that since dr is an invariant of the group, so is $\mathcal{C}_{rr}$. $\mathcal{C}_{rr}$ is called the *group normal*. For nonlinear actions, the flow (11.17) is still G-invariant, since as pointed out in Theorem 7, $\frac{\partial}{\partial r}$ is the *invariant derivative*. See [280] and our discussion in Section 3.3 below. In fact, as we will see the evolution given by (11.17) is in a certain precise sense the simplest possible non-trivial G-invariant flow.

We have just formulated the invariant geometric heat flow in terms of concepts intrinsic to the group itself, i.e., based on the group arc-length. For different reasons, which we will explain shortly, it is useful to formulate the group velocity $\mathcal{C}_{rr}$ in terms of Euclidean notions such as the Euclidean normal and Euclidean curvature. In order to do this, we need to calculate

$$< \mathcal{C}_{rr}, \vec{\mathcal{N}} >,$$

where $\vec{\mathcal{N}}$ is the Euclidean unit (inward) normal, and $< \cdot, \cdot >$ is the standard inner product in $\mathbf{R}^2$. In this way, we will be able to decompose the group normal $\mathcal{C}_{rr}$ into its Euclidean unit normal $\vec{\mathcal{N}}$ and Euclidean unit tangential $\vec{\mathcal{T}}$ components, and to re-write the flow (11.17) as

$$\frac{\partial \mathcal{C}}{\partial t} = \alpha \vec{\mathcal{T}} + \beta \vec{\mathcal{N}}.$$
(11.18)

In order to calculate α and β, assume for the moment that the curve $\mathcal{C}$ is parametrized by the Euclidean arc-length v. Then,

$$\frac{\partial^2 \mathcal{C}}{\partial r^2} = \frac{1}{g^2}\frac{\partial^2 \mathcal{C}}{\partial v^2} - \frac{g_v}{g^3}\frac{\partial \mathcal{C}}{\partial v},$$
(11.19)

where g is the group metric defined in Section 3.1. (In this case, g is a function of v.) Now, using the relations

$$\mathcal{C}_{vv} = \kappa \vec{\mathcal{N}}, \quad \mathcal{C}_v = \vec{\mathcal{T}},$$

we obtain

$$\alpha = -\frac{g_v}{g^3}, \quad \beta = \frac{\kappa}{g^2}.$$
(11.20)

In general, $g(v)$ in (11.20) is written as a function of κ and its derivatives (see Section 3.5).

The importance of the formulation (11.18) can be seen from the following:

Lemma 3 ([95]) *Let β be a geometric quantity for a curve, i.e., a function whose definition is independent of a particular parametrization. Then a family of curves which evolves according to*

$$C_t = \alpha \vec{T} + \beta \vec{N}$$

can be converted into the solution of

$$C_t = \bar{\alpha} \vec{T} + \bar{\beta} \vec{N}$$

for any continuous function $\bar{\alpha}$, by changing the space parametrization of the original solution. Since β is a geometric function, $\beta = \bar{\beta}$ when the same point in the (geometric) curve is considered.

In particular, the above lemma shows that $\text{Img}[C(p,t)] = \text{Img}[\hat{C}(w,t)]$, where $C(p,t)$ and $\hat{C}(w,t)$ are the solutions of

$$C_t = \alpha \vec{T} + \beta \vec{N},$$

and

$$\hat{C}_t = \bar{\beta} \vec{N},$$

respectively. For proofs of the lemma, see [95, 323].

Therefore, assuming that the normal component β of $\vec{v}$ (the curve evolution velocity) does not depend on the curve parametrization, we can consider the evolution equation

$$\frac{\partial C}{\partial t} = \beta \vec{N}, \tag{11.21}$$

where $\beta = <\vec{v}, \vec{N}>$, i.e., the projection of the velocity vector in the Euclidean normal direction. Since C_{rr} is a geometric quantity, equation (11.18) can be reduced to (11.21).

The formulation (11.17) gives a very intuitive formulation of the invariant geometric heat flow. On the other hand, the formulation given by equation (11.18), together with (11.20), gives an Euclidean-type flow which also allows us to simplify the flow using the result of Lemma 3. This type of analysis is crucial, since it allows one to understand the partial differential equation underlying the flow and to study its essential properties (such as short and long term existence, convergence, etc.). This will be a key technique when we study affine invariant flows in Section 3.5. Finally, reduction of equation (11.17) to (11.21) allows one to numerically implement the flow on computer.

In fact, there is now available an efficient numerical algorithm due to Osher and Sethian [285, 339] to do this.

The flow given by (11.17) is non-linear, since the group arc-length r is a function of time. This flow gives the invariant geometric heat-type flow of the group, and provides the invariant direction of the deformation. For subgroups of the full projective group $SL(\mathbb{R}, 3)$, we show in Theorem 8 below that the most general invariant evolutions are obtained if the group curvature χ and its derivatives (with respect to arc-length) are incorporated into the flow:

$$
\begin{aligned}
\frac{\partial C(p,t)}{\partial t} &= \Psi(\chi, \frac{\partial \chi}{\partial r}, \ldots, \frac{\partial^n \chi}{\partial r^n}) \frac{\partial^2 C(p,t)}{\partial r^2}, \\
C(p,0) &= C_0(p),
\end{aligned}
\tag{11.22}
$$

where $\Psi(\cdot)$ is a given function. (We discuss the existence of possible solutions of (11.22) in [327].) Since the group arc-length and group curvature are the basic invariants of the group transformations, it is natural to formulate (11.22) as the most general geometric invariant flow.

Since we have expressed the flow (11.17) in terms of Euclidean properties (equations (11.18), (11.20)), we can do the same for the general flow (11.22). All what we have to do is to express χ as a function of κ and it derivatives. This is done by expressing the curve components $x(p)$ and $y(p)$ as a function of κ, and then computing χ.

3.3. Uniqueness of Invariant Heat Flows

In this section, we give a fundamental result, which elucidates in what sense our invariant heat-type equations (11.17) are unique. We use here the action of the projective group $SL(\mathbb{R}, 3)$ on $\mathbb{R}^2$ as defined in Section 2.2.1. We will first note that locally, we may express a solution of (11.17) as the graph of a function $y = u(x,t)$.

Lemma 4 *Locally, the evolution (11.17) is equivalent to*

$$
\frac{\partial u}{\partial t} = \frac{1}{g^2} \frac{\partial^2 u}{\partial x^2},
$$

where g is the G-invariant metric ($g = dr/dx$).

Proof. Indeed, locally the equation

$$
C_t = C_{rr},
$$

becomes

$$
x_t = x_{rr}, \quad y_t = y_{rr}.
$$

Now $y(r, t) = u(x(r, t), t)$, so

$$y_t = u_x x_t + u_t, \quad y_{rr} = u_{xx} x_r^2 + u_x x_{rr}.$$

Thus,

$$u_t = y_t - u_x x_t = y_{rr} - u_x x_{rr} = x_r^2 u_{xx}.$$

Therefore the evolution equation (11.17) reduces to

$$u_t = g^{-2} u_{xx},$$

since $dr = g \, dx$. $\square$

We can now state the following fundamental result:

Theorem 9 *Let G be a subgroup of the projective group $SL(\mathbb{R}, 3)$. Let $dr = g \, dp$ denote the G-invariant arc-length and χ the G-invariant curvature. Then*

1. *Every differential invariant of G is a function*

$$I\left(\chi, \frac{d\chi}{dr}, \frac{d^2 \chi}{dr^2}, \ldots, \frac{d^n \chi}{dr^n}\right)$$

   of χ and its derivatives with respect to arc length.
2. *Every G-invariant evolution equation has the form*

$$\frac{\partial u}{\partial t} = \frac{1}{g^2} \frac{\partial^2 u}{\partial x^2} I, \tag{11.23}$$

   where I is a differential invariant for G.

We are particularly interested in the following subgroups of the full projective group: Euclidean, similarity, special affine, affine, full projective. (See our discussion below for the precise results.)

Corollary 1 *Let G denote the similarity, special affine, affine, or full projective group. (see remark on the Euclidean group below). Then there is, up to a constant factor, a unique G-invariant evolution equation of lowest order, namely*

$$\frac{\partial u}{\partial t} = \frac{c}{g^2} \frac{\partial^2 u}{\partial x^2},$$

where c is a constant.

Remark. Part 1 of the Theorem 9 (suitably interpreted) does not require G to be a subgroup of the projective group; however for part 2 and the corollary this is essential (see surfaces section for an extension). One can, of course, classify the differential invariants, invariant arc-lengths, invariant evolution equations, etc., for any group of transformations in the plane, but the interconnections are more complicated. See Lie [210] and Olver [280] for the details of the complete classification of all groups in the plane and their differential invariants.

Remark. The uniqueness of the Euclidean and affine flows (see the next section), was also proven in [7], using a completely different approach. In contrast with the results here presented, the ones in [7] were proven independently for each group, and when considering a new group, a new analysis had to be carried out. Our result is a general one, and can be applied to any subgroup. Also, with the geometric approach presented here, we believe that the result is clear and intuitive. Note also that in the affine case (see next section for the specific equation), it is enough to ask for the "lowest order" in order to obtain the heat flow. All other requirements in [7] are then unnecessary, and they are just properties of the flow as proven in [328, 323].

Proof of Theorem.
Part 1 follows immediately from Theorem 8, and the definitions of dr and χ. (Note by Theorem 7 for a subgroup of $SL(\mathbb{R}, 3)$ acting on $\mathbb{R}^2$, we have each differential invariant of order k is in fact unique.)
As for part 2, let

$$\mathbf{v} = \xi(x, u)\partial_x + \varphi(x, u)\partial_u$$

be an infinitesimal generator of G, and Let pr $\mathbf{v}$ denote its prolongation to the jet space. Since dr is (by definition) an invariant one-form, we have

$$\mathbf{v}(dr) = [\,\mathrm{pr}\,\mathbf{v}(g) + gD_x\xi\,]dx,$$

which vanishes if and only if

$$\mathrm{pr}\,\mathbf{v}(g) = -gD_x\xi = -g(\xi_x + u_x\xi_u). \tag{11.24}$$

Applying pr $\mathbf{v}$ to the evolution equation (11.23), and using condition (11.24), we have (since ξ and φ do not depend on t)

$$\begin{aligned}
\mathrm{pr}\,\mathbf{v}[u_t - g^{-2}u_{xx}I] &= (\varphi_u - u_x\xi_u)u_t - 2g^{-2}(\xi_x + u_x\xi_u)u_{xx}I - \\
&\quad - g^{-2}\mathrm{pr}\,\mathbf{v}[u_{xx}]I - g^{-2}u_{xx}\mathrm{pr}\,\mathbf{v}[I]. \tag{11.25}
\end{aligned}$$

If G is to be a symmetry group, this must vanish on solutions of the equation; thus, in the first term, we replace u_t by $g^{-2}u_{xx}I$. Now, since G was assumed to be a subgroup of the projective group, which is the symmetry group of the second order ordinary differential equation $u_{xx} = 0$, we have pr $\mathbf{v}[u_{xx}]$

is a multiple of u_{xx}; in fact, inspection of the general prolongation formula for pr $\mathbf{v}$ (see Theorem 5) shows that in this case

$$\text{pr } \mathbf{v}[u_{xx}] = (\varphi_u - 2\xi_x - 3\xi_u u_x)u_{xx}. \tag{11.26}$$

(The terms in pr $\mathbf{v}[u_{xx}]$ which do not depend on u_{xx} must add up to zero, owing to our assumption on $\mathbf{v}$.) Substituting (11.26) into (11.25) and combining terms, we find

$$\text{pr } \mathbf{v}[u_t - g^{-2}u_{xx}I] = g^{-2}u_{xx}\text{pr } \mathbf{v}[I],$$

which vanishes if and only if pr $\mathbf{v}[I] = 0$, a condition which must hold for each infinitesimal generator of G. But this is just the infinitesimal condition that I be a differential invariant of G, and the theorem follows. $\square$

The Corollary follows from the fact that, for the listed subgroups, the invariant arc length r depends on lower order derivatives of u than the invariant curvature χ. (This fact holds for most (but *not* all) subgroups of the projective group; one exception is the group consisting of translations in x, u, and isotropic scalings $(x, u) \mapsto (\lambda x, \lambda u)$.) For the Euclidean group, it is interesting to note that the simplest nontrivial flow is given by (constant motion)

$$u_t = c\sqrt{1 + u_x^2}, \quad c \text{ a constant.}$$

(Here $g = \sqrt{1 + u_x^2}$.) In this case the curvature (the ordinary planar curvature κ) has order 2. This equation is obtained for the invariant function $I = 1/\kappa$. The Euclidean geometric heat equation (see next Section) is indeed given by the flow in the Corollary. The orders are indicated in the following table:

Group	Arc Length	Curvature
Euclidean	1	2
Similarity	2	3
Special Affine	2	4
Affine	4	5
Projective	5	7

The explicit formulas are given in the following table:

Group	ArcLength	Curvature
Euclidean	$\sqrt{1+u_x^2}\,dx$	$\dfrac{u_{xx}}{(1+u_x^2)^{3/2}}$
Similarity	$\dfrac{u_{xx}\,dx}{(1+u_x^2)}$	$\dfrac{(1+u_x^2)u_{xxx}-3u_xu_{xx}^2}{u_{xx}^2}$
Special Affine	$(u_{xx})^{1/3}dx$	$\dfrac{P_4}{(u_{xx})^{8/3}}$
Affine	$\dfrac{\sqrt{P_4}}{u_{xx}}\,dx$	$\dfrac{P_5}{(P_4)^{3/2}}$
Projective	$\dfrac{(P_5)^{1/3}}{u_{xx}}\,dx$	$\dfrac{P_7}{(P_5)^{8/3}}$

Here

$$P_4 = 3u_{xx}u_{xxxx}-5u_{xxx}^2,$$
$$P_5 = 9u_{xx}^2u_{xxxxx}-45u_{xx}u_{xxx}u_{xxxx}+40u_{xxx}^3,$$
$$P_7 = \frac{1}{3}u_{xx}^2[6P_5D_x^2P_5-7(D_xP_5)^2]+2u_{xx}u_{xxx}P_5D_xP_5$$
$$-(9u_{xx}u_{xxxx}-7u_{xxx}^2)P_5^2.$$

Some of this invariants will be specifically derived below and used to present the corresponding invariant flows of the viewing transformations.

3.4. Euclidean Invariant Flows

We now show how to use the general theory presented above, for the computation of the invariant heat flows corresponding to the Euclidean group. In the next section, we will discuss the affine group.
Recall from our discussion in Section 2.2.1 that a general Euclidean transformation in the plane is given by

$$\tilde{\mathcal{X}} = R\mathcal{X} + V,$$

where $\mathcal{X} \in \mathbf{R}^2$, R is a 2×2 rotation matrix, and V is a 2×1 translation vector. The Euclidean transformations constitute a group, and give some of the basic shape deformations which appear in computer vision applications. We proceed to find, based on the above developed method, an Euclidean invariant geometric heat equation. From (11.17), we obtain that the *Euclidean geometric heat flow* is given by

$$\mathcal{C}_t = \mathcal{C}_{vv}, \qquad (11.27)$$
$$\mathcal{C}(p,0) = \mathcal{C}_0(p).$$

(Recall that v is the Euclidean arc-length.) The Euclidean metric is defined by equation (11.15), and if the curve is already parametrized by arc-length, then of course $g_{euc}(v) \equiv 1$. Therefore, using equation (11.20) we obtain

$$\alpha_{euc} = 0, \quad \beta_{euc} = \kappa.$$

Then, the "Euclidean type" equation equivalent to (11.27) is (see equation (11.18))

$$\mathcal{C}_t = \kappa \vec{\mathcal{N}}. \tag{11.28}$$

Equation (11.28) has a large research literature devoted to it. Gage and Hamilton [124] proved that any smooth, embedded convex curve converges to a round point when deforming according to it. Grayson [139] proved that any non-convex embedded curve converges to a convex one. Hence, any simple curve converges to a round point when evolving according to the Euclidean geometric heat equation. The flow is also known as the *Euclidean shortening flow*, since the Euclidean perimeter shrinks as fast as possible when the curve evolves according to (11.28); see [139]. Equation (11.28) was also found to be very important for image enhancement applications [9, 326], and was introduced into the theory of shape in computer vision by [180, 182, 181]. In order to proof that the flow indeed smoothes the curve, results from [18, 124, 139] can be used.

3.5. Affine Invariant Flows

In this section, we present the affine flow corresponding to equation (11.17) as first developed in [328, 322, 327]. We first make some remarks about classical affine differential geometry.

3.5.1. *Sketch of Affine Differential Geometry*
An affine transformation transforms disks into ellipses, and rectangles into parallelograms. Recall from Section 2.2.1 that a general *affine transformation* in the plane ($\mathbf{R}^2$) is defined by

$$\tilde{\mathcal{X}} = A\mathcal{X} + B, \tag{11.29}$$

where $\mathcal{X} \in \mathbf{R}^2$ is a vector, $A \in \mathrm{GL}_2^+(\mathbf{R})$ (the group of invertible real 2×2 matrices with positive determinant) is the affine matrix, and $B \in \mathbf{R}^2$ is a translation vector. As we have seen (11.29) form a real Lie group A(2), called the *group of proper affine motions*. We will also consider the case of when we restrict $A \in \mathrm{SL}(2, \mathbf{R})$ (i.e., the determinant of A is 1), in which case (11.29) gives us the *group of special affine motions*, $\mathrm{A}_{sp}(2)$.

In the case of Euclidean motions (in which case A in (11.29) is a rotation matrix), we have that the Euclidean curvature κ of a given plane curve is a differential invariant of the transformation. In the case of general affine transformations, in order to keep the invariance property, a new definition of curvature is necessary. In this section, this *affine* curvature is presented [35, 143, 328]. See [35] for general properties of affine differential geometry. We should note that when we say "affine invariant" in the sequel, we will mean with respect to $A_{sp}(2)$ (the group of special affine motions). However, all our "invariants" will be *relative* invariants with respect to the full affine group, in the sense that the transformed quantities will always differ by some function of the determinant of the transforming matrix. See [86, 280] for the precise definition of "relative invariant." We will consider flows which are "absolutely" invariant with respect to the full affine group in Section 3.7. As above, let $C : S^1 \to \mathbf{R}^2$ be an embedded curve with parameter p (where S^1 denotes the unit circle). We now make the invariant re-parametrization of $C(p)$ by defining a new parameter s such that

$$[C_s, C_{ss}] = 1, \tag{11.30}$$

where $[\mathcal{X}, \mathcal{Y}]$ stands for the determinant of the 2×2 matrix whose columns are given by the vectors $\mathcal{X}$, $\mathcal{Y} \in \mathbf{R}^2$. This relation is invariant under proper affine transformations, and the parameter s is the *affine arc-length*. Setting

$$g_{aff}(p) := [C_p, C_{pp}]^{1/3}, \tag{11.31}$$

the parameter s is explicitly given by

$$s(p) = \int_0^p g_{aff}(\xi) d\xi. \tag{11.32}$$

Note, we have assumed (of course) that g_{aff} (the *affine metric*) is different from zero at each point of the curve, i.e., the curve has no inflection points. In general, affine differential geometry is defined just for convex curves [35, 143]. In Section 3.5.2, we will show how to overcome this problem for the evolution of non-convex curves.

By differentiating (11.30) we obtain that the two vectors C_s and C_{sss} are linearly dependent and so there exists μ such that

$$C_{sss} + \mu C_s = 0. \tag{11.33}$$

The last equation and (11.30) imply

$$\mu = [C_{ss}, C_{sss}], \tag{11.34}$$

and μ is the *affine curvature*, i.e., the simplest non-trivial differential affine invariant function of the curve C; see [35]. Moreover, one can easily show [328] that ds, C_s, C_{ss}, μ, and the area enclosed by a closed curve, are (absolute) invariants of the group $A_{sp}(2)$ of special affine motions and relative invariants of the group $A(2)$ of proper affine motions.

3.5.2. Affine Geometric Heat Equation

With s the *affine* arc-length, the affine-invariant geometric heat flow is given by [328, 327]

$$\begin{cases} \mathcal{C}_t = \mathcal{C}_{ss}, \\ \mathcal{C}(p,0) = \mathcal{C}_0(p). \end{cases} \tag{11.35}$$

Since s is only defined for convex curves, the flow (11.35) is defined *a priori* for such curves only. However, in fact the evolution can be extended to the non-convex case, in the following natural manner. Observe that if $\mathcal{C}$ is parametrized by the Euclidean arc-length, then

$$g_{aff}(v) \quad = \quad [\mathcal{C}_v, \mathcal{C}_{vv}]^{1/3} = [\vec{\mathcal{T}}, \kappa \vec{\mathcal{N}}]^{1/3} = \kappa^{1/3},$$

and we obtain

$$\alpha_{aff} = -\frac{(\kappa^{1/3})_v}{\kappa}, \quad \beta_{aff} = \kappa^{1/3}.$$

Now one can easily compute that

$$\mathcal{C}_{ss} = \kappa^{1/3} \vec{\mathcal{N}} + \text{tangential component}.$$

(See [328, 327].) Hence, using Lemma 3, we obtain that the following flow is geometrically equivalent to the affine invariant flow (11.35):

$$\mathcal{C}_t = \kappa^{1/3} \vec{\mathcal{N}}. \tag{11.36}$$

Note that the flow (11.36) is affine invariant flow, and is also *well-defined for non-convex curves*. (We should also observe here that inflection points are affine invariant.) We should also add that recently Alvarez et al. [7] derived (11.36) using a completely different approach.

In summary, despite the fact that we cannot define the basic differential invariants of affine differential geometry on non-convex curves, nevertheless an affine invariant heat-type flow can be formulated. This is possible due to the possibility to "ignore" the tangential component of the deformation velocity, together with the invariant property of inflection points. One can see that while $\mathcal{C}_{ss}$ contains three derivatives, its normal component contains only two. This allows one to write the geometric affine heat flow as a function of κ.

The key results in this theory are the following [19, 328, 322, 327]:

Theorem 10 ([328]) *A smooth convex curve evolving according to the affine geometric heat flow remains smooth and convex.*

Theorem 11 ([328]) *A convex curve evolving according to the geometric heat flow converges to an elliptical point.*

Theorem 12 ([19, 322, 327]) *Let $C(\cdot, 0) : S^1 \to \mathbf{R}^2$ be a C^2 embedded curve in the plane. Then there exists a unique one parameter family of C^2 curves $C : S^1 \times [0, T) \to \mathbf{R}^2$ for some $T > 0$, satisfying the affine heat equation*

$$C_t = \kappa^{1/3} \mathcal{N}.$$

Moreover, there is a $t_0 < T$ such that $C(\cdot, t)$ is convex for all $t_0 < t < T$.

Theorem 13 *Any given C^2 embedded plane curve converges to an elliptical point when evolving according to (11.36) .*

Moreover, in [19] we show how to extend (11.36) to Lipschitz initial curves, and in particular, polygons. This eliminates the necessity of the viscosity framework [60, 72, 97] as proposed in [7], being also a stronger result.
In [323], it is formally proven that the affine geometric flow (11.35) (or its geometric analogue (11.36)) smoothes the curve. For example, it is shown that the total curvature and the number of inflection points decrease with time (scale-parameter). Figure 11.1 gives an example of this flow.

3.6. Projective Invariant Flows

We would like to make some remarks about projective invariant flows. The projective maps constitute the most general geometric transformations on planar shapes (or planar curves). Projective invariant flows have been considered by [100, 99, 326, 325, 327, 334]. Recall from Section 2.2.1 that the projective group acts on $\mathbf{R}^2$ via linear fractional transformations. Let w denote the projective arc-length. Then

$$C_t = C_{ww} \tag{11.37}$$

is the projective flow. The flow is more complicated than the Euclidean and affine evolutions, because of the higher derivatives involved. Explicitly,

$$dw = g_{pro} dp,$$

where the projective metric is given by,

$$g_{pro}(p) = \left(R(p) - \frac{3}{2} \frac{\partial Q(p)}{\partial p} \right)^{1/3}, \tag{11.38}$$

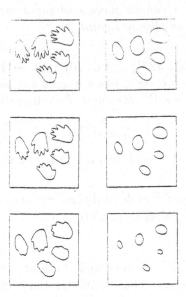

Figure 11.1. Example of the affine geometric heat flow (from [323]). The hands are related during all the evolution by the same affine transformation.

where

$$Q(p) = q_2(p) - q_1^2(p) - \frac{\partial q_1(p)}{\partial p}, \quad R(p) = -3q_1(p)q_2(p) + 2q_1^3(p) - \frac{\partial^2 q_1(p)}{\partial p^2},$$

and

$$q_1(p) = \frac{1}{3}\frac{[C_{ppp}, C_p]}{[C_p, C_{pp}]}, \quad q_2(p) = \frac{1}{3}\frac{[C_{pp}, C_{ppp}]}{[C_p, C_{pp}]}.$$

Assuming that the curve is parametrized by the Euclidean arc-length v, we obtain

$$q_1(v) = -\frac{1}{3}\frac{\kappa_v}{\kappa}, \quad q_2(v) = \frac{1}{3}\kappa^2,$$

and $Q(v)$, $R(v)$ and $g_{pro}(v)$ can be computed. Clearly, we must assume that g_{pro} is well-defined and non-zero on the curve for the projective flow to be defined.

We would like to discuss why singularities should arise in this flow. First, we must define the real projective space $\mathbf{RP}^2$. This is the space of lines through the origin in $\mathbf{R}^3$. Two nonzero points $p_1, p_2 \in \mathbf{R}^3$ determine the same point $p \in \mathbf{RP}^2$ if and only if they are scalar multiples of each other $p_1 = \lambda p_2$, $\lambda \neq 0$. We will refer to the coordinates of $p_1 \in \mathbf{R}^3$, (x^1, x^2, x^3), as *homogeneous coordinates* of p. Notice this induces a canonical projection

$$\pi : \mathbf{R}^3 \backslash \{0\} \to \mathbf{RP}^2.$$

Coordinate charts on $\mathbf{RP}^2$ are constructed by considering all lines with a given component, say x^i, nonzero; the coordinates are then provided by ratios x^k/x^i, $k \neq i$, which amounts to the choice of canonical representative of such a line given by normalizing its i-th component to unity. Thus we may embed $\mathbb{R}^2$ as an open subset of $\mathbf{RP}^2$ via

$$\iota : \mathbb{R}^2 \hookrightarrow \mathbf{RP}^2, \quad (x, y) \mapsto (x, y, 1).$$

Projective space $\mathbf{RP}^2$ may then be regarded as the completion of $\mathbb{R}^2$ by adjoining all "directions at infinity." Thus we have following diagram:

$$
\begin{array}{ccc}
 & & \mathbf{R}^3 \backslash \{0\} \\
 & & \downarrow \pi \\
\mathbf{R}^2 & \xrightarrow{\iota} & \mathbf{P}^2
\end{array}
.
$$

Now suppose, we are given the flow in $\mathbf{R}^2$

$$C_t = C_{ww}. \tag{11.39}$$

Via ι and π we may lift this to a flow in $\mathbf{R}^3$, say

$$\hat{C}_t = \hat{C}_{ww}, \tag{11.40}$$

which is invariant with respect to scaling. Under certain conditions, one can conclude, using results from the theory of reaction-diffusion equations [345], short term existence for this flow [100]. Suppose we consider initial curves which differ by a scaling:

$$\hat{C}_1(0, w) = \lambda(w)\hat{C}_2(0, w).$$

Then via (11.40), the resulting flows will differ by precisely the same scaling:

$$\hat{C}_1(t, w) = \lambda(w)\hat{C}_2(t, w).$$

By the uniqueness of the solutions of differential equations, this means that the flow (11.39) in $\mathbf{R}^2$ is induced by (11.40).

Given this, even assuming that (11.40) remains smooth, singularities could develop in (11.39) in the following two ways: First, since $\mathbf{R}^2 \subset \mathbf{P}^2$ and the flow (11.39) is induced by the projectively invariant flow via ι, points may go off to infinity, and so the flow in $\mathbb{R}^2$ can be singular. Secondly, the projection π may give cuspidal singular points in (11.39). Thus the projective invariant case of (11.17) does not have the nice smoothing properties of the Euclidean and affine models.

3.7. Similarity and Full Affine Invariant Flows

In this section, we consider flows which are invariant relative to the scale-invariant versions of the Euclidean and affine groups, namely the similarity and full affine groups as defined in Section 2.2.1. We begin with the heat flow for the similarity group (rotations, translations, and isotropic scalings). This flow was first presented and analyzed in [324]. We assume for the remainder of this section that our curves are strictly convex ($\kappa > 0$). Accordingly, let $\mathcal{C}$ be a smooth strictly convex plane curve, p the curve parameter, and as above, let $\vec{N}, \vec{T}$, and v denote the Euclidean unit normal, unit tangent, and Euclidean arc-length, respectively. Let

$$\sigma := \frac{\partial v}{\partial p}$$

be the speed of parametrization, so that

$$\mathcal{C}_p = \sigma \vec{T}, \quad \mathcal{C}_{pp} = \sigma_p \vec{T} + \sigma^2 \kappa \vec{N}.$$

Then clearly,

$$\mathcal{C}_p \cdot \mathcal{C}_p = \sigma^2,$$
$$[\mathcal{C}_p, \mathcal{C}_{pp}] = \sigma^3 \kappa.$$

For the similarity group (in order to make the Euclidean evolution scale-invariant), we take a parametrization p such that

$$\mathcal{C}_p \cdot \mathcal{C}_p = [\mathcal{C}_p, \mathcal{C}_{pp}],$$

which implies that

$$\sigma = 1/\kappa.$$

Therefore the similarity group invariant arc-length is the standard angle parameter θ, since

$$\frac{d\theta}{dv} = \kappa,$$

where v is the Euclidean arc-length. (Note that $\vec{T} = [\cos\theta, \sin\theta]^T$.) Thus the *similarity normal* is $\mathcal{C}_{\theta\theta}$, and the *similarity invariant flow* is

$$\mathcal{C}_t = \mathcal{C}_{\theta\theta}. \tag{11.41}$$

Projecting the similarity normal into the Euclidean normal direction, the following flow is obtained

$$\mathcal{C}_t = \frac{1}{\kappa}\vec{N}, \tag{11.42}$$

and both (11.41) and (11.42) are geometrically equivalent flows.

Instead of looking at the flow given by (11.42) (which may develop singularities), we reverse the direction of the flow, and look at the expanding flow given by

$$\frac{\partial \mathcal{C}}{\partial t} = -\frac{1}{\kappa}\vec{\mathcal{N}}. \tag{11.43}$$

For completeness, we state the following results for the flow (11.43) (the proofs are given in [324]; see also [124, 181]):

Lemma 5 ([324]) *The following evolution equations are obtained for a curve evolving according to (11.43):*

1. *Evolution of Euclidean metric m (dp = mdv, where v is the Euclidean arc-length):*
$$m_t = m.$$

2. *Evolution of Euclidean tangent $\vec{T}$:*
$$\vec{T}_t = -\frac{\kappa_v}{\kappa^2}\vec{\mathcal{N}}.$$

3. *Evolution of Euclidean normal $\vec{\mathcal{N}}$:*
$$\vec{\mathcal{N}}_t = \frac{\kappa_v}{\kappa^2}\vec{T}.$$

4. *Evolution of Euclidean perimeter $\mathbf{P}$:*
$$\mathbf{P}_t = \mathbf{P}.$$

5. *Evolution of area $\mathbf{A}$:*
$$\mathbf{A}_t = \oint \frac{1}{\kappa}dv.$$

6. *Evolution of Euclidean curvature κ:*
$$\kappa_t = -\left(\frac{1}{\kappa}\right)_{vv} - \kappa.$$

7. *Evolution of tangential angle θ:*
$$\theta_t = \frac{\kappa_v}{\kappa^2}.$$

Theorem 14 ([324])

1. *A simple convex curve remains simple and convex when evolving according to the similarity invariant flow (11.43).*

2. *The solution to (11.43) exists (and is smooth) for all $0 \le t < \infty$.*

Lemma 6 ([324]) *Changing the curve parameter from p to θ, we obtain that the radius of curvature r, $r := 1/\kappa$, evolves according to*

$$r_t = r_{\theta\theta} + r. \tag{11.44}$$

Theorem 15 ([324]) *A simple (smooth) convex curve converges to a disk when evolving according to (11.43).*

It is important to note that in contrast with (11.43), (11.42) can deform a curve towards singularities. Since (11.43) can be seen as a smoothing process (heat flow), the inverse equation can be seen as an enhancement process. The importance of this for image processing, as well as the extension of the theory to non-convex curves, is currently under investigation.

Using a similar argument, one may show the invariant equation for the full affine flow $(GL(\mathbb{R}, 2) \times_s \mathbb{R}^2)$ is given by

$$C_t = \frac{C_{ss}}{\mu}. \tag{11.45}$$

As for the similarity flow, this will develop singularities. The backward flow (add a minus sign to the right-hand side of (11.45)), can be shown to asymptotically converge to an ellipse.

When the heat flow can develop singularities, as in the scale-invariant cases described above, one can use the most general flow given by (11.22), i.e., to multiply the velocity by functions of the group curvature and its derivatives. We are currently investigating these more general flows and their possible smoothing properties.

4. Geometric Heat Flows Without Shrinkage

In the previous sections, we derived intrinsic geometric versions of the (non-geometric) classical heat flow (or Gaussian filtering). Using this geometric methodology, we now will solve the problem of shrinkage to which we alluded above. This theory is developed in [324], to which we refer the interested reader for all the details and relevant references.

A curve deforming according to the classical heat flow shrinks. This is due to the fact that the Gaussian filter also affects low frequencies of the curve coordinate functions [277]. Oliensis [277] proposed to change the Gaussian kernel to a filter which is closer to the ideal low pass filter. This way, low frequencies are less affected, and less shrinkage is obtained. With this approach, which is also non-intrinsic, the semi-group property holds just approximately. Note that in [24, 396] (see also [7]), it was proved that filtering with a Gaussian kernel is the unique linear operation for which the causality criterion holds. In fact, the approach presented in [277], which is closely related to wavelet approaches [241, 252], violates this important principle.

Lowe [235] proposes to estimate the amount of shrinkage and to compensate for it. The estimate is based on the amount of smoothing (variance/time) and the curvature. This approach, which only reduces the shrinkage problem, is again non-intrinsic, since it is based on Gaussian filtering, and works only for small rates of change. The semi-group property is violated as well.

Horn and Weldon [158] also investigated the shrinkage problem, but only for convex curves. In their approach, the curve is represented by its extended circular image, which is the radius of curvature of the given curve as a function of the curve orientation. The scale-space is obtained by filtering this representation.

We now show how to solve the shrinkage problem with a variation of the geometric flows described above. The resulting flows will keep all the basic properties of scale-spaces, while preserving area (length) and performing geometric smoothing at the same time [324].

4.1. Area Preserving Euclidean Flow

We now solve the shrinkage problem with the Euclidean geometric heat flow following ideas of Gage [124].

Consider the evolution (11.21) given above. When a closed curve evolves according to (11.21), it is easy to prove that the enclosed area $\mathbf{A}$ evolves according to

$$\frac{\partial \mathbf{A}}{\partial t} = -\oint \beta dv. \tag{11.46}$$

Therefore, in the case of the Euclidean geometric heat flow (11.28) we obtain $(\beta = \kappa)$

$$\frac{\partial \mathbf{A}}{\partial t} = -2\pi, \tag{11.47}$$

and the area decreases. Moreover

$$\mathbf{A}(t) = \mathbf{A}_0 - 2\pi t,$$

where $\mathbf{A}_0$ is the area enclosed by the initial curve $\mathcal{C}_0$. As pointed out in [124, 124, 139], curves evolving according to (11.28) can be normalized in order to keep constant area. The normalization process is given by a change of the time scale, from t to τ, such that a new curve is obtained via

$$\tilde{C}(\tau) := \psi(t)\,C(t),\tag{11.48}$$

where $\psi(t)$ represents the normalization factor (time scaling). (The equation can be normalized so that the point $\mathcal{P}$ to which $C(t)$ shrinks is taken as the origin.) In the Euclidean case, $\psi(t)$ is selected such that

$$\psi^2(t) = \frac{\partial \tau}{\partial t}.\tag{11.49}$$

The new time scale τ must be chosen to obtain $\tilde{\mathbf{A}}_\tau \equiv 0$. Define the collapse time T, such that $\lim_{t \to T} \mathbf{A}(t) \equiv 0$. Then,

$$T = \frac{\mathbf{A}_0}{2\pi}.$$

Let

$$\tau(t) = -T\,\ln(T - t).\tag{11.50}$$

Then, since the area of $\tilde{C}$ and C are related by the square of the normalization factor $\psi(t) = \left(\frac{\partial \tau}{\partial t}\right)^{1/2}$, $\tilde{\mathbf{A}}_\tau \equiv 0$ for the time scaling given by (11.50). The evolution of $\tilde{C}$ is obtained from the evolution of C and the time scaling given by (11.50). Taking partial derivatives in (11.48) we have

$$
\begin{aligned}
\frac{\partial \tilde{C}}{\partial \tau} &= \frac{\partial t}{\partial \tau}\frac{\partial \tilde{C}}{\partial t}\\
&= \psi^{-2}(\psi_t C + \psi C_t)\\
&= \psi^{-2}\psi_t C + \psi^{-1}\kappa\vec{\mathcal{N}}\\
&= \psi^{-2}\psi_t C + \tilde{\kappa}\vec{\mathcal{N}}\\
&= \psi^{-3}\psi_t \tilde{C} + \tilde{\kappa}\vec{\mathcal{N}}.
\end{aligned}
$$

From Lemma 3, the flow above is geometrically equivalent to

$$\frac{\partial \tilde{C}}{\partial \tau} = \psi^{-3}\psi_t < \tilde{C}, \vec{\mathcal{N}} > \vec{\mathcal{N}} + \tilde{\kappa}\vec{\mathcal{N}}.\tag{11.51}$$

Define the *support function* ρ as

$$\rho := -<C, \vec{\mathcal{N}}>.$$

Then, it is easy to show that

$$\mathbf{A} = \frac{1}{2}\oint \rho\,dv.$$

Therefore, applying the general area evolution equation (11.46) to the flow (11.51), together with the constraint $\tilde{\mathbf{A}}_\tau \equiv 0$, we obtain

$$\frac{\partial \tilde{\mathcal{C}}}{\partial \tau}(p, \tau) = \tilde{\kappa}\vec{\mathcal{N}} - \frac{\pi\tilde{\rho}}{\tilde{\mathbf{A}}(\tau)}\vec{\mathcal{N}}. \tag{11.52}$$

Note that the flow (11.52) exists for all $0 \leq \tau < \infty$. Since $\tilde{\mathbf{A}}_\tau \equiv 0$ when $\tilde{\mathcal{C}}$ evolves according to (11.52), the enclosed area $\tilde{\mathbf{A}}(\tau)$ in (11.46) can be replaced by $\mathbf{A}_0$, obtaining

$$\frac{\partial \tilde{\mathcal{C}}}{\partial \tau}(p, \tau) = \left(\tilde{\kappa} - \frac{\pi\tilde{\rho}}{\mathbf{A}_0}\right)\vec{\mathcal{N}}, \tag{11.53}$$

which gives a local, area preserving, flow.

Since $\mathcal{C}$ and $\tilde{\mathcal{C}}$ are related by dilations, the flows (11.28) and (11.53) have the same geometric properties [124, 124, 139, 324]. The properties of this flow can also be obtained directly from the general results on characterization of evolution equations given in [7]. In particular, since a curve evolving according to the Euclidean heat flow satisfies all the required properties of a multi-scale representation, so does the normalized flow. An example of this flow is presented in figure 11.2.

4.2. Area Preserving Affine Flow

From the general evolution equation for areas (11.46), and the flow (11.36), we have that when a curve evolves according to the affine geometric heat flow, the enclosed area evolves according to

$$\mathbf{A}_t = -\oint \kappa^{1/3} dv. \tag{11.54}$$

Following [35], we define the *affine perimeter* $\mathbf{L}$ as

$$\mathbf{L} := \oint [\mathcal{C}_p, \mathcal{C}_{pp}]^{1/3} dp.$$

Then it is easy to show that [328]

$$\mathbf{L} = \oint \kappa^{1/3} dv.$$

Therefore,

$$\mathbf{A}_t = -\mathbf{L}. \tag{11.55}$$

As in the Euclidean case, we define a normalized curve

$$\tilde{\mathcal{C}}(\tau) := \psi(t)\mathcal{C}(t), \tag{11.56}$$

Figure 11.2. Example of the area preserving Euclidean geometric heat flow (from [324]).

such that when $\mathcal{C}$ evolves according to (11.36), $\tilde{\mathcal{C}}$ encloses a constant area. In this case, the time scaling is chosen such that [324]

$$\frac{\partial \tau}{\partial t} = \psi^{4/3}. \tag{11.57}$$

(We see from the Euclidean and affine examples that in general, the exponent λ in $\frac{\partial \tau}{\partial t} = \psi^\lambda$ is chosen such that $\tilde{\beta} = \psi^{1-\lambda}\beta$.) Taking partial derivatives in (11.56), using the relations (11.46), (11.55), and (11.57), and constraining $\mathbf{A}_\tau \equiv 0$, we obtain the following geometric affine invariant, area preserving, flow:

$$\frac{\partial \tilde{\mathcal{C}}}{\partial \tau} = \left(\tilde{\kappa}^{1/3} - \frac{\tilde{\rho}\,\tilde{L}}{2\tilde{\mathbf{A}}(\tau)} \right) \vec{\mathcal{N}}. \tag{11.58}$$

Since $\tilde{\mathbf{A}}_\tau \equiv 0$, $\tilde{\mathbf{A}}(\tau)$ in (11.58) can be replaced by $\mathbf{A}_0$ to obtain

$$\frac{\partial \tilde{\mathcal{C}}}{\partial \tau} = \left(\tilde{\kappa}^{1/3} - \frac{\tilde{\rho}\,\tilde{\mathbf{L}}}{2\mathbf{A}_0} \right) \vec{\mathcal{N}}. \tag{11.59}$$

Note that in contrast with the Euclidean area preserving flow given by equation (11.53), the affine one is not local. This is due to the fact that the rate of area change in the Euclidean case is constant, but in the affine case it depends on the affine perimeter.

As in the Euclidean case, the flow (11.59) satisfies the same geometric properties as the affine geometric heat flow (11.36). Therefore, it defines a geometric, affine invariant, area preserving multi-scale representation.

4.3. Length Preserving Flows

Similar techniques to those presented in previous sections, can be used in order to keep fixed other curve characteristics, e.g., the Euclidean length $\mathbf{P}$ [324]. In this case, when $\mathcal{C}$ evolves according to the general geometric flow

$$\frac{\partial \mathcal{C}}{\partial t} = \beta \vec{\mathcal{N}},$$

and

$$\tilde{\mathcal{C}}(\tau) := \psi(t)\,\mathcal{C}(t), \tag{11.60}$$

we obtain the following length preserving geometric flow:

$$\frac{\partial \tilde{\mathcal{C}}}{\partial \tau}(p,\tau) = \left(\tilde{\beta} - \frac{\oint \tilde{\beta}\tilde{\kappa}}{\mathbf{P}_0} \tilde{\rho} \right) \vec{\mathcal{N}}. \tag{11.61}$$

The computation of (11.61) is performed again taking partial derivatives in (11.60), and using the relations (see for example [124])

$$\mathbf{P}_t = -\oint \beta\kappa dv,$$

$$\mathbf{P} = \oint \kappa\rho dv,$$

together with the constraint

$$\tilde{\mathbf{P}}_\tau \equiv 0.$$

The following flows are the corresponding length preserving Euclidean, affine, and similarity heat flows respectively:

$$\frac{\partial \tilde{\mathcal{C}}}{\partial \tau}(p,\tau) = \left(\tilde{\kappa} - \frac{\oint \tilde{\kappa}^2}{\mathbf{P}_0} \tilde{\rho} \right) \vec{\mathcal{N}}, \tag{11.62}$$

$$\frac{\partial \tilde{\mathcal{C}}}{\partial \tau}(p,\tau) = \left(\tilde{\kappa}^{1/3} - \frac{\oint \tilde{\kappa}^{4/3}}{\mathbf{P}_0} \tilde{\rho} \right) \vec{\mathcal{N}}, \tag{11.63}$$

$$\frac{\partial \tilde{\mathcal{C}}}{\partial \tau}(p,\tau) = \left(-\tilde{\kappa}^{-1} + \tilde{\rho} \right) \vec{\mathcal{N}}. \tag{11.64}$$

Note that in the similarity case, the flow is completely local. Another local, length preserving flow may be obtained for the Euclidean constant motion given by

$$\mathcal{C}_t = \vec{\mathcal{N}}. \tag{11.65}$$

This flow, obtained taking $r \equiv v$ and $\Psi(\chi) = \chi^{-1}$ in (11.22), models the classical Huygens principle or morphological dilation with a disk [320] (of course, it is not a geometric heat flow of the form (11.17)). In this case, the rate of change of length is constant and the length preserving flow is given by

$$\frac{\partial \tilde{\mathcal{C}}}{\partial \tau}(p,\tau) = \left(1 - \frac{2\pi \tilde{\rho}}{\mathbf{P}_0} \right) \vec{\mathcal{N}}. \tag{11.66}$$

Note that a smooth initial curve evolving according to the Euclidean constant motion (11.65), as well as to the flow given by (11.42), can develop singularities [7, 320]. In this case, the physically correct weak solution of the flow is the viscosity (or *entropy* [345]) one [7, 320]. See [320, 324] for examples of this flow.

5. Invariant Geometric Surface Flows

The results in Section 3.3 have been recently generalized. We can now write down the general form that any G-invariant evolution in n-independent and one dependent variable must have. For details see [282]. Thus for $n = 1$, we get the family of all possible invariant curve evolutions in the plane under a given transformation group G, and for $n = 2$ the family of all possible invariant surface evolutions under a given transformation group G. We let

$$\omega = g\,dx^1 \wedge \ldots \wedge dx^n,$$

denote a G-invariant n-form with respect to the transformation group G acting on $\mathbf{R}^n$. Let $E(g)$ denote the variational derivative of g. We only consider forms such that $E(g) \neq 0$. We call such a g a G-invariant volume function.

Theorem 16 ([282]) *Notation as above. Then every G-invariant evolution equation has the form*

$$u_t = \frac{g}{E(g)} I, \tag{11.67}$$

where I is a differential invariant.

Theorem 17 ([282]) *Suppose G is a connected transformation group, and $g dx$ a G-invariant n-form such that $E(g) \neq 0$. Then $E(g)$ is invariant if and only if G is volume preserving.*

It is now trivial to give the simplest possible invariant surface evolution. This gives for example the surface version of the affine shortening flow for curves. This equation was also derived using completely different methods by Alvarez et al. [7] (see corresponding chapter). Note than in contrast with the large number of properties needed in [7] to derive the flow below, only two are needed in our approach: affine invariance and "simplest as possible." We define the *(special) affine group* on $\mathbf{R}^3$ as the group of transformations generated by $SL_3(\mathbf{R})$ (the group of 3×3 matrices with determinant 1) and translations.

Let $\mathcal{S}$ be a smooth strictly convex surface in $\mathbf{R}^3$, which we write locally as the graph (x, y, u). Then one can compute that the Gaussian curvature is given by

$$\kappa = \frac{u_{xx} u_{yy} - u_{xy}^2}{(1 + u_x^2 + u_y^2)^2}.$$

From [35], the affine invariant metric is given by

$$g = \kappa^{1/4} \sqrt{1 + u_x^2 + u_y^2}.$$

(One can also write

$$g = \kappa^{1/4} \sqrt{\det g_{ij}},$$

where g_{ij} are the coefficients of the first fundamental form.)
We now have:

Corollary 2 ([282]) *Notation as above. Then*

$$u_t = c \kappa^{1/4} \sqrt{1 + u_x^2 + u_y^2}, \tag{11.68}$$

(for c a constant) is the simplest affine invariant surface flow. This corresponds to the global evolution

$$\mathcal{S}_t = c \kappa^{1/4} \vec{\mathcal{N}}, \tag{11.69}$$

where $\vec{\mathcal{N}}$ denotes the inward normal.

We will call the evolution

$$\mathcal{S}_t = \kappa^{1/4} \vec{\mathcal{N}}, \tag{11.70}$$

the *affine surface flow*. Note that it is the surface analogue of the affine heat equation.

6. Conclusions

In this chapter, we have outlined the theory of differential invariants, and applied them to invariant flows that appear in computer vision. Using this theory, we have defined G-invariant heat-type flows where G is a subgroup of the projective group $SL(\mathbb{R}, 3)$. As we have indicated, these flows, first described in [328, 322, 325, 327, 324], are the simplest possible. In certain cases, such diffusions have been used to define new geometrically invariant scale-spaces. They have also been employed for various problems in image processing and computer vision. See [7, 9, 110, 180, 182, 323, 326, 327] and the references therein. We have also discussed area and length preserving versions of these equations in which there is no shrinkage. See [324] for full details.

In summary, in addition to novel results as the classification and uniqueness of the geometric heat flows, this chapter gives a complete picture of the relevant theory of differential invariants and geometric invariant curve flows. Extension of this theory to surface flows was presented as well. Details and proofs can be found in [282].

ON OPTIMAL CONTROL METHODS IN COMPUTER VISION AND IMAGE PROCESSING

Benjamin Kimia

Brown University
Laboratory for Engineering Man-Machine Systems
Providence, Rhode Island 02912, USA

and

Allen Tannenbaum

University of Minnesota
Department of Electrical Engineering
Minneapolis, MN 55455, USA

and

Steven Zucker

McGill University
Department of Electrical Engineering
Montreal, Quebec, Canada
and
Isaac Newton Institute for Mathematical Sciences
University of Cambridge, Cambridge, U.K.

Abstract. In this chapter, we discuss the employment of methods from optimal control for problems in computer vision and image processing. The underlying principle will be that of dynamic programming and the associated Hamilton-Jacobi equation which allows a unified approach to tackle a number of different issues in vision. In particular, we will consider problems concerning shape theory, morphology, optical flow, nonlinear scale-spaces, and shape-from-shading. We will also discuss some numerical issues concerned with computer implementations of this methodology.

1. Introduction

A number of problems in computer vision and image processing have been found to be amenable to treatment using the calculus of variations. At the heart of the classical theory of the calculus of variations are the Euler-Lagrange equations, which are "necessary conditions" that the solution must satisfy. The calculus of variations has found a modern incarnation in optimal control theory where one finds the extensive use of the principle of dynamic programming and the associated Hamilton-Jacobi-Bellman equation. (This will be discussed in detail below.)

The classical calculus of variations is built upon smooth functions, which has been a major source of problems for computer vision applications, because many functions are inherently discontinuous. Perhaps the most common examples arise when one object occludes another. This chapter is built upon the observation that the viscosity theory of Crandall-Lions [73] has recently made it possible to extend Hamilton-Jacobi theory to singular functions, thereby substantially extending the applicability to practical computer vision problems. (An alternative approach for such an extension can be based on nonsmooth analysis as well. See Clarke [64].)

Our goal in this chapter is to introduce the general computer vision audience to these techniques through several illustrative problems: shape theory, morphology, optical flow, nonlinear scale-spaces, and shape-from-shading. As such the chapter will have a tutorial flavor. (We should mention that there are several other important problems in vision which may also be treated through such optimal control techniques which we don't consider here, e.g., image segmentation (see [264, 202] and the references therein), and "snakes" or active contours (see [174, 34] and the references therein).)

In contrast to more classical techniques, we will indicate how such problems can be formulated as generalized solutions within Hamilton-Jacobi theory. Moreover, since much of the motivation for developing these generalized techniques comes from applications in control theory, some of this background is included as motivation. This not only permits the recovery of solutions involving singularities, but also opens up a wider perspective on the problems in computer vision by revealing connections with variational theory and systems and control.

The form of the chapter is as follows. In Section 2, we discuss the notion of "generalized solution" which opens up the use of nonsmooth functions in our infinitesimal analysis of a number of vision problems. In Section 3, we outline Hamilton-Jacobi theory and the resulting notion of viscosity solution of the Hamilton-Jacobi equation. Here our point of view is based on the dynamic programming principle of Richard Bellman. In Section 4, we give our first application of this theory, namely to the problem of optical flow. We then continue with the shape-from-shading problem in Section 5. In Section 6, we use these ideas for a computational theory of shape which leads to the reaction-diffusion scale-space. Section 7 is then devoted to morphology, and finally in Section 8, we review the Osher-Sethian theory for the numerical implementation of curve evolution laws via a Hamilton-Jacobi formulation.

2. Generalized or "Weak" Solutions of Conservation Equations

There is an obvious conceptual difficulty that must be addressed when considering discontinuous solutions of partial differential equations—the derivatives are simply not well defined in the usual sense. However, there is another perspective that does lead one to suspect that a proper resolution of this difficulty exists, and that is through the notion of *conservation laws*. In particular, conservation laws are typically written in an integral form, which naturally leads to the notion of "weak solution" which we will formally define in this section. The idea is as follows. Suppose we were studying the flow of a fluid through a container at a point. An integral conservation law would certainly hold in a region around this point to describe the flow (the sum of material flowing in equals the sum of that which remains minus that which flows out). That is, the conservation law would involve terms regarding the change in amount of material, and the flux of material. If the flow were smooth, we could immediately write the integral conservation law in differential form:

$$u_t + g(u)_x = 0, \quad u(x,0) = u_0. \tag{12.1}$$

which is called a hyperbolic conservation law. Now, if a singularity existed in the flow at this point, an integral conservation law would still hold in a neighborhood around the point; the remarkable mathematical result is that one can characterize a generalized or weak form of a solution to the differential hyperbolic conservation law that is still well-formed mathematically.

Suppose that u is a strong classical solution of (12.1). Let C_0^1 denote the set of continuously differentiable functions with compact support on $\mathbb{R} \times \mathbb{R}^+$. (Here $\mathbb{R}^+$ denotes the set of real numbers ≥ 0.) Let $\psi \in C_0^1$. We suppose that the support of ψ is contained in the rectangle $R = [x_1, x_2] \times [0, T]$ with $|x_1|, |x_2|, T$ chosen sufficiently large so that

$$\psi(x_1, t) = \psi(x_2, t) = \psi(x, T) = 0, \quad \forall t \in \mathbb{R}^+, \, x \in \mathbb{R}.$$

Then multiplying (12.1) by ψ and integrating, we see that

$$\int_{-\infty}^{\infty} \int_{0}^{\infty} (u_t + g(u)_x)\psi \, dx \, dt = \iint_{R} (u_t + g(u)_x)\psi \, dx \, dt = 0.$$

The trick is now to integrate by parts:

$$\int_{x_1}^{x_2} \int_0^T u_t \psi \, dx \, dt = \int_{x_1}^{x_2} [u(x,T)\psi(x,T) - u(x,0)\psi(x,0)] \, dx$$
$$- \int_{x_1}^{x_2} \int_0^T u\psi_t \, dx \, dt$$
$$= \int_{-\infty}^{\infty} -u_0(x)\psi(x,0) \, dx - \int_{x_1}^{x_2} \int_0^T u\psi_t \, dx \, dt.$$

Similarly,

$$\int_{x_1}^{x_2} \int_0^T g(u)_x \psi \, dx \, dt = \int_0^T [g(u(x_2,t))\psi(x_2,t) - g(u(x_1,t))\psi(x_1,t)] \, dt$$
$$- \int_0^T \int_{x_1}^{x_2} g(u)\psi_x \, dx \, dt$$
$$= - \int_0^T \int_{x_1}^{x_2} g(u)\psi_x \, dx \, dt.$$

Hence, we derive the *weak form* of the conservation law (12.1):

$$\int_{-\infty}^{\infty} \int_0^{\infty} (u\psi_t + g(u)\psi_x) \, dx \, dt + \int_0^{\infty} u_0\psi \, dx = 0. \qquad (12.2)$$

The point is that (12.2) makes sense for $u(x,t)$ bounded and measurable (with bounded measurable initial data u_0), for all $\psi \in C_0^1$. Thus such functions $u(x,t)$ which satisfy (12.2) for all $\psi \in C_0^1$ are called *weak solutions*. Finally, it is easy to show that if a continuously differentiable function u satisfies (12.2) for all $\psi \in C_0^1$, then u is a classical solution of (12.1).

3. Hamilton-Jacobi Equation

In this section, we review the connection of Hamilton-Jacobi theory with optimal control in order to motivate the definition of "viscosity solution." This will be important for developing techniques to select the physically plausible solution from among an infinity of generalized or weak solutions. The discussion we give below can be applied more generally to the Hamilton-Jacobi-Bellman equation, but we are only interested in outlining the main ideas here. We refer the interested reader to the excellent new book of Fleming-Soner [103] for an extensive treatment of viscosity theory in the optimal control framework. See also the original work of Lions on this topic [232].

Let us begin by studying the calculus of variations problem on the fixed interval $[t, t_1]$. Set

$$U := L^{\infty}([t, t_1]; \mathbb{R}^n).$$

Consider the dynamical system

$$\frac{dx}{ds} = u(s), \quad u \in U \tag{12.3}$$

that is $x(s)$ is the *state variable* and $u(s)$ is the *control.* We require the boundary conditions $x(t) = x$, and $x(t_1) \in \mathcal{Q}$ where $\mathcal{Q} \subset \mathbb{R}^n$ is closed. (We should note that one can take more general functions $g(s, x(s), u(s))$ on the right-hand side of (12.3). We have chosen just to consider the simplest classical case in order to motivate our treatment of the Hamilton-Jacobi equation via control principles.) Then the problem is to minimize the integral (*cost functional*)

$$J := \int_t^{t_1} L(s, x(s), u(s)) ds + \phi(x(t_1)), \tag{12.4}$$

subject to the constraint (12.3) over all Lipschitz continuous curves $x : [t, t_1] \to \mathbb{R}^n$ which satisfy the endpoint conditions.

The associated Hamiltonian function is

$$H(t, x, p) := \max_{v \in \mathbb{R}^n} \{ -v \cdot p - L(t, x, v) \}. \tag{12.5}$$

The dual formula is

$$L(t, x, v) := \max_{p \in \mathbb{R}^n} \{ -v \cdot p - H(t, x, p) \}. \tag{12.6}$$

The extremal points where the maximum is taken on in (12.5) and (12.6) are related by

$$p = -L_v, \quad v = -H_p \tag{12.7}$$

(the Legendre transformation). This duality between L and H is basically that of the Lagrangian and Hamiltonian formulations of classical mechanics where v is velocity and p is momentum.

Define the *value function* or the *optimal return*

$$V(t, x) := \inf_{u \in U} J(t, x; u). \tag{12.8}$$

Then for continuously differentiable V, one may show using dynamic programming that V satisfies the Hamilton-Jacobi equation

$$-\frac{\partial V}{\partial t} + H(t, x, \nabla_x V) = 0. \tag{12.9}$$

(Note that ∇_x denotes the gradient computed with respect to the space variables x.) Assuming that $\mathcal{Q} = \mathbb{R}^n$, we get the boundary condition

$$V(t_1, x) = \phi(x). \tag{12.10}$$

(See a sketch of the derivation of equation (12.9) below.)
One can also show that one gets the same Hamilton-Jacobi equation as
(12.9) for the following *exit time problem*: Minimize

$$J := \int_t^{t_e} L(s, x(s), \dot{x}(s))ds + \Phi(t_e, x(t_e)),$$

where t_e is the exit time of the "state trajectory" $(s, x(s))$ from $[t_0, t_1] \times X$,
where $X \in \mathbb{R}^n$ is closed, and Φ is a function such that

$$\Phi(t, x) := \begin{cases} f(t, x) & \text{if } (t, x) \in [t_0, t_1) \times \mathbb{R}^n , \\ \phi(x) & \text{if } (t, x) \in \{t_1\} \times \mathbb{R}^n . \end{cases} \qquad (12.11)$$

Remark. The problems considered above are called *finite horizon problems*.
There is a related class of optimization problems called *infinite horizon*.
Suppose that we have a dynamical system of the form

$$\frac{dx}{ds} = g(x(s), u(s)),$$
$$x(0) = x_0.$$

The cost in this case is given by a functional of the form

$$J(x, u) := \int_0^t e^{-\alpha s} L(x(s), u(s))ds + e^{-\alpha t} \phi(x(t)).$$

A Hamilton-Jacobi theory with associated value functions can be developed
in this setting as well.

One needs to introduce *generalized solutions* to the Hamilton-Jacobi
equation (12.9) in case V is not differentiable. A standard example [103]
for the exit time problem may be constructed as follows: We let the
"Lagrangian"

$$L(t, x, v) := 1 + \frac{v^2}{4},$$

$\Phi \equiv 0$, $X := [-1, 1]$, $[t_0, t_1] = [0, 1]$. So we want to minimize

$$\int_t^{t_e} \{1 + \frac{\dot{x}(s))^2}{4}\}ds.$$

The optimal control in this case for given initial condition (t, x) can be
computed to be

$$u_{opt} := \begin{cases} -2 & \text{if } x \leq -t, \\ 0 & \text{if } |x| < t, \\ 2 & \text{if } x \geq t. \end{cases}$$

The corresponding value function is

$$V(t, x) := \begin{cases} 1 - t & \text{if } |x| \leq -t, \\ 1 - |x| & \text{if } |x| \geq t. \end{cases}$$

V is not differentiable for $t = |x|$. Note however that V satisfies the corresponding Hamilton-Jacobi equation

$$-\frac{\partial V(t, x)}{\partial t} + \left(\frac{\partial V(t, x)}{\partial x}\right)^2 - 1 = 0$$

except when $t = |x|$.

In fact, this illustrates the general fact that the value function satisfies the Hamilton-Jacobi at all points where it is differentiable. Let us call $\mathcal{V}$ a *generalized solution* of the Hamilton-Jacobi equation (12.9) if $\mathcal{V}$ is locally Lipschitz and satisfies (12.9) almost everywhere. Then one may show that the value function is a generalized solution in this sense. (See our discussion in the next section.) The problem is that typically there are many generalized solutions. We therefore want to pick out one "natural" solution. We show how to do this next.

3.1. Viscosity Solutions

The above discussion motivates the introduction of *viscosity solutions*. Viscosity solutions were introduced by Crandall and Lions [73]. For another approach to the generalized solutions of Hamilton-Jacobi using techniques from *nonsmooth analysis*, see [64] and the references therein.

We set $C^\infty = C^\infty(\mathbb{R}^n \times (0, \infty))$. Consider the "approximation" of the Hamilton-Jacobi equation (12.9) given by

$$-\frac{\partial V_\epsilon}{\partial t} - \epsilon \Delta V_\epsilon + H(t, x, \nabla_x V_\epsilon) = 0. \tag{12.12}$$

Let V_ϵ be a classical solution of (12.12), and let $\omega \in C^\infty$. Suppose that (x_ϵ, t_ϵ) is a local maximum of $V_\epsilon - \omega$. Then via the maximum principle,

$$\frac{\partial V_\epsilon}{\partial t}(x_\epsilon, t_\epsilon) = \frac{\partial \omega}{\partial t}(x_\epsilon, t_\epsilon),$$
$$\nabla_x V_\epsilon(x_\epsilon, t_\epsilon) = \nabla_x \omega(x_\epsilon, t_\epsilon),$$
$$\Delta V_\epsilon(x_\epsilon, t_\epsilon) \leq \Delta \omega(x_\epsilon, t_\epsilon),$$

where Δ denotes the Laplacian operator. Thus we see from (12.12) that for all $\omega \in C^\infty$, we have

$$-\frac{\partial \omega}{\partial t}(x_\epsilon, t_\epsilon) - \epsilon \Delta \omega(x_\epsilon, t_\epsilon) + H(x_\epsilon, t_\epsilon, \nabla_x \omega(x_\epsilon, t_\epsilon)) \leq 0. \tag{12.13}$$

Similarly, playing the same game for a local minimum, we get that for $\omega \in C^\infty$,

$$-\frac{\partial \omega}{\partial t}(x_\epsilon, t_\epsilon) - \epsilon \Delta \omega(x_\epsilon, t_\epsilon) + H(x_\epsilon, t_\epsilon, \nabla_x \omega(x_\epsilon, t_\epsilon)) \geq 0. \tag{12.14}$$

Now if V_ϵ converges uniformly on compact subsets to $V : \mathbb{R}^n \times (0, \infty) \to \mathbb{R}$ continuous, and if we take $\omega \in C^\infty$, and assume that $V - \omega$ has a strict local maximum (x_o, t_o), then there exists (x_ϵ, t_ϵ) which converges to (x_o, t_o) such that each (x_ϵ, t_ϵ) is a local maximum point of $V_\epsilon - \omega$. Hence (12.13) is satisfied, and we can pass to the limit. We therefore make the following definition:

Definition. Let $V : \mathbb{R}^n \times (0, \infty) \to \mathbb{R}$ be continuous.

1. V is a *viscosity subsolution* of (12.9) if for any given $\omega \in C^\infty$, at each local maximum point of $V - \omega$, say (x_o, t_o), we have

$$-\frac{\partial \omega}{\partial t}(x_o, t_o) + H(x_o, t_o, \nabla_x \omega(x_o, t_o)) \leq 0.$$

2. V is a *viscosity supersolution* of (12.9) if for any given $\omega \in C^\infty$, at each local minimum point of $V - \omega$, say (x_o, t_o), we have

$$-\frac{\partial \omega}{\partial t}(x_o, t_o) + H(x_o, t_o, \nabla_x \omega(x_o, t_o)) \geq 0.$$

3. V is a *viscosity solution* of (12.9) if it is both a viscosity subsolution and supersolution.

A beautiful fact is that the value function defined by (12.8) is the unique **viscosity solution** of (12.9). For the eikonal equation (the "prairie fire" model of computer vision) this has an especially neat form.

3.2. Eikonal Equation

It was Barles [26] who first made explicit the connection between entropy theory of hyperbolic conservation laws and the viscosity theory of the Hamilton-Jacobi theory. Consider the eikonal equation

$$-\frac{\partial V}{\partial t} + H\left(\frac{\partial V}{\partial x}\right) = 0, \tag{12.15}$$
$$V(x, 0) = V_o(x),$$

where the Hamiltonian is given by

$$H(p) := \sqrt{1 + p^2}.$$

Then one can show in this case that the value function which gives the unique viscosity solution of (12.15) is given by the Lax-Oleinik formula (see [103, 9])

$$V(t, x) = \min_y \left[V_o(y) + tH^* \left(\frac{x - y}{t} \right) \right]. \tag{12.16}$$

Here H^* is the *conjugate function* of H defined by

$$H^*(w) = \max_u [uw - H(u)].$$

This is closely related to the entropy condition introduced by Sethian in relation to the prairie fire model of flame propagation. We work in $\mathbb{R}^2$ for simplicity. Let

$$C = \{(x, y) \in \mathbb{R}^2 : y = \phi(x)\},$$

where ϕ is any continuous function. We assume that the particles below C are burnt, and the region above C is filled with a combustible fluid. We assume that the front propagates in the normal direction with constant speed one. Sethian describes the motion of the flame front even when singularities develop by introducing the following entropy condition:

"Once a particle is burnt, it remains burnt."

Let

$$y = V(x, t), \quad t > 0$$

denote the position of the prairie fire when evolving according to this law. Then Barles [26] noticed the following key fact:

Proposition 1 *Notation as above. Then the position of the prairie fire is given by $y = V(x, t)$, where V is the (unique) viscosity solution of the eikonal equation (12.15) with boundary condition*

$$V(x, 0) = \phi(x).$$

Proof. The proof is a straightforward calculation, which we include to stress the interrelationships between ideas. One may show using the above description of the prairie fire evolution that the front is given by

$$V(x, t) = \max_{|z - x| \le t} \{ \phi(z) + \sqrt{t^2 - |z - x|^2} \}.$$

It is then easy to check that this is the Lax-Oleinik (value function) formula (12.16) given above. $\square$

Because of this proposition, we will identify the entropy and viscosity solutions in the sequel.

3.3. Derivation of Hamilton-Jacobi Equation

In this section, we give a sketch of how the Hamilton-Jacobi equation may be derived. The key is the following fundamental observation of Richard Bellman:

Principle of Optimality:

From any point on an optimal trajectory, the remaining trajectory is optimal for the corresponding problem initiated at that point.

The Principle of Optimality is the basis of *dynamic programming*. Let us see how from the principle of optimality, we may derive the Hamilton-Jacobi equation. We consider the optimal control problem (12.4) subject to the constraint (12.3). Recall that the value function $V(t, x)$ is the optimal value of the cost functional starting at x at time t. We assume that V is continuously differentiable. Suppose that we know the function for $t + \delta$ where $\delta > 0$ is small. We want to work backward to t. Assume that between times t and $t+\delta$ a fixed u is applied. This yields approximately a contribution of $L(t, x, u)\delta$ from the integral term of the cost functional. Moreover, it transfers the state approximately to $x + u\delta$ at time $t + \delta$. The value function (i.e., the optimal return) is known from the point. Thus

$$J \approx L(t, x, u)\delta + V(t + \delta, x + u\delta),$$

assuming an optimal path from $t + \delta$ to the terminal point t_1. The value function of x at t is thus the infimum of this last expression with respect to all possible choices of u, that is

$$V(t, x) = \inf_{u \in U}[L(t, x, u)\delta + V(t + \delta, x + u\delta)].$$

Letting $\delta \to 0$ (and assuming the V is sufficiently smooth), we see that

$$V(t + \delta, x + u\delta) \approx V(t, x) + \frac{\partial V}{\partial t}\delta + \nabla_x V(t, x)u\delta,$$

from which we get

$$V(t, x) = \inf_{u \in U}[L(t, x, u)\delta + V(t, x) + \frac{\partial V(t, x)}{\partial t}\delta + \nabla_x V(t, x)u\delta].$$

Since $V(t, x)$ and $\frac{\partial V(t, x)}{\partial t}\delta$ are independent of u, they can be taken out of the infimum, and so cancelling the $V(t, x)$ from both sides of the equation, dividing by δ and taking the limit as $\delta \to 0$, we get the resulting Hamilton-Jacobi as required.

4. Optical Flow

Optical flowis the vector field of apparent velocities of light patterns in an image; see [151, 346] and the references therein. There are a number of approaches to motion-field estimation which are based on trying to compute the intensity flow. One of the most popular approaches in this area is due to Horn and Schunck[157] who base their method on the assumption that the image intensities of scene points are preserved over time together with a smoothness constraint. Their basic premise is that optical flow cannot be computed locally, since only one measurement is available at any point, while the optical flow vector field has two components. An additional constraint must be found to uniquely determine the vector field. They observed that the distribution of velocities on the image domain is not random. Rather, each velocity vector in a domain is often constrained to have only a *slight* variation from its neighboring velocities.

Specifically, Horn and Schunck showed that the constant intensity assumption, namely, that points retain their intensity values from one frame to another, leads to a zero total derivative,

$$E_t + E_x u + E_y v = 0, \qquad (12.17)$$

where $E = E(x, y, t)$ is image intensity, and u, v are horizontal and vertical components of the optical flow vector, respectively. Written differently,

$$E_t + \nabla E \cdot (u, v) = 0, \qquad (12.18)$$

(where the gradient ∇ is computed with respect to the variables x, y). This indicates that only the normal component of the velocity, say V_n, can be computed:

$$
\begin{aligned}
V_n &= \frac{-E_t}{\|\nabla E\|^2} \nabla E \\
&= \frac{-E_t}{E_x^2 + E_y^2} (E_x, E_y).
\end{aligned}
\qquad (12.19)
$$

That only the normal velocity can be computed is known is one form of the *aperture problem*; see [151]. Since this is a single equation involving two unknowns (u, v), a second constraint is sought. Horn and Schunck noted that each velocity is constrained by the velocities of its neighbors which can be expressed as a smoothness constraint. Smoothness can be the magnitude of the gradient of velocities, for example. They argued that since equation (12.17) cannot hold exactly due to noise, and since smoothness of the flow is required, a linear combination of the two, namely, how close $E_t + E_x u + E_y v$ is to zero, and how smooth the flow is, must be minimized. Therefore, the optical flow vector field is obtained as the solution to the following optimization problem

$$\min_{(u,v)} \int \int [(E_x u + E_y v + E_t)^2 + \alpha^2 (u_x^2 + u_y^2 + v_x^2 + v_y^2)] \, dx \, dy. \qquad (12.20)$$

4.1. Euler-Lagrange equation for optical flow

Clearly, there are a number of problems with this approach, the main one being that the optical flow errors are high near discontinuities, since smoothness is violated at these points. Thus a natural proposal is to achieve *piecewise-smoothing* by moving away from the domain of smooth solutions and into the domain of generalized functions.

Smoothness, as an additional constraint, should constrain the large number of possible solutions to one. However, linearly combining the two constraints of smoothness and equation (12.17) is to trade-off "smoothness" and "fit" in ways that are not always desirable. Rather, equation (12.17) must *always* hold; among all of its possible solutions, we should then pick the smoothest one. This reduces to a constrained minimization problem:

$$\begin{cases} \min_{(u,v)} \int \int [u_x^2 + u_y^2 + v_x^2 + v_y^2]\, dx\, dy, \\ \text{subject to} \quad E_t + E_x u + E_y v = 0. \end{cases} \tag{12.21}$$

To solve this problem, we use the classical method of Lagrange multipliers [131] for a function of two variables. Let us consider the general problem

$$\begin{cases} \min_{(u,v)} \mathcal{J}(u,v) \\ \text{subject to} \quad \mathcal{G}(u,v) = 0. \end{cases} \tag{12.22}$$

where

$$\mathcal{J}(u,v) = \int \int L(x, y, u, v, u_x, u_y, v_x, v_y)\, dx\, dx, \tag{12.23}$$

for

$$L(x, y, u, v, u_x, u_y, v_x, v_y) = u_x^2 + u_y^2 + v_x^2 + v_y^2, \tag{12.24}$$

and where

$$\mathcal{J}(u,v) = E_t + E_x u + E_y v. \tag{12.25}$$

This problem can be solved by solving the unconstrained problem

$$\min_{(u,v,\lambda)} \{\mathcal{J}(u,v) + \lambda \mathcal{G}(u,v)\}. \tag{12.26}$$

Computing the first variation, we see that

$$\begin{cases} F_u - \frac{\partial F_{u_x}}{\partial x} - \frac{\partial F_{u_y}}{\partial y} + \lambda \mathcal{G}_u = 0, \\ F_v - \frac{\partial F_{v_x}}{\partial x} - \frac{\partial F_{v_y}}{\partial y} + \lambda \mathcal{G}_v = 0, \\ E_t + E_x u + E_y v = 0. \end{cases} \tag{12.27}$$

In our case, we get that

$$\begin{cases} -2u_{xx} - 2u_{yy} + \lambda E_x = 0, \\ -2v_{xx} - 2v_{yy} + \lambda E_y = 0, \\ E_t + E_x u + E_y v = 0. \end{cases} \tag{12.28}$$

which upon eliminating λ reduces to

$$\begin{cases} \frac{\Delta u}{\Delta v} = \frac{E_x}{E_y}, \\ E_t + E_x u + E_y v = 0. \end{cases} \tag{12.29}$$

Substituting $v = -\frac{E_t + E_x u}{E_y}$ from the second equation into the first, we obtain

$$\begin{aligned} E_y \Delta u - E_x \Delta v &= 0, \\ E_y \Delta u - E_x \Delta [-\frac{E_t + E_x u}{E_y}] &= 0, \\ E_y \Delta u + E_x \Delta [\frac{E_t}{E_y}] + E_x \Delta [\frac{E_x}{E_y} u] &= 0, \\ E_y \Delta u + E_x \Delta [\frac{E_t}{E_y}] + E_x \Delta [\frac{E_x}{E_y}] u + E_x [\frac{E_x}{E_y}]_x u_x + E_x [\frac{E_x}{E_y}]_y u_y + E_x \frac{E_x}{E_y} \Delta u &= 0, \end{aligned} \tag{12.30}$$

Therefore u satisfies the following PDE,

$$[E_x^2 + E_y^2] \Delta u + E_x E_y [\frac{E_x}{E_y}]_x u_x + E_x E_y [\frac{E_x}{E_y}]_y u_y + E_x E_y \Delta [\frac{E_x}{E_y}] u + E_x E_y \Delta [\frac{E_t}{E_y}] = 0. \tag{12.31}$$

We now assume that

$$E_x E_y \Delta [\frac{E_x}{E_y}] \leq 0.$$

Then (12.31) becomes a degenerate elliptic, proper equation to which we can apply the theory of weak solutions as in [72].

Remark. The authors are presently studying other variational formulations of optical flow for use in feedback control problems (controlled active vision). In fact, making the connection with the geometric heat equation (see Section 6.4 below), Arun Kumar [204] is studying the following type of problem:

$$\begin{cases} \min_{(u,v)} \int \int \sqrt{u_x^2 + u_y^2 + v_x^2 + v_y^2} \, dx \, dy, \\ \text{subject to} \quad E_t + E_x u + E_y v = 0, \end{cases} \tag{12.32}$$

as well as the analogue of (12.20) with the α parameter. The Euler-Lagrange equation in this case will be nonlinear, and closely related to the geometric heat equation.

5. Shape-from-Shading

In this section, we discuss a Hamilton-Jacobi approach to the famous shape-from-shading problem following [311, 232]. The *shape-from-shading* problem in computer vision may be formulated as follows. Suppose that we have a function of two variables $z(x, y)$ describing the surface of an object. We can take the function $z(x, y)$ to be defined on the open connected bounded, smooth subdomain $\Omega \subset \mathbb{R}^2$. The shaded image of the surface is given as a *brightness distribution*

$$0 < E(x,y) \leq 1$$

at each pixel. The brightness is determined by a *reflectance map* or shading rule $R(\mathcal{N})$ which characterizes the surface properties and which provides the connection between the surface orientation and the image ($\mathcal{N}(x,y)$ denotes the unit surface normal). Then the shape-from-shading problem consists in recovering the depth $z(x,y)$ from $E(x,y)$. This leads to the equation

$$E(x,y) = R(\mathcal{N}(x,y)),$$

which may be interpreted as a first-order nonlinear partial differential equation whose unknown is z. For a rather complete set of references on this problem, we refer the interested reader to [156]. See also the interesting new optimal control approach in [91], and a level set approach in [183]. More precisely, let

$$p := \frac{\partial z}{\partial x}, \quad q := \frac{\partial z}{\partial y},$$

the components of the surface gradient. The unit normal is given then by

$$\mathcal{N}(x,y) = \frac{1}{\sqrt{1+p^2+q^2}} \begin{bmatrix} -p \\ -q \\ 1 \end{bmatrix}.$$

The most studied case of shape-from-shading is the *Lambertian* case, in which we have diffuse reflection properties and uniform illumination. Here $E(x,y)$ is proportional to the cosine of the angle between the normal $\mathcal{N}$ and the light source direction d. We will assume for simplicity that we have a vertical light source. Then we get the equation

$$E(x,y) = \mathcal{N}(x,y), \quad (x,y) \in \Omega. \tag{12.33}$$

We will take the boundary condition

$$z(x,y) = 0 \quad (x,y) \in \partial\Omega. \tag{12.34}$$

(Note we assume that $0 < E \leq 1$ on $\overline{\Omega}$.) Set

$$u(x,y) := \sqrt{\frac{1}{E(x,y)^2} - 1}.$$

Then, we may write (12.33) and (12.34) equivalently as

$$\begin{aligned} \sqrt{p^2+q^2} &= u(x,y), \quad (x,y) \in \Omega, \\ z(x,y) &= 0, \quad (x,y) \in \partial\Omega, \end{aligned} \tag{12.35}$$

where

$$p = \frac{\partial z}{\partial x}, \quad q = \frac{\partial z}{\partial y}.$$

Let us now explicitly put the shape-from-shading equation (12.35) into the Hamilton-Jacobi framework to which the above generalized solution theory applies. Define the Hamiltonian to be

$$H((x,y), t, (p, q)) := \sqrt{p^2 + q^2} - u(x, y), \tag{12.36}$$

so that the Hamilton-Jacobi equation becomes

$$-\frac{\partial z}{\partial t} + H((x, y), t, (\frac{\partial z}{\partial x}, \frac{\partial z}{\partial y})) = 0.$$

Since $z(x, y)$ is independent of t, we get (12.35).
The viscosity solution theory for Hamilton-Jacobi equations of the type (12.35) is worked out in the book of Lions [232], once more based on ideas from optimal control. Define for $\psi, \eta \in \overline{\Omega}$, the functional

$$L(\psi, \eta) := \inf\{\int_0^1 z(x(s), y(s)) ds : (x(0), y(0)) = \psi, \quad (x(1), y(1)) = \eta\}$$

where the infimum is taken over all curves $(x(s), y(s)) \in C^1([0, 1], \overline{\Omega})$, with $\dot{x}(s)^2 + \dot{y}(s)^2 \leq 1$.
We now assume that $0 < E \leq 1$ on $\overline{\Omega}$, $E(\psi_o) = 1$, and $E(x, y) < 1$ on $\overline{\Omega}\backslash\{\psi_o\}$ for some $\psi_o \in \Omega$. (The discussion below easily generalizes for a finite set of points where E may be 1.)
We now state the following result from [232], Chapter 5 to which we refer the reader for the proof.

Theorem 1 *Let $\alpha \in \mathbb{R}$. Then the unique viscosity solution of (12.35) which satisfies the boundary condition $z(\psi_o) = \alpha$ is given by the following value function:*

$$z(\psi) = \min\{\inf_{\eta \in \partial\Omega} L(\psi, \eta), \alpha + L(\psi, \psi_o)\}. \tag{12.37}$$

Further, the compatibility condition

$$|\alpha| \leq \inf_{\eta \in \partial\omega} L(\psi_o, \eta) \tag{12.38}$$

holds.
Conversely, if (12.38) holds, the function $z(\psi)$ given by (12.37) is a viscosity solution of (12.35) with $z(\psi_o) = \alpha$.

Thus the viscosity solution of the corresponding Hamilton-Jacobi equation will give the solution to the shape-from-shading problem in this context.

6. Computational Theory of Shape

When we look at a planar shape we usually give it some interpretation, such as a conglomerate of elementary parts or as a basic shape with some structured protuberances. Very early on in computer vision research, people wanted to understand and model mathematically the way we make such "gestalt" decisions, that seem to be crucial in the process of associating meaning to the picture we see. Planar shapes are described in the computer in various ways, using either contour or region description, each of them making different types of information explicit. While all complete descriptions are obviously equivalent we would like to have algorithms to choose the description that is readily suited for the task at hand and algorithms that use the description of choice to come up with the "right" interpretations. Blum [40] argued that a very useful concept in analyzing shape is the notion of the shape skeleton, a planar graph, made of interconnected curve portions retaining some of the characteristics of a shape, and perhaps enabling its decomposition into meaningful parts. The skeleton of a planar shape is formally defined as the set of points, whose minimal distance to the boundary is attained at two or more different points on it. This concept is perhaps best understood via the so-called "prairie fire" model : assume that the shape under consideration is set on fire simultaneously along its boundary and the fire propagates inward. The skeleton will be made of the points the fire quenches, and it can be shown that complete reconstruction of the original shape is possible given the skeleton and the so-called quenching function, the distance of each skeleton point to the outside world. Below we will give the differential equation describing such a prairie fire propagation which as we have already shown is the classical eikonal equation. However, it is easy to see that the skeleton is defined by the shock fronts that this equation generates. Recently an extension of such a shape analysis process was found very useful in providing reasonable interpretations to planar shapes.

Thus the theme of this part of the chapter will be a certain approach for representing shape based on shape analysis via boundary evolution equations. As we will see, such evolution equations again solve certain variational problems, and so fit into the optimal control (calculus of variations) paradigm being discussed here. The approach given here is based on [178, 180, 181, 182]. (See also the nice results of [239].) One of the main ideas in this work, is to consider an explicit treatment of singularities, which is founded on a series of conservation laws. In this way we consolidate the two major approaches to shape–the one based on boundary and the other on interior information–by a simultaneous representation of both. Most importantly, a natural hierarchy of "parts" emerges as shapes evolve. Thus this approach, which is inherently non-linear, stands in contrast to earlier linear (e.g., heat equation) scale-space models, and initiates the direct handling of discontinuous events.

A very interesting fact is that the differential equations which appear in the vision problem have also been used in the modelling of evolving phase boundaries as well as in the curve shortening problem [139, 123, 125]. (More will be said about this below.) The subject has much mathematical depth and is rapidly developing, and so we will only be able to touch on a number of the more relevant aspects to vision here.

6.1. Evolution Equations

We will be considering the families of embedded closed curves $C : S^1 \times [0, T) \to \mathbb{R}^2$ evolving according to functions of the curvature. (Since this discussion is informal, we will be rather loose with the hypotheses we will be putting on the families we consider here. See [16, 182] for the formal mathematical treatments.)

We will be considering families of closed curves $C : S^1 \times [0, t_f) \to \mathbb{R}^2$ evolving according to functions of the curvature. More precisely, the general deformation of a curve in the plane of interest in vision problems may be given by

$$\frac{\partial C}{\partial t} = \alpha(s, t)\mathcal{T} + \hat{\beta}(s, t)\mathcal{N} \tag{12.39}$$

where $\mathcal{N}$ is the (inward) normal, $\mathcal{T}$ is tangent, and $\alpha, \hat{\beta}$ are smooth functions. Notice that since we are only interested in shape we may take $\alpha = 0$. (Changing α only changes the curves parametrization, and not its shape.) Furthermore, as is typical in this area, we constrain the deformations to be determined by the local geometry of the curve, i.e., $\hat{\beta}(s, t) = \beta(\kappa)$ where $\kappa(s, t)$ denotes the Gaussian curvature of the curve $C(s, t)$. Thus we are led to the following equation:

$$\frac{\partial C}{\partial t} = \beta(\kappa)\mathcal{N}. \tag{12.40}$$

In the mathematics literature, a number of cases for the function β have been explored. For example, there has been a great deal of work in connection with the geometric heat equation in which $\beta(\kappa) = \kappa$. In this case, the isoperimetric ratio L^2/A of the curve approaches 4π as the enclosed area approaches 0, see [125, 139]. In the problems of interest to us in computer vision, the function of interest is

$$\beta(\kappa) = a\kappa + 1 \tag{12.41}$$

where $a \geq 0$. The case $a = 0$, i.e., $\beta \equiv 1$ is very important, and here equation (12.40) becomes an equation that has been studied in relation to problems in geometric optics [21], flame propagation [337], and shape morphology [40], as well as shape decomposition. Indeed, this is the differential equation for the prairie fire model described above. The κ part gives a diffusive effect, while the constant part gives a wave (hyperbolic) effect which tends to create singularities and break a shape into its constituent parts. This point will be elaborated on below.

Remark. In [182], we study the very general evolution equation

$$\frac{\partial C}{\partial t} = (\beta_0 \kappa + \beta_1)\mathcal{N}, \tag{12.42}$$

for constants $\beta_0 \geq 0$ and β_1. The resulting scale-space for various ratios of β_0 and β_1, we call the *reaction-diffusion scale-space*. Mathematical results connected to this scale-space may be found in [181].

Let us first discuss some general properties of equation (12.40). We first introduce some standard notation. Set

$$\rho(s,t) := \left\| \frac{\partial C}{\partial s} \right\| = [x_s^2 + y_s^2]^{1/2}, \tag{12.43}$$

denote the length along the curve. The arc-length parameter **s** is then defined as

$$\mathbf{s}(s,t) := \int_0^s \rho(\zeta,t)d\zeta. \tag{12.44}$$

Let the positive orientation of a curve be defined so that the interior is to the left when traversing the curve. The *tangent, curvature, normal, orientation* and *length* are defined in the standard way. We will take the normal to be pointing inwards, where the inward or outward is determined by the interior, or equivalently by the orientation of the curve. We then have that

$$\mathcal{T} := \frac{\partial C}{\partial s} = \frac{1}{\rho}\frac{\partial C}{\partial s}, \tag{12.45}$$

$$\kappa := \frac{1}{\rho}\left\| \frac{\partial \mathcal{T}}{\partial s} \right\| \tag{12.46}$$

$$\mathcal{N} := \frac{1}{\kappa\rho}\frac{\partial \mathcal{T}}{\partial s} \tag{12.47}$$

$$L(t) := \int_0^{2\pi} \rho(s,t)ds. \tag{12.48}$$

Finally, we let

$$\overline{\kappa}(t) := \int_{2\pi}^{0} |\kappa(s,t)| \rho(s,t) ds \qquad (12.49)$$

denote the total absolute curvature.

The behavior of the classical solutions equation (12.40) can be rather thoroughly analyzed and one can prove useful results (from the applied point of view) of the kind [16, 178, 180, 181, 182]. Let $C(s,t)$ be a classical solution of (12.40) for $t \in [0,t')$ and $\kappa\beta(\kappa) \le M$ for all $\kappa \in \mathbb{R}$ (regarding β as a function of κ). Then,

$$L(t) \le L(0)e^{Mt}. \qquad (12.50)$$

In case $\beta(\kappa) = a\kappa + 1$, we get

$$L(t) \le \min(L(0) + 2\pi t, L(0)e^{\frac{t}{4a}}). \qquad (12.51)$$

Let $C(s,t)$ be a classical solution of (12.40) for $t \in [0,t')$. Suppose that $\kappa\beta(\kappa) \le M$, and $\beta_\kappa \le 0$. Then

$$\overline{\kappa}(t) \le \overline{\kappa}(0). \qquad (12.52)$$

Moreover, if $[0,t')$ is an interval on which a classical solution exists, then one may also show that

$$d_H(C_t, C_0) \le \sqrt{(4at)}, \qquad (12.53)$$

where d_H denotes the Hausdorff metric on compact subsets of $\mathbb{R}^2$. The above facts easily imply that

$$\lim_{t \to t'} C_t = C^* \qquad (12.54)$$

in the Hausdorff metric, and the curve C^* regarded as a mapping $C^* \to \mathbb{R}^2$ is Hölder continuous with exponent $1/2$. In vision, a major theme are the weak solutions of equations of the type (12.40) which we will now describe.

6.2. Conservation Laws

In this section, we concentrate on an interesting special case of β, namely the case in which $\beta \equiv 1$. Here in the classical manner we will derive a *hyperbolic conservation law*.

With $\beta \equiv 1$, we can write the equation (12.40) as

$$x_t = \frac{y_s}{(x_s^2 + y_s^2)^{1/2}} \qquad (12.55)$$

$$y_t = -\frac{x_s}{(x_s^2 + y_s^2)^{1/2}}. \tag{12.56}$$

Note we are taking $C(s,t) = C_t(s) = (x(s,t), y(s,t))$, with initial curve $C(s,0) =: C_0(s)$ for $0 \le s \le S$. (Thus $(x(s,t), y(s,t))$ are the position coordinates of the curve $C(s,t)$.)

Now let us see how from equations (12.55) and (12.56) we can derive a hyperbolic conservation law. Indeed, as long as C_t stays smooth and non-self-intersecting, by virtue of the implicit function theorem, we can express the front in the form

$$y = U(t, x). \tag{12.57}$$

(Note that C_t is the graph of (12.57).) One can then verify that U satisfies the Hamilton-Jacobi equation

$$\frac{\partial U}{\partial t} - (1 + U_x^2)^{1/2} = 0. \tag{12.58}$$

Set

$$u := \frac{\partial U}{\partial x}. \tag{12.59}$$

Then differentiating (12.58) with respect to x we see that

$$u_t - \left((1 + u^2)^{1/2}\right)_x = 0. \tag{12.60}$$

Equation (12.59) is in the form of a "hyperbolic conservation law" which has a huge classical and modern literature devoted to it; see [345] and the references therein. We shall not go into these laws in depth here, but would like to give the interested reader some of the physical motivation behind such PDE's. Notice that (12.58) is the classical eikonal equation which we discussed above in Section 3.2.

Explicitly, a *hyperbolic conservation law* is given by the hyperbolic PDE (i.e. a wave-type equation)

$$u_t + F(u)_x = 0. \tag{12.61}$$

The conservation law that (12.61) is expressing mathematically may be formulated as follows. Suppose material is distributed along a line with coordinate x. Suppose moreover the distribution satisfies the physical conservation law that the rate of change of the amount of material in a fixed interval equals the flux of the material through the boundaries. If u then denotes the density, and $F(u)$ the flux, then mathematically this conservation law may be expressed as

$$\frac{d}{dt} \int_x^{x+\Delta x} u \, dx = F(u(x)) - F(u(x + \Delta x)). \tag{12.62}$$

Letting $\Delta x \to 0$ gives (12.61), the required conservation law.

Using this interpretation for the moving front given by (12.55,12.56), we see that it is the slope that is conserved. We should also note that such hyperbolic conservation laws appear in gas dynamics related to the Riemann problem [345].

A very important property of such hyperbolic conservation laws (from the point of view of the vision problem we are studying, and other applications [345]) is the fact that one may get discontinuous solutions for (12.61), even in the presence of smooth initial data. Once such discontinuities or *shocks* develop, one must be careful in defining precisely what one means by "solution" to (12.61). This is precisely where the notion of viscosity solution arises and as we have seen above in Section 3.2, the viscosity solution coincides with the entropy condition imposed by Sethian [337], interpreting the hyperbolic evolution law in the prairie-fire sense that "once a particle is burnt, it remains burnt." (See [345] for an extensive discussion on classical entropy conditions.)

Geometrically, it is very easy to see how discontinuities develop for the system (12.55,12.56). Indeed, assuming that C_t remains smooth it is easy to compute that the Gaussian curvature satisfies the following evolution equation:

$$\kappa_t = -\kappa^2. \tag{12.63}$$

Now we can explicitly solve (12.63) to find that

$$\kappa(s,t) = \frac{\kappa(s,0)}{1 + t\kappa(s,0)}. \tag{12.64}$$

Notice then that if $C = C_0$ the initial curve is anywhere concave, i.e. has Gaussian curvature negative at any point, $\kappa(s,t)$ will blow up in finite time, and the resulting curve will develop corners, i.e., we get a **shock.**

The entropy condition prevents a non-self-intersecting curve from burning into a self-intersecting one. It is very important to note that such entropy (weak) solutions may be obtained as limits of (classical) solutions of the hyperbolic equation (12.61) perturbed by certain viscosity terms [345]. A similar effect ("artificial viscosity") occurs in the discretization of the system. This means that essentially (when everything converges) we pick up such an entropy condition in the digital implementation of the hyperbolic conservation law.

It is precisely the formation of such singularities that allow us to decompose a given figure into its parts. The various possibilities are defined by a classification of the shocks; see [178].

6.3. Diffusion Equation and Image Smoothing

We now would like to discuss the notion of scale-space and the connection to diffusion equations. See the recent thesis [104] for an extensive list of references on this topic.

One of the pillars of image processing is Gaussian smoothing in which a Gaussian operator is convolved against an image given by a gray-scale map $\Phi_0 : \mathbb{R}^2 \to \mathbb{R}$. The semigroup property for the Gaussian implies that a "pyramid" of images can be created, such that each image in the pyramid has less structure than its predecessor. The resulting family of images is called a "scale-space;" see Witkin and Koenderink [386, 187, 190]. In classical computer vision, the (linear) scale-space of images $\Phi(x, y; t)$ obeys the linear heat equation

$$\Phi_t = \Delta\Phi := \Phi_{xx} + \Phi_{yy},$$

where (x, y) are the position coordinates of the image, and t is the scale parameter for the diffusion.

The advantage of image smoothing is noise removal; the problem is that it also removes important structure, e.g., by smoothing across edges. Technically the root of the problem is that images of objects are discontinuous: imagine a dark cube against a light background. The image would consist of several distinct regions, corresponding to the separate (visible) faces of the cube or the background. We seek a smoothing operation that does not cross between these regions, but rather stays within them. This is one of the key motivations for the research of nonlinear scale-spaces driven by nonlinear versions of the linear heat equation pioneered by Perona and Malik [293]. We will now discuss a geometric nonlinear version of the classical heat equation to handle such issues.

6.4. Euclidean Curve Shortening

The geometric heat equation

$$\frac{\partial C}{\partial t} = \kappa \mathcal{N}, \tag{12.65}$$

has a number of properties which make it very useful in image processing, and in particular, the basis of a nonlinear scale-space for shape representation [6, 5, 178, 180, 182, 257].

First of all it is a *Euclidean curve shortening flow*, in the sense that the Euclidean perimeter shrinks as quickly as possible when the curve evolves according to (12.65) [123, 125, 139]. Since the argument for this involves the calculus of variations, let us briefly sketch it.

Let $C = C(\eta, t)$ be a smooth family of closed curves where t parametrizes the family and η the given curve, say $0 \leq \eta \leq 1$. (Note we assume that

$C(0,t) = C(1,t)$ and similarly for the first derivatives.) Consider the length functional

$$L(t) := \int_0^1 \|\frac{\partial C}{\partial \eta}\| d\eta.$$

Then differentiating (taking the "first variation"), and using integration by parts, we see that

$$
\begin{aligned}
L'(t) &= \int_0^1 \frac{\langle \frac{\partial C}{\partial \eta}, \frac{\partial^2 C}{\partial \eta \partial t} \rangle}{\|\frac{\partial C}{\partial \eta}\|} d\eta \\
&= -\int_0^1 \langle \frac{\partial C}{\partial t}, \frac{1}{\|\frac{\partial C}{\partial \eta}\|} \frac{\partial}{\partial \eta} \left[\frac{\frac{\partial C}{\partial \eta}}{\|\frac{\partial C}{\partial \eta}\|} \right] \|\frac{\partial C}{\partial \eta}\| \rangle d\eta.
\end{aligned}
$$

(Note that we multiplied and divided by $\|\frac{\partial C}{\partial \eta}\|$ in the latter integral.) But noticing now that

$$\|\frac{\partial C}{\partial \eta}\| d\eta = ds$$

is arc-length, we get easily form the definition of curvature that the last integral is

$$-\int_0^{L(t)} \langle \frac{\partial C}{\partial t}, \kappa \mathcal{N} \rangle ds$$

that is, we see

$$L'(t) = -\int_0^{L(t)} \langle \frac{\partial C}{\partial t}, \kappa \mathcal{N} \rangle ds.$$

Thus the direction in which $L(t)$ is decreasing most rapidly is when

$$\frac{\partial C}{\partial t} = \kappa \mathcal{N}.$$

A much deeper fact is that closed curves converge to "round" points when evolving according to (12.65); see [125, 139].

6.5. Connection with Classical Heat Equation

We will see in Section 8 that the evolution of the curve according to (12.65) may be recovered as the level set of a surface evolving according to

$$\Phi_t = \kappa \|\nabla \Phi\|, \qquad (12.66)$$

where

$$\kappa = \frac{\Phi_{yy} \Phi_x^2 - 2\Phi_x \Phi_y \Phi_{xy} + \Phi_{xx} \Phi_y^2}{(\Phi_x^2 + \Phi_y^2)^{3/2}}.$$

(This is in fact the curvature of the curve C regarded as the level set of the corresponding evolution; see Section 8 and [6, 5, 9, 285].)

Now as above, the classical heat equation in the plane is given by

$$\Phi_t = \Delta\Phi,$$

where Δ denotes the Laplacian operator. As we have argued, for reasons of smoothing and edge detection, we want a diffusion which is only made in the direction of the edge. This would keep the location of the edge, while smoothing on the sides of the edge.

Thus we want a diffusion equation which does not diffuse in the direction of the gradient $\nabla\Phi$. This can be easily computed as (see [9])

$$\Phi_t = \nabla\Phi - \frac{1}{\|\nabla\Phi\|^2}\langle\nabla^2\Phi(\nabla\Phi),\nabla\Phi\rangle.$$

($\nabla^2\Phi$ denotes the Hessian of ∇.) This equation may be written as

$$\Phi_t = \|\nabla\Phi\|\mathrm{div}\frac{\nabla\Phi}{\|\nabla\Phi\|},$$

which is exactly the heat equation (12.66).

Remark. The preceding theory of curve evolution is only Euclidean invariant. Very recently, an affine invariant approach has been developed in [6, 328, 323, 327]. Affine invariants are very important in computer vision, e.g., in problems involving object recognition. It turns out the the unique affine invariant evolution is

$$\frac{\partial C}{\partial t} = \kappa^{1/3}\mathcal{N}. \tag{12.67}$$

This evolution has a number of remarkable properties which we only state here (proofs may be found in [328, 323, 327]. First of all, the evolution (12.67) shrinks the affine perimeter as quickly as possible, and converges to an ellipse. Moreover, image processing smoothing algorithms turn out to be more stable when implemented according to (12.67) rather than (12.65).

7. Mathematical Morphology

In this section, we demonstrate how the standard algebraic view of mathematical morphology is complemented with a geometric, differential view, which is captured by a Hamilton-Jacobi equation. The basic idea is to give a continuous implementation of morphology following [44, 5, 320, 20]. Mathematical morphology is an image analysis methodology which was introduced in the late sixties and is taking on increasing importance; see [250, 336, 148] and the references therein. It addresses applications in various domains: medicine, biology, MR imagery, radar and remote sensing images, quality control, science of materials, fingerprints, and many others. Its approach to image analysis problems is algebraic in that binary images are considered as sets which are then transformed– the spatial domain— via *morphological transformations,* whose definitions are usually based on *structuring elements,* i.e., particular shapes that are translated in the images and used as probes. In one of the simplest possible cases, dilating a planar shape S with respect to a disk of given radius r, the transformed shape is the union of all disks whose center is in S. Alternatively, the boundary of the transformed shape is obtained by moving each point along the normal by a constant r, by Huygens' principle. The question posed in this section is whether for a class of structuring elements, e.g., an elliptical structuring element, algebraic unions of sets can be translated to geometric deformations, thereby generalizing the Huygens' principle.

Three observations underlie a geometric PDE view of mathematical morphology: 1) certain structuring elements (convex) are scalable in that a sequence of repeated operations is equivalent to a single operation, but with a larger structuring element of the same shape; 2) To determine the outcome of the operation, it is sufficient to consider how the boundary is modified; 3) The modifications of the boundary are such that we can move each point along the normal by certain amount, which is dependent on the structuring element. Taken together, these observations (when the size of the structuring element shrinks to zero) assert that mathematical morphology operations with a convex structuring element are captured by a differential deformation of the boundary along the normal. We now expand on each of these observations.

First, for a class of structuring elements $\hat{\mathcal{B}}$, n repeated dilations (or erosions) with a structuring element $\mathcal{B} \in \hat{\mathcal{B}}$, is equivalent to a single dilation with a structuring element of the same shape, but which is "scaled up" n times. Let us denote this scale by λ and the scaled structuring element as $\mathcal{B}(\lambda)$. For example, 10 dilations with a circle of radius of 1, $\mathcal{B}(1)$, is exactly a single dilation with a circle of radius 10, $\mathcal{B}(10)$. The class of structuring elements $\hat{\mathcal{B}}$ for which this property holds is exactly the set of *convex* structuring elements [250], Theorem 1.5.1, pp. 22–23. Now, conversely, given a single operation with a convex structuring element, the operation can be decomposed as a sequence of n operations with the same structuring element but of "size" $1/n^{\text{th}}$ the original. In the limit, as $n \to \infty$, this notion of size constitutes the "time" axis for the differential deformation.

Second, the outcome of mathematical morphology operations with a shape S, can be determined by knowledge of its boundary, ∂S. Thus, the mapping from the shape S to S' by morphological operations can be reduced to operations taking ∂S to $\partial S'$. As such, appropriate geometric evolution of the boundary can model algebraic mathematical morphology operations on shapes as sets.

Third, for a sufficiently small but finite structuring element, mathematical morphology operations along the smooth portions of the boundary can be viewed as a deformation along the normal by a certain amount. Formally, let $C_0(s) = (x_0, y_0)$ denote the initial curve, and $C(s, t)$ denote the curve evolving by a mathematical morphology operation, say dilation, by some structuring element $B(t)$ of size t. Then the evolution of the boundary along the normal is given by

$$C(\sigma, t + \lambda) - C(\sigma, t) = \Gamma(\sigma, t, \lambda)\mathcal{N}, \qquad (12.68)$$

where $\mathcal{N}$ is the normal to the boundary of the shape, σ is the length parameter along the boundary, and Γ is the distance along $\mathcal{N}$ that a point on the boundary would move in a dilation by a structuring element of size λ. Note that $\Gamma(s, t, 0) = 0$. Then, in the limiting case as $\lambda \to \infty$,

$$\frac{\partial C}{\partial t} = \lim_{\lambda \to 0} \frac{C(s, t + \lambda) - C(s, t)}{\lambda} \qquad (12.69)$$

$$= \lim_{\lambda \to 0} \frac{\Gamma(s, t, \lambda)}{\lambda - 0} \mathcal{N} \qquad (12.70)$$

$$= \lim_{\lambda \to 0} \frac{\Gamma(s, t, \lambda) - \Gamma(s, t, 0)}{\lambda} \mathcal{N} \qquad (12.71)$$

$$= \frac{\partial \Gamma(s, t, 0)}{\partial \lambda} \mathcal{N}. \qquad (12.72)$$

Therefore, the mathematical morphology operation of dilation with a structuring element of size t is governed by the partial differential equation

$$\begin{cases} \frac{\partial C}{\partial t} = \beta(\sigma, t)\mathcal{N} \\ C(\sigma, 0) = C_0(\sigma), \end{cases} \qquad (12.73)$$

where $\beta(s, t) = \frac{\partial \Gamma(s, t, 0)}{\partial \lambda}$, and $C_0(\sigma)$ is the boundary of the original shape. Note that at singular points, the deformation is handled by notions of shock, rarefaction wave, and entropy in the context of the "shape from deformation" framework, as introduced in [178, 182, 179, 181].

Having shown that mathematical morphology operations can be modeled as geometric evolution governed by a partial differential equation as in equation (12.73), we now examine the dependence of β on the shape of the structuring element B and the original shape S.

Theorem 2 *The amount of differential deformation, β, of a shape S at a point P due to dilation (erosion), is the maximal (minimal) projection of the structuring element B onto the normal $\mathcal{N}$ of the boundary at P.*

Corollary 1 *The differential deformation, β, at a point P of the boundary, does not depend on the shape of the boundary at point P.*

Direct simulations of equation (12.73) are not robust. Observe that shapes undergoing morphological operations change their topology in that they merge and split, leading to problems in administrating new formed boundaries and old annihilated ones. In addition, singularities (or shocks) form often, as shapes evolve, and boundary points in the vicinity of the singularity bunch-up and merge, leading to numerical errors. Typically, a reparametrization is required, but at expense of adding smoothing to the evolution. Singularities, however, are the salient features of shape and must be handled and represented explicitly. Furthermore, numerical simulations of boundaries are prone to a buildup of error. One way to contain these errors is to represent a curve as the level set of a surface. Note that the evolving surface does handle topological events and, since the grid is fixed, no resampling is needed. There are more subtle arguments for re-posing the curve evolution problem as a surface evolution problem. (See also Section 8 below.)

Let $z = \phi_0(x, y)$ denote the initial surface, which is obtained from the initial curve by placing a "tent" on it, e.g. its distance transform. Let $z = \phi(x, y, t)$ denote the evolved surface. Now if the surface evolves by

$$\left\{ \begin{array}{l} \phi_t + \beta(\phi_x^2 + \phi_y^2)^{1/2} = 0 \\ \phi(x, y, 0) = \phi_0(x, y), \end{array} \right. \tag{12.74}$$

then the curve evolves by equation (12.73). In the case of evolution designed to implement mathematical morphology operations, as we will show in Theorem 2, for any convex structuring element, the amount of deformation is only dependent on the orientation at the point of deformation, $\beta = \beta(\theta)$. In addition, note that

$$\left\{ \begin{array}{l} \cos\theta = -\dfrac{\phi_y}{\sqrt{\phi_x^2 + \phi_y^2}} \\ \sin\theta = \dfrac{\phi_x}{\sqrt{\phi_x^2 + \phi_y^2}}, \end{array} \right. \tag{12.75}$$

so that θ depends on (ϕ_x, ϕ_y), namely, only the first derivatives of ϕ. Expressed differently, equation (12.74) can be rewritten as

$$\left\{ \begin{array}{l} \phi_t + \mathcal{H}(\phi_x, \phi_y) = 0 \\ \phi(x, y, 0) = \phi_0(x, y), \end{array} \right. \tag{12.76}$$

where $\mathcal{H}(\phi_x, \phi_y) = \beta(\phi_x, \phi_y)(\phi_x^2 + \phi_y^2)^{1/2}$ is the Hamiltonian associated with this Hamilton-Jacobi equation.

8. Numerical Implementations

We now briefly discuss the powerful numerical algorithms developed by Osher and Sethian for curve evolution. Their work in turn is based on writing the approximations in *conservation form* and applying the *Godunov method* [208]. For more details, see [285, 339, 340].
Let $C(p,t) : S^1 \times [0, \tau) \to \mathbb{R}^2$ be a family of curves satisfying the following evolution equation:

$$\frac{\partial C}{\partial t} = \beta(\kappa)\mathcal{N}. \tag{12.77}$$

There are a number of problems which must be solved when implementing curve evolution equations such as (12.77) on computer. For example, singularities may develop. We have already seen for $\beta \equiv 1$ in (12.77), even a smooth initial curve can develop singularities. The question is how to continue the evolution after the singularities appear. As we previously discussed, a natural way is to choose the solution which agrees with the Huygens principle [338], or as Sethian observed if the front is viewed as a burning flame, this solution is based on the principle that *once a particle is burnt, it stays burnt* [339]. The importance of this solution in shape analysis was previously pointed out and analyzed in [178, 180, 181]. We have already indicated above that, from all the *weak* solutions corresponding to (12.77), the one derived from the Huygens principle is *unique*, and can be obtained via the entropy condition constraint.
In any numerical algorithm, we have the key requirements of accuracy and stability. The numerical algorithm must approximate the evolution equation, and it must be robust. Sethian [339] showed that a simple, Lagrangian, difference approximation, requires an impractically small time step in order to achieve stability. The basic problem with Lagrangian formulations is that the marker particles on the evolving curve come very close during the evolution.
The algorithm proposed by Osher and Sethian [285, 338, 339, 340] provides a reliable numerical solution for curve (and hypersurface) evolution. It is based on the Hamilton-Jacobi theory that we discussed above, and thus fits the theme of this chapter. Indeed, first the curve is embedded in a two dimensional surface, and then the equations of motion are solved using a combination of straightforward discretization, and numerical techniques derived from hyperbolic conservation laws [347].
The embedding step is done in the following manner: The curve $C(p,t)$ is represented by the zero level set of a smooth and Lipschitz continuous

function $\Phi : \mathbb{R}^2 \times [0, \tau) \to \mathbb{R}$. Assume that Φ is negative in the interior and positive in the exterior of the zero level set. We consider the zero level set, defined by

$$\{X(t) \in \mathbb{R}^2 : \Phi(X, t) = 0\}. \qquad (12.78)$$

We have to find an evolution equation of Φ, such that the evolving curve $C(t)$ is given by the evolving zero level $X(t)$, i.e.,

$$C(t) \equiv X(t). \qquad (12.79)$$

By differentiating (12.78) with respect to t we obtain:

$$\nabla\Phi(X, t) \cdot X_t + \Phi_t(X, t) = 0. \qquad (12.80)$$

Note that for the zero level, the following relation holds:

$$\frac{\nabla\Phi}{\|\nabla\Phi\|} = -\mathcal{N}. \qquad (12.81)$$

In this equation, the left side uses terms of the surface Φ, while the right side is related to the curve C. The combination of equations (12.77) to (12.81) gives

$$\Phi_t = \beta(\kappa) \|\nabla\Phi\|, \qquad (12.82)$$

and the curve C, evolving according to (12.77), is obtained by the zero level set of the function Φ, which evolves according to (12.82). Sethian [339] called this scheme an *Eulerian formulation* for front propagation, because it is written in terms of a fixed coordinate system.

The second step of the algorithm consists of the discretization of the equation (12.82). If singularities cannot develop during the evolution, as in the geometric heat equation flow (12.66), a straightforward discretization can be performed [285]. If singularities can develop, as in the case of $\beta = 1$, a special discretization must be implemented. In this case, the implementation of the evolution of Φ is based on a *monotone* and *conservative* numerical algorithm, derived from the theory of hyperbolic conservation laws [208, 285, 347]. For a large class of functions β of this type, this numerical scheme automatically obeys the entropy condition, i.e., the condition derived from the Huygens principle [340, 347]. For velocity functions such as $\beta = a\kappa + 1$, a combination of both methods is used [285]. It is important to note that the discretization of the evolution equations is performed on a fixed *rectangular grid* [339]. This rectangular grid can be associated with the *pixel* grid of digital images, making this discretization method natural for image processing. Since the evolving curve is given by the level set of the function Φ, we have to find this level set (Φ is discrete now). This is done using a very simple contour finding algorithm described in [340].

We describe now the numerical algorithm just for the equation

$$\frac{\partial C}{\partial t} = (a\kappa + 1)\mathcal{N}. \tag{12.83}$$

If the curve C is a level set of a function Φ, then its curvature κ can be computed as:

$$\kappa = \frac{\Phi_{yy}\Phi_x^2 - 2\Phi_x\Phi_y\Phi_{xy} + \Phi_{xx}\Phi_y^2}{(\Phi_x^2 + \Phi_y^2)^{3/2}}.$$

Since

$$\| \nabla\Phi \| = (\Phi_x^2 + \Phi_y^2)^{1/2},$$

and in our case

$$\beta = (a\kappa + 1),$$

we obtain from equation (12.82)

$$\Phi_t = a\kappa\|\nabla\Phi\| + \|\nabla\Phi\|. \tag{12.84}$$

8.1. Conservation Form

We give some details now about the numerical implementations of equations of the form (12.84). These are based on the generalizations of numerical schemes for a single conservation law. Thus consider the eikonal equation once again,

$$\phi_t = H(\nabla_x\phi), \tag{12.85}$$

where as above,

$$H(p) := \sqrt{1 + p^2}.$$

The idea is to use methods from conservation laws applied to this case. Explicitly, we differentiate (12.85) with respect to x, and derive

$$\psi_t = [H(\psi)]_x. \tag{12.86}$$

There turns out to be a natural requirement that we can impose on an approximation scheme to an equation of the form (12.86) which guarantees that we do not converge to "non-solutions,' that is functions which are not weak solutions of the original equation or weak solutions which do not satisfy the entropy condition.

We say that an algorithm for the approximation of (12.86) is in *conservation form* if it can be written as

$$\psi_j^{n+1} = \psi_j^n - \frac{\Delta t}{\Delta x}[g(u_{j-p}^n, u_{j-p+1}^n, \ldots, u_{j+q}^n) - g(u_{j-p-1}^n, u_{j-p}^n, \ldots, u_{j+q-1}^n)],$$

(12.87)

for some *numerical flux function* g of $p + q + 1$ arguments. The function g must be Lipschitz and satisfy the (consistency) condition that $g(u, \ldots, u) = -H(u)$. Further, we say that the approximation procedure (12.87) is *monotone* if the right hand side of (12.87) is a non-decreasing function of its arguments. Then one can prove that a scheme which can be written in monotone conservation form will automatically enforce the entropy condition. (See e.g, [208].)

There are several first order monotone schemes for such equations, the two most famous being the Lax-Friedrichs and the Godunov method [208]. The Godunov method is *upwind* (the discretization points in the upwind or upstream direction [350]), the correct direction from which the characteristic information propagates. This turns out to be a crucial property for boundaries [285].

In [285], a new upwind scheme is proposed for Hamiltonian functions which satisfy

$$H(p) = h(p^2), \quad h'(p) < 0.$$

In this technique, we use the flux function

$$g_{HJ}(\psi_j^n, \psi_{j+1}^n) := h((\min(\psi_j^n, 0))^2 + (\max(\psi_{j+1}^n, 0))^2).$$

This numerical scheme has the advantage of generalizing to several dimensions. If we now return to the one dimensional eikonal equation

$$\phi_t = \sqrt{1 + \phi_x^2},$$

the entropy weak solution can therefore be approximated by

$$\phi_j^{n+1} = \phi_j^n - \Delta t g_{HJ}(D_-\phi_j^n, D_+\phi_j^n),$$

where

$$D_-\phi_j^n := (\phi_j^n - \phi_{j-1}^n)/\Delta x, \quad D_+\phi_j^n := (\phi_{j+1}^n - \phi_j^n)/\Delta x,$$

(just integrate (12.87) with respect to x for the flux function g_{HJ}).
Higher order schemes for the numerical implementation of equations such as (12.82) are described in [285] which we leave to the interested reader.

9. Conclusions

The calculus of variations and the resulting Euler-Lagrange and Hamilton-Jacobi equations have proven to be a powerful methodology for numerous feedback control problems. In this chapter, we have tried to indicate through a number of illustrative problems, how these techniques may be also very useful in image processing and computer vision. An important direction we believe in the future will the the consideration of variational symmetries in these laws which should have a very strong connection to the differential invariant theory going on in the vision community as well.

Acknowledgement

This work was supported in part by grants from the National Science Foundation DMS-8811084, ECS-9122106, and IRI 9225139, by the Air Force Office of Scientific Research AFOSR-90-0024, F49620-93-1-0344, and F49620-94-1-00S8DEF, by the Army Research Office DAAL03-91-G-0019 and DAAH04-93-G-0332, by the Natural Sciences and Engineering Research Council of Canada, and by the SUN Microsystems Foundation. S.W.Z. was supported by Fellowships from the SERC, Churchill College, Cambridge, and the Canadian Institute for Advanced Research. The authors would like to thank Professor Ofer Zeitouni of the Technion for a judicious reading of the manuscript.

NONLINEAR SCALE-SPACE

Luc M. J. Florack

Alfons H. Salden

Bart M. ter Haar Romeny

Jan J. Koenderink

Max A. Viergever

Utrecht University, 3D Computer Vision Research Group
Heidelberglaan 100, 3584 CX Utrecht, the Netherlands

Abstract. We propose a generalisation of scale-space theory for scalar images. The starting point is the assumption that the conventional Gaussian model constitutes an *unbiased* (that is, task independent), multiscale image representation. A generalised scale-space is then considered as a conventional scale-space "in disguise", representing the data in a format that is more convenient for *specific* applications. Although formally equivalent to the conventional representation (at least locally), a generalised representation may be more apt for a dedicated task. In particular, it may potentially solve the so-called "localisation problem" of linear scale-space. Several models based on nonlinear diffusion have emerged by the desire to deal with this problem. The proposed theory provides a unifying framework for a variety of such models that can be related to conventional scale-space in a one-to-one way.

Our defining constraint for a generalised scale-space is the requirement of *equivalence*: it should formally correspond to a transformation of linear scale-space. The key idea is a *metric transform* that preserves the intrinsic properties of the spatial domain. This allows one to regard a generalised scale-space as a strategy for *reading out* a single data representation. The equivalence constraint is shown to yield a particular class of (linear or nonlinear) diffusion equations. Conventional, linear scale-space is a convenient representative of the equivalence class for "universal" purposes. The emphasis is on *nonlinear scale-spaces*, although the principle of equivalence can be used within the linear context as well. Examples are included to illustrate the theory both in the linear as well as in the nonlinear sector.

1. Introduction

A description of image structure requires a decision on the level(s) of *inner scale* or *resolution* on which the structure of interest manifests itself. If there is no a priori preference for a given scale, one must therefore treat all scales in a similar way. For a general, "front-end" system this is a natural way of handling input data. This plain observation has led to the construct of a *scale-space* [386, 187, 24, 213, 109, 115]. This is a 1-parameter family of images representing a given input image on a continuum of scales.

The defining equation that governs the behaviour of the input image upon "blurring" (i.e. increasing scale or decreasing resolution), is the *isotropic diffusion equation*. The evolution parameter in this parabolic equation is related to the inner scale, and the input image serves as its initial condition. This clearly reveals the process of scaling in the above sense as a *continuous semigroup operation*. On the basis of this scale-space concept one can formulate any local image property in terms of *differential invariants*, leaving scale as an intrinsically free parameter [109, 358, 110, 111].

Whereas the isotropic diffusion equation concisely captures the notion of isotropic scale, various nonlinear diffusion equations have been proposed in the literature to solve what has been called the "localisation problem" of linear scale-space. The problem is due to the fact that, in a linear representation, the parametrisation of resolution is independent of image structure. Indeed, for any *particular* application this may not be convenient, but it makes perfect sense for a *general* representation of image data (cf. the role of the "sensorium" in [196]).

Nevertheless, in a particular application one may want to consider local neighborhoods with a spatially varying emphasis on certain preferred scales, depending on whether there is something "interesting" or not. The key word is of course "interesting"; it expresses an *a priori bias* of a data-interpreting routine. Since the bias goes with the application, it makes sense to express it as a task dependent *mapping* of a single representation of the *de facto* data. If we are willing to consider linear scale-space as a standard representation of input data common to a variety of different routines, each of which is designed to support a particular application, then the question arises of how to reformat this into an "intermediate" representation that provides a more convenient input for each such application.

For example, one application of interest is image segmentation. In order to facilitate a segmentation task, one may want to resolve regions with pronounced gradients with a higher resolution (and hence with an "improved localisation") than flat regions. The plain observation that large gradients are more likely to occur at boundaries of meaningful segments is an instance of an a priori bias that can be built into the representation. This is the idea behind nonlinear diffusion, proposed in the literature, showing promising results with regard to enhancement and suppression of certain (primarily gradient-based) structures [67, 142, 293, 273, 258, 341, 381, 380, 316, 57, 9, 5, 286].

However, we feel that there are some unsolved problems concerning previous work on nonlinear diffusion. Firstly we argue that the incorporation of nonlinearities in a *primary* stage is undesirable for a general purpose vision system, since it obscures modularity. We would like to consider conventional scale-space as a basic representation of the initial data, waiting to be read out (but not written to!) by a gamut of high-level routines. Each such routine may be dedicated to solve a particular task, and may be preoccupied with a particular interpretation of the "evidence". In this picture, we regard a generalised scale-space as an intermediate representation, locally equivalent to linear scale-space, expressing a particular, biased way in which some dedicated routine reads out a common set of input data. The bias is formulated as a specific *metric transform*, one that can effectively be realized by means of a spatial transformation on a *flat* image domain.

We will emphasise nonlinear generalisations. By construction, it will be seen that a nonlinear scale-space can be described by a nonlinear diffusion equation. But note that the above interpretation of a nonlinear scale-space is quite different from contemporary methods aimed at solving a nonlinear initial value problem, using the high-resolution data as the initial condition. Moreover, nonlinear diffusion equations are often designed to converge towards stationary solutions, and the evolution parameter often enters as a mere iteration dummy, instead of a free scale parameter. Finally, the approach based on nonlinear diffusion equations, although aesthetically attractive for its concise way of representing a defining task by means of a single equation, obscures the fundamental role of *scale*. We would like to preserve the multiscale character of the representation (because the interpretation of images *does* generally require a multiscale approach [195]), and therefore consider nonlinear diffusion equations in the context of nonlinear scale-spaces. From these considerations it is clear that comparisons with existing models of nonlinear diffusion are to be made with some care. The chapter is organised as follows. Section 2 contains the main theory on nonlinear scale-space. In section 2.1 we outline the general approach for constructing nonlinear scale-spaces. In section 2.2 we establish the notations and conventions used in the remainder of this chapter. Section 2.3 is meant to review conventional, linear scale-space theory, while section 2.4 is devoted to the actual generalisation. The examples in section 3 illustrate the relevance of our work in the context of conventional scale-space theory and previous work on nonlinear diffusion. We end the chapter by a discussion and some concluding remarks in section 4. An appendix is added to explain the covariant formalism used in this chapter.

2. Nonlinear Scale-Spaces

2.1. General Approach

In order to define a nonlinear scale-space, we proceed as follows. Our basic requirement is that a nonlinear scale-space should be a manifestation of linear scale-space "in disguise". To be more precise, we require the existence of a (at least locally defined) spatial transformation and scale reparametrisation that maps one into the other. Such a transformation may depend on the image data, and expresses an a priori bias towards certain greyvalue structures (e.g. "edges"), or explicit spatial locations (as in a foveal system). Contrary to a mere coordinate reparametrisation, such a transformation has an operational significance; it affects the way the linear data are resampled when read out by a nonlinear routine (say, one sample per volume element *after* the transformation). Although it formally looks the same as a coordinate transformation when described in any *given* coordinate system, it is intended to describe a genuine image deformation.

The abovementioned image transformations are to be distinguished from coordinate reparametrisations, which serve no other purpose than to switch from one *descriptive* coordinate system to another. As soon as one introduces dummy "overlay" coordinates, one has to face the problem of how to make sure that essential results do not depend on the choice of these coordinates. There are at least two common ways to achieve this. One is to agree on an unambiguous coordinate convention[1], or *gauge condition*, the other is to use the *covariant formalism*, which guarantees that relations, expressed in any given coordinate system, are form invariant under general coordinate transformations. We will henceforth use the covariant formalism, leaving it to the user to decide on the coordinates (if necessary). When using the covariant formalism, coordinate invariance is a trivial consequence of the formal notation (*manifest invariance*).

In   order   to distinguish operationally significant transformations relating different scale-space representations from physically meaningless transforma- tions that merely relate different coordinate parametrisations of a given scale-space the former will be referred to as *active transformations* and the latter as *passive transformations*. Invariance under passive transformations will thus

---

[1]Note that in the literature one frequently encounters a restriction to Cartesian coordinate frames. In that case one is left with a residual gauge degree of freedom: the group of Cartesian coordinate transformations. Therefore one essentially ends up with the same problem, be it of a simpler form: either one needs an additional convention to single out an unambiguous frame, or one has to resort to a covariant formalism that treats all admissible systems on equal footing.

be a trivial consequence of the covariant formalism. For a discussion of active and passive viewpoint, the reader is referred to [374, appendix C.1], [301, section 11.3] and [275, 276]. See also Figure 13.1 and Example 1 below.

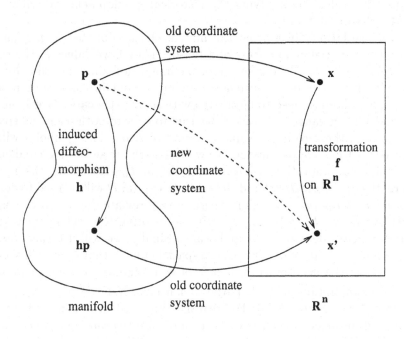

Active view: **f** induces **h**

Passive view: **f** induces change from old to new coordinate system

Figure 13.1. The active and passive view of transformation (adapted from [275]).

2.2. Notation and Conventions

We rely on the reader's familiarity with the covariant formalism, and some basic notions such as Riemannian metric spaces, covariant differentiation, etc. In order to make this chapter self-contained, we have added an appendix in which we explain the basic concepts underlying a coordinate independent description of image structure; see appendix 1. The reader may find it helpful to consult a more detailed explanation in e.g. [212, 348, 254, 206, 209, 332]. We will introduce *tensors* by their "classical" notation. For mathematically inclined readers familiar with modern terminology this may appear somewhat old-fashioned. However, it does have the advantage that we do not have to raise the abstraction beyond an elementary level. Moreover, the classical representation of tensors, being based on coordinates, lends itself excellently for computer implementations. In order to appreciate contemporary views, the reader may find it instructive to learn the "translation rules" by studying the historical development, described e.g. in [348, chapter 4, vol. I].

We will use latin indices from the middle of the alphabet (i, j, k, l, m, n) to label the spatial components of a tensor. They have values in the range $1, \ldots, d$, if d denotes the dimension of the image domain. Any value for d is allowed; all we require is *isotropy* of the image domain (so spatiotemporal images, for example, need to be treated with some extra care). Greek indices (μ, ν) within the range $0, \ldots, d$ will be used to refer to both scale and spatial components, the zero index value referring to scale. The symbol s will be reserved for the *scale* parameter. We will furthermore agree on a condensed, self-explanatory notation using ∂_i and ∂_s (or ∂_0) to denote the partial differential operators $\partial/\partial x^i$ and $\partial/\partial s$, etc. A capital D will be used to denote a *covariant derivative*. We will furthermore abbreviate the n-th order image derivative $D_{i_1} \ldots D_{i_n} \phi$ by $\phi_{i_1 \ldots i_n}$ (with ϕ denoting the scalar image value at an implicitly defined point and scale). Finally, we will use the *Einstein summation convention* with respect to spatial indices, that is, we will silently assume that the occurrence of a pair of equal indices in a given product implies a summation over all spatial dimensions: $X_i Y^i \overset{\text{def}}{=} \sum_{i=1}^d X_i Y^i$ (this is also called a *contraction*). For reasons explained in appendix 1 we will distinguish between *covariant* or *lower* and *contravariant* or *upper* indices. The Einstein convention applies to pairs consisting of one lower and one upper index.

2.3. Covariant Description of Linear Scale-Space

Consider the linear, isotropic diffusion equation

$$\Delta\phi = \partial_s\phi . \tag{13.1}$$

This equation can be regarded as the generating equation of a *linear scale-space* $\phi : \mathbb{R}^d \times \mathbb{R}^+ \to \mathbb{R}$ associated with a given image $\psi : \mathbb{R}^d \to \mathbb{R}$, which serves as the initial condition $\phi(\mathbf{x}; s_0)$ to (13.1). This initial condition need not be a function in the conventional sense: *distributions* are allowed as well. In fact it is better to think of an image as a distribution rather than a conventional function. After all, any linear operation performed on an image amounts to a mapping of some "test function" (that is, a filter) to a real or complex number. Because such an action invariably involves both raw image data as well as a correlating test function, it is appropriate to model an image as a *regular tempered distribution* [191, 333, 114], at least if the test functions are sufficiently smooth. We will not bother about boundary conditions; instead, we will silently assume that the spatial image domain is all of $\mathbb{R}^d$ (this assumption is not important for the present discussion). Of fundamental interest are all derivatives of the Dirac distribution $\delta(\mathbf{x})$, that is the *conventional differential operators*. When inserted as the initial condition to the diffusion equation these infinitesimal operators are "blown up" so as to yield a complete family of *scaled differential operators* [198, 109, 110, 115], appropriate for differentiation (in the exact sense) at any given spatial scale. In particular, diffusion of the "zero-scale" Dirac distribution $\delta(\mathbf{x})$ yields the normalised Gaussian as the basic (zeroth order) scale-space kernel.

The goal of this section is to find a parametrisation of (13.1) which holds in any spatial coordinate system. At this point we will not reparametrise s, since we want to retain a clear notion of *scale* throughout. As a proper measure of scale we may take Gaussian width σ, related to s according to $2s = \sigma^2$ (possibly after a prior shift of s).

In the covariant formalism the formulation of scale-space invariant under arbitrary reparametrisations of the independent, spatial variables requires the introduction of a *metric affinity* on the basis of which we can define a covariant derivative. This allows us to rewrite (13.1) in covariant form, simply by replacing ordinary derivatives by covariant derivatives (see appendix 1):

$$D_i D^i \phi = \partial_s \phi \,, \tag{13.2}$$

or, by writing out the affinity terms, using η_{ij} to denote the covariant components of the metric tensor:

$$\frac{1}{\sqrt{\eta}} \partial_i \left\{ \eta^{ij} \sqrt{\eta} \, \partial_j \phi \right\} = \partial_s \phi \,, \tag{13.3}$$

(we have used the fact that $\Gamma_{ij}^j = \left\{ \begin{matrix} j \\ i\,j \end{matrix} \right\} = \partial_i \ln \sqrt{\eta}$, with $\eta = \det \eta_{ij}$).

Note that (13.2) expresses a *conservation law*: if we introduce the $(d+1)$-dimensional "current" J^μ ($\mu = 0, \dots, d$) by

$$J^\mu = (J^0; J^i) = (\phi; -D^i\phi) \,, \tag{13.4}$$

then we have (identifying D_0 with ∂_s)

$$D_\mu J^\mu = 0 \,. \tag{13.5}$$

The conserved scalar quantity is the overall (or, equivalently, the average) intensity; this follows by a spatial integration of (13.5) over some enclosing spatial region $\Omega \subset \mathbb{R}^d$, bounded by $\partial\Omega$ (with outward oriented surface element $d\Omega_i$, say):

$$\int_\Omega dx \sqrt{\eta(\mathbf{x})} \, D_\mu J^\mu(\mathbf{x}; s) = 0 \quad \Leftrightarrow$$

$$\frac{d}{ds} \int_\Omega dx \sqrt{\eta(\mathbf{x})} \, \phi(\mathbf{x}; s) = - \oint_{\partial\Omega} d\Omega_i \sqrt{\eta(\mathbf{x})} \, J^i(\mathbf{x}; s) . \qquad (13.6)$$

Equation (13.6) states that the increase of overall (or average) intensity within Ω equals the inflow through its bounding surface $\partial\Omega$. In the (hypothetical) case that the boundary encloses the support region of the image, the overall (or average) intensity within the region will remain constant throughout the scaling process. Note that $\sqrt{\eta(\mathbf{x})} \, dx$ and $\sqrt{\eta(\mathbf{x})} \, d\Omega_i$ are covariant measures.

By its very construction, equation (13.2) describes exactly the same diffusion process as equation (13.1). All that we have gained is its explicit coordinate representation for arbitrary coordinate systems. As such, it is merely a matter of academic interest: introducing general coordinates to parametrise the Laplacian operator in a covariant way (thus making the dummy role of the coordinates explicit). But if we forget about its construction, and in particular, about the assumptions on its underlying, flat metric, then we do actually have a generalisation beyond aesthetics. There are at least two ways of generalising the metric:

- We may consider an *arbitrary Riemannian metric* g_{ij} instead of the *flat* metric η_{ij} (cf. definition 10 in appendix 1). This means that we allow for an intrinsically curved image domain (nonzero Riemann tensor).
- We may allow for an *image induced metric* $g_{ij}[\phi]$. In that case the covariant metric tensor becomes a functional of the image ϕ. This reflects itself in an image dependent metric affinity, which in turn introduces a nonlinearity in (13.2) through the image induced covariant Laplacian.

Both ways of extending the domain of validity of the basic diffusion equation essentially boil down to a nontrivial substitution for the Riemannian metric $\eta_{ij} \rightarrow g_{ij}$.

In the first case, the nontriviality resides in the fact that there generally do not exist coordinates in which expression (13.2) reduces to its familiar Cartesian representation (cf. (13.55) in appendix 1). Put differently, if the image domain is flat, then one can always make the affinity Γ^k_{ij} vanish *globally* upon a suitable coordinate transformation. If, on the other hand, the image domain is intrinsically curved, then although one can transform away a nonzero affinity in any given point, one cannot make its derivatives vanish simultaneously[2]. We will not pursue this idea any further here.

---

[2]This is much like removing the sensation of gravity (that is Γ^k_{ij}) locally, by making a free fall: this does not necessarily remove an overall gravitational field.

We will consider the second case. In a linear scale-space representation, the absence of a priori knowledge of image structure is implicit in taking an image independent metric. The a priori equivalence of local neighborhoods is implicit in an identically vanishing Riemann tensor[3] (its Cartesian components vanish identically, and hence they vanish in any coordinate system). If we do not want to alter the intrinsic properties of the base space, then we should be careful about the choice of an image dependent metric. We then have to insist on a vanishing Riemann tensor. Equivalently, we have to assume that, at least *formally*, the image dependent covariant metric tensor $g_{ij}[\phi]$ is related to the flat metric η_{ij} by some active coordinate transformation (that is, the formal manifestation of a genuine diffeomorphism of the underlying spatial manifold in a given coordinate system).

Using a general, image dependent Riemannian metric may seem somewhat awkward on first encounter, but giving it some thought it may be appreciated that this idea is, at least qualitatively, similar to various models of nonlinear diffusion proposed in the literature. Another example of a greyvalue dependent metric is the "grey weighted distance transform", which is actually a conformal mapping used to account for matter occlusion [315]. In the next subsection we propose a method for generalising (13.2) to the generating equation of a nonlinear scale-space.

2.4. Nonlinear Scale-Space

Starting point for our construction of a nonlinear scale-space will be the generalisation of (13.2) discussed in the previous section, with the flat metric η_{ij} replaced by an image dependent metric $g_{ij}[\phi]$. We parametrise the base manifold by a coordinate "overlay". The insignificance of such a coordinatisation is manifest in the covariant formalism. We then transform the covariant metric by an active, image dependent, spatial coordinate transformation (it won't be necessary to actually find this transformation in closed form, as long as it can be guaranteed to exist). The mapping of elementary volume cells by this active transformation directly relates to a resampling and read-out strategy for conventional scale-space.

Note that even within the context of linear scale-space, one can carry out active transformations, in which case they must be independent of the measurement data. Nevertheless they can be nontrivial; an instance of this is the log-polar coordinate system used in models of mammalian retinas, which corresponds to a flat, rotationally invariant, conformal metric (see

---

[3]More generally, we might have taken any umbilical image domain instead of a flat one, corresponding to a Riemann tensor of the form $R_{ijkl} = \kappa(\eta_{ik}\eta_{jl} - \eta_{il}\eta_{jk})$ with constant "Riemannian curvature" κ.

also Example 3 in section 3).

Such a coordinate system has physical significance; for each level of scale one can imagine an elementary volume cell to correspond to a *receptive field*, causing a linear decrease of resolution as a function of eccentricity (indeed, an operational consequence of the log-polar transformation). The log-polar parlance does not at all prevent us from using *any coordinate system whatsoever* for the description of such a foveal system (although one usually conforms to a likewise log-polar, descriptive coordinate system, for obvious reasons). The only distinction with the nonlinear case is that, in the latter case, the active coordinate transformations depend on the data.

Under a coordinate transformation $\mathbf{x} \to \tilde{\mathbf{x}}$ the covariant and contravariant components of the metrics are related by

$$g^{ij} = \frac{\partial \tilde{x}^i}{\partial x^k} \frac{\partial \tilde{x}^j}{\partial x^l} \eta^{kl} \,, \quad g_{ij} = \frac{\partial x^k}{\partial \tilde{x}^i} \frac{\partial x^l}{\partial \tilde{x}^j} \eta_{kl} \,. \tag{13.7}$$

The idea is now to replace the metric η_{ij} (or η^{ij}) by a general, image dependent tensor $g_{ij}[\phi]$ ($g^{ij}[\phi]$, respectively) in such a way that one can formally interpret it as an image dependent local coordinate transformation. An arbitrary, symmetric, positive definite tensor g^{ij} (g_{ij}, respectively) can always be expressed by the following metric transformation:

$$g^{ij} = E_k^i E_l^j \eta^{kl} \,, \quad g_{ij} = F_i^k F_j^l \eta_{kl} \,, \tag{13.8}$$

with a suitable choice of E_j^i (with $F_k^i E_j^k = E_k^i F_j^k = \delta_j^i$). Vice versa, any choice of E_j^i will yield such a tensor, provided its determinant is nonvanishing. In order not to introduce an explicit, image independent spatial bias, we must assume that E_j^i does not depend *explicitly* on $\mathbf{x}$ (recall that this may well be the case at the linear level, so we can disregard it here), but only *implicitly* as a functional of the image intensity function $\phi(\mathbf{x}; s)$. Likewise, to prevent an explicit resolution bias, it should depend on spatially dimensionless image derivatives $\sqrt{s}^n \phi_{i_1 \ldots i_n}$ only, that is, derivatives with respect to the "natural coordinates" $\xi^i = x^i / \sqrt{s}$.

In view of the above discussion, we shall need a restrictive *integrability condition* for the admissible choices of E_j^i. By construction, the Riemann tensor R_{jkl}^i associated with (13.7) (see appendix 1, Definition 12, with formal substitution $\eta_{ij} \to g_{ij}$) vanishes identically.

We may formulate the integrability condition by requiring that the Riemann tensor that formally corresponds to (13.8) vanishes as well:

$$R_{jkl}^i[E] \equiv 0 \,. \tag{13.9}$$

If this integrability condition is satisfied, we may write the associated (locally well-defined) active coordinate transformation as the transformation whose Jacobian is just the nonsingular E^i_j:

$$d\tilde{x}^i[\phi] = E^i_j[\phi]\, dx^j \ . \tag{13.10}$$

The implicit assumption of nonsingularity can be made explicit by writing the tensor E^i_j in exponential form, say

$$E^i_j[\phi] = \exp^i_j(\mathbf{T}[\phi]) \ . \tag{13.11}$$

The exponential, indexed by the mixed pair (i,j), is defined by its power series in terms of the mixed tensor $T^i_j[\phi]$ as usual. We will only consider *local* functionals, that is functionals that can be expressed as mere functions of image derivatives at a fixed point up to some order n:

$$E^i_j[\phi](\mathbf{x};s) = E^i_j(\phi(\mathbf{x};s), \sqrt{s}\,\phi_i(\mathbf{x};s),\ldots,\sqrt{s}^n\,\phi_{i_1\ldots i_n}(\mathbf{x};s)) \ . \tag{13.12}$$

To avoid confusion, we will write $\mathcal{D}_i$ instead of D_i to denote the formal, nonlinear derivative in terms of the formal metric $g_{ij}[\phi]$ (that is, same definition as the covariant derivative D_i, but with covariant metric η_{ij} formally replaced by $g_{ij}[\phi]$; note that $\mathcal{D}_i$ commutes with $g_{ij}[\phi]$). For the moment we will disregard active transformations of the scalar parameter s:

$$\mathcal{D}_i\mathcal{D}^i\phi = \partial_s\phi \ . \tag{13.13}$$

The formal analogue of (13.5) is

$$\mathcal{D}_\mu\mathcal{J}^\mu[\phi] = 0 \ . \tag{13.14}$$

Here, the generalised, nonlinear "gradient" operator is defined by $\mathcal{D}_\mu = (\partial_s; \mathcal{D}_i)$, and the $(d+1)$-dimensional "current" by $\mathcal{J}^\mu[\phi] = (\phi; -\mathcal{D}^i\phi) = (\phi; -g^{ij}[\phi]\mathcal{D}_j\phi)$. In the linear limit (i.e. $g_{ij} \to \eta_{ij}$, or $T^i_j \to 0$), the current is linear in ϕ, its spatial part reduces to minus the image gradient, while the scale part is just ϕ itself. Note that the scale component of the current, $\mathcal{J}^0[\phi] = \phi$, remains unaltered in the generalised case.
The d-current $\mathcal{J}^i[\phi] = -\mathcal{D}^i\phi$ controls the scale evolution of the image. In the linear case it is proportional to the image gradient, which is a natural choice for an unbiased representation. In the nonlinear case, the rate of change $\delta\mathcal{J}^i[\phi]/\delta\phi$ is no longer insensitive to the data ϕ, thus introducing a bias; the way the current responds to an increment $\delta\phi$ of input data depends on the local "state" of ϕ. This is just the idea behind a nonlinear scale-space: to give a multiscale representation emphasising "relevant" over "irrelevant" data, relevance being defined operationally by virtue of the nonlinearity.
In order to unconfound active and passive coordinate transformations in (13.13), we have to split the formal operator $\mathcal{D}_i$ into the true covariant

operator D_i (the one that accounts for covariance) and the nonlinearity (the *essential* modification introduced by the active transformation). Using (13.7) we can rewrite (13.13) into covariant form

$$D_i \left(A^{ij}[\phi] \, D_j \phi \right) = B[\phi] \partial_s \phi \, . \tag{13.15}$$

Here, we have defined

$$A^{ij}[\phi] \overset{\text{def}}{=} \sqrt{\frac{g[\phi]}{\eta}} \, g^{ij}[\phi] = \frac{1}{\det E[\phi]} E^i_k[\phi] E^j_l[\phi] \eta^{kl} \, , \tag{13.16}$$

$$B[\phi] \overset{\text{def}}{=} \sqrt{\frac{g[\phi]}{\eta}} = \frac{1}{\det E[\phi]} \, . \tag{13.17}$$

Equation (13.14) or (13.15) corresponds to a conservation law, since (by regularity of $E^i_j[\phi]$) there exists a reparametrisation $s \mapsto \tilde{s}[\phi]$, such that

$$B[\phi] \partial_s \phi = \partial_{\tilde{s}} \phi \, , \tag{13.18}$$

in terms of which we have

$$D_\mu J^\mu[\phi] = 0 \, , \tag{13.19}$$

with $J^\mu[\phi] = (J^0[\phi]; J^i[\phi]) = (\phi; -A^{ij}[\phi] D_j \phi)$ (and $D_0 = \partial_{\tilde{s}}$). Note that the active transformation of scale is determined by (13.11):

$$d\tilde{s}[\phi] = \exp(\operatorname{tr} \mathbf{T}[\phi]) \, ds \, . \tag{13.20}$$

(By (13.8) we have $\sqrt{g[\phi]} = \exp(-\operatorname{tr} \mathbf{T}[\phi]) \sqrt{\eta}$.) One can now either integrate (13.14) with a formal measure $\sqrt{g[\phi]}(\mathbf{x}; s) \, d\mathbf{x}$, or (13.19) with the usual covariant measure $\sqrt{\eta(\mathbf{x})} \, d\mathbf{x}$, in order to obtain the familiar form of the conservation law in integral form, as we did in (13.6).

In the exponential notation of (13.11), the nonlinearity becomes

$$A^{ij}[\phi] = \exp(-\operatorname{tr} \mathbf{T}[\phi]) \exp^i_k(\mathbf{T}[\phi]) \exp^j_l(\mathbf{T}[\phi]) \eta^{kl} \, , \tag{13.21}$$

$$B[\phi] = \exp(-\operatorname{tr} \mathbf{T}[\phi]) \, . \tag{13.22}$$

The nonlinearity disappears in the linearising limit, as it should:

$$\lim_{g_{ij} \to \eta_{ij}} A^{ij}[\phi] = \eta^{ij} \, , \quad \lim_{g_{ij} \to \eta_{ij}} B[\phi] = 1 \, . \tag{13.23}$$

It should be appreciated that, despite the virtually arbitrary transformation freedom, the nonlinearities (13.16) and (13.17) are of a rather restricted form. In particular, one cannot simply pick *any* two-tensor $A^{ij}[\phi]$ and scalar $B[\phi]$ because of the integrability condition (13.9).

We end this section with the remark that we have not yet specified any specific form of the nonlinearity; this should be motivated by applications. Rather, we have proposed a general equivalence class of scale-spaces, with conventional linear scale-space as a universal representative. The equivalence relation is the existence of an active transformation that links the members of the class. By construction, the diffusion equation that governs nonlinear scale-space is of a rather restricted nature. It allows us to appreciate nonlinear scale-space in the context of a multiresolution framework: it is a reformatted version of linear scale-space. In this precise sense it is indeed just a linear scale-space "in disguise".

3. Examples

The following example illustrates the fundamental difference between an active and a passive transformation.

Example 1 (Active versus Passive Transformations)
Consider the following formal coordinate transformation in two dimensions:

$$\begin{cases} \tilde{x} &= \lambda x \\ \tilde{y} &= y \,. \end{cases} \tag{13.24}$$

It may serve to represent either a passive or an active transformation. How to tell the difference?
Let us first consider the trivial case, taking (13.24) to represent a passive transformation. Recall that this is nothing but a mere reparametrisation of coordinates. Both (x, y) and $(\tilde{x}, \tilde{y})$ refer to the *same* point in the spatial image domain, P say. Let us assume that the metric is given in the $(\tilde{x}, \tilde{y})$-system by

$$d\ell^2 = d\tilde{x}^2 + d\tilde{y}^2 \,. \tag{13.25}$$

Then in the (x, y)-system it is obviously given by

$$d\ell^2 = \lambda^2 dx^2 + dy^2 \,. \tag{13.26}$$

Isotropic diffusion of a scalar image can be described in either system. If

$$\tilde{\phi}_{\tilde{x}\tilde{x}} + \tilde{\phi}_{\tilde{y}\tilde{y}} = \tilde{\phi}_s \tag{13.27}$$

holds in $(\tilde{x}, \tilde{y})$-coordinates, then we have

$$\lambda^{-2}\phi_{xx} + \phi_{yy} = \phi_s \tag{13.28}$$

in (x, y)-coordinates, where ϕ and $\tilde{\phi}$ are related by the scalar transformation law

$$\tilde{\phi}(\tilde{x}, \tilde{y}) \stackrel{\text{def}}{=} \phi(\lambda x, y) \,. \tag{13.29}$$

Note that although $\tilde{\phi}$ and ϕ have *different functional* forms, this transformation law states that the greyvalue in a *given point* P is not affected by the passive transformation: after all, nothing has happened on the physical level. The scalar transformation law (13.29) guarantees that (13.27) and (13.28) are physically identical. One should always keep in mind that this is what is meant with the sloppy notation $\tilde{\phi} = \phi$ in classical tensor calculus (cf. (13.59) in appendix 1)!

Next, let us consider the case where (13.24) represents an active transformation. This means that it tells us how the *base manifold itself* is mapped in terms of a *given parametrisation* (or coordinate chart). In other words, (x, y) and $(\tilde{x}, \tilde{y})$ now refer to *physically distinct* points in the spatial image domain, say P and $\tilde{P}$, respectively. Every physical attribute attached to the base manifold, notably the image greyvalue, is dragged along from P to $\tilde{P}$. This dragging of points therefore manifests itself in an *image deformation*[4] $\phi \mapsto \tilde{\phi}$, defined by

$$\tilde{\phi}(\tilde{x}, \tilde{y}) \stackrel{\text{def}}{=} \phi(x, y) . \tag{13.30}$$

Equation (13.30) is *not* the same as the scalar transformation law (13.29); in the present case, $\tilde{\phi}$ and ϕ are *functionally the same* when expressed in the coordinates of corresponding mapped and original points $\tilde{P} : (\tilde{x}, \tilde{y})$ and $P : (x, y)$, respectively. A caveat: it is implicitly understood that one uses (13.30) whenever we are dealing with active transformations. The very idea behind this migration of points is to arrange it such that the metric assumes the familiar form

$$d\ell^2 = d\tilde{x}^2 + d\tilde{y}^2 . \tag{13.31}$$

This implies that, in terms of (x, y)

$$d\ell^2 = \lambda^2 dx^2 + dy^2 . \tag{13.32}$$

In turn this implies that *isotropic* diffusion of the *warped* image $\tilde{\phi}$ (disregarding a reparametrisation of s),

$$\tilde{\phi}_{\tilde{x}\tilde{x}} + \tilde{\phi}_{\tilde{y}\tilde{y}} = \tilde{\phi}_s , \tag{13.33}$$

amounts to *non-isotropic* diffusion of the *original* image, by virtue of (13.30):

$$\lambda^{-2}\phi_{xx} + \phi_{yy} = \phi_s . \tag{13.34}$$

---

[4]In modern terminology the deformed image induced by the diffeomorphism on the base manifold is usually called the "carry-along" of the original image. Similar terminology applies to the naturally induced metric tensor and other geometric objects that live on the manifold.

The latter equation, although formally identical to (13.28), shows the (physically significant) effect of an active transformation when applied to the image.

Of course, the example above is only of academic interest. Alvarez et al. [9] proposed a similar scaling expressed in terms of a local frame (v, w), for which the coordinate axes are locally tangential to the isophote and gradient direction, respectively. Replacing x by v and y by w in (13.34), and then taking the singular, "geometric" limit $\lambda \to \infty$ (after reparametrising $t = \lambda^2 s$), one obtains an interesting result:

$$\phi_{vv} = \phi_t . \tag{13.35}$$

Note that ϕ_{vv} can be parsed into first and second order derivatives with respect to the Cartesian coordinates x and y:

$$\phi_{vv} = \left(\phi_x^2 + \phi_y^2\right)^{-1} \left(\phi_x^2 \phi_{yy} - 2\phi_x \phi_y \phi_{xy} + \phi_y^2 \phi_{xx}\right) . \tag{13.36}$$

The partial derivatives on the r.h.s. can be calculated at any scale by means of convolution with corresponding derivatives of the normalised Gaussian Green's function, most efficiently in the Fourier domain. See Figure 13.2, 13.3, and 13.4.

One may wonder how our nonlinear scale-space model relates to other models of nonlinear diffusion. A classical example of such a model has become known as the Perona and Malik equation [293] (but see also [67, 142]).

Example 2 (Nonlinear Scale-Space and Perona-Malik Diffusion)
An equation proposed by Perona and Malik is given by

$$D_i \left(C^{ij}[\phi] D_j \phi\right) = \partial_s \phi , \tag{13.37}$$

in which the nonlinearity is given by

$$C^{ij}[\phi] = c[\phi] \eta^{ij} , \tag{13.38}$$

$$c[\phi] = \exp\left(-\frac{1}{2}\lambda^2 D_i \phi \eta^{ij} D_j \phi\right) . \tag{13.39}$$

Note that this choice corresponds to a *conformal transformation* of the metric.

Recall our criterion for a nonlinear scale-space: it should formally correspond to a spatial transformation of linear scale-space. Equation (13.37) does not satisfy this criterion: one generally cannot obtain a nontrivial, conformal metric by means of a coordinate transformation. This is only possible for a very restricted class of metrics. To see this, suppose that

$$g_{ij}(\mathbf{x}) = e^{-2\tau(\mathbf{x})} \eta_{ij} , \tag{13.40}$$

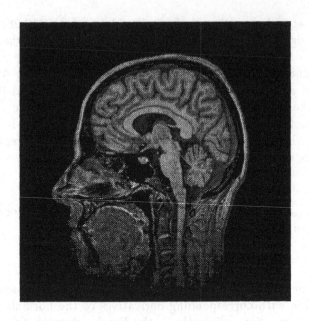

Figure 13.2. Magnetic resonance image of a brain (256 × 256 pixels) used as initial image for the generation of a scale-space.

for some unknown scalar function $\tau(\mathbf{x})$, and flat metric tensor η_{ij} (cf. (13.8)). The corresponding Riemann tensor (see Definition 12 in appendix 1, with the upper index lowered to first position) is given by

$$R_{ijkl} = \quad \frac{1}{2}e^{-2\tau}\left[(d-2)\left\{\eta_{ik}Y_{jl} - \eta_{il}Y_{jk} + (\eta \leftrightarrow Y)\right\}\right.$$
$$\left.+2X\left\{\eta_{ik}\eta_{jl} - \eta_{il}\eta_{jk}\right\}\right], \tag{13.41}$$

with

$$Y_{ij} = D_iD_j\tau + D_i\tau D_j\tau, \tag{13.42}$$

and

$$X = \left(1 - \frac{d-1}{3}\right)\eta^{ij}D_iD_j\tau + (d-2)\left(\frac{d-1}{6} - 1\right)\eta^{ij}D_i\tau D_j\tau. \tag{13.43}$$

The proof of this is tedious but straightforward: insert (13.40) into Definition 12 of Appendix 1 and follow the definitions. The following trick is helpful: first assume that η_{ij} is an overall constant—such a choice exists—and work your way through the successive steps (using ordinary instead of covariant derivatives). Finally, replace all ordinary derivatives by covariant derivatives. By covariance the result will then be correct for arbitrary coordinate systems.

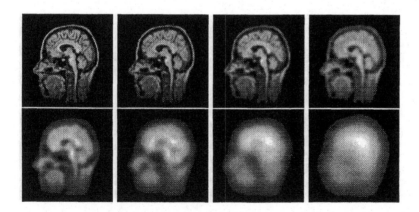

Figure 13.3. 8 Levels from a conventional linear scale-space corresponding to $\phi_s = \phi_{xx} + \phi_{yy}$. Scale levels: $1.00, 1.54, 2.36, 3.62, 5.55, 8.52, 13.1, 20.1$ pixels, respectively. The underlying isotropic diffusion equation has been solved by straightforward multiplication with the appropriate Gaussian Green's function in Fourier space.

A necessary and sufficient condition for the existence of a coordinate transformation that maps the flat metric η_{ij} to the conformal metric g_{ij} is the vanishing of the Riemann tensor. This will be the case if and only if

$$(d - 2)Y_{ij} + X\eta_{ij} = 0 \,. \tag{13.44}$$

The reader may verify that this puts a severe restriction on the function τ. In fact, the solution is completely determined by (13.44), up to a pair of integration constants; there is certainly no freedom for any image dependency. Therefore the Perona-Malik equation (13.37) does *not*, by our definition, correspond to a nonlinear scale-space (but recall that it was not intended to do so: it was meant as a—less restrictive—generalisation of greyvalue diffusion). A spin-off of this exercise is an interesting case of a generalised *linear* scale-space: see Example 3.

Apart from illustrating the general procedure for generalising the scale-space paradigm in a simple context, the following example shows that there are also nontrivial, *linear* representations of scale-space.

Example 3 (Foveal Scale-Space: Log-Polar Transform)
Recall that our construction of a nonlinear scale-space is based on a spatial transformation whose Jacobian does not *explicitly* depend on the spatial coordinates **x**. This was done in order to avoid an a priori bias with respect to spatial location *at the nonlinear level*. This is not really a restriction, since such an explicit spatial bias has nothing to do with a nonlinearity; it may be taken into account at a linear stage. For a real-time operating

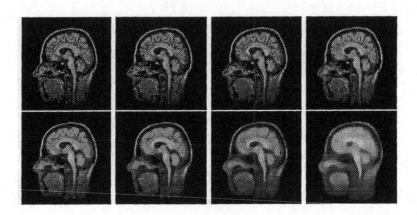

Figure 13.4. 8 Levels from the nonlinear diffusion scheme $\phi_t = \phi_{vv}$, as proposed by Alvarez et al. Iterations: $2, 4, 8, 16, 32, 64, 128, 256$, respectively. The underlying nonlinear diffusion equation has been solved by forward Euler iteration. To this end, ϕ_{vv} was calculated at each iteration by evaluating the r.h.s. of (13.36) at an inner scale of $\sigma = 0.8$ pixel units, and the Euler step size was taken to be $\Delta t = 0.2$ (square pixel units).

system it is in fact quite natural to actually enforce an a priori spatial bias; its finite processing capacity simply prevents otherwise. For such a system it makes sense to focus on (a neighborhood of) a definite spatial location.

We may follow the arguments that led to our construction of a nonlinear scale-space also within this linear case: start with (13.2), and perform an active, in this case *image independent* transformation, which we now allow to be a function of the radial separation r from a given "foveal point" (rotational symmetry). Clearly, this will destroy spatial shift invariance, but this is easily repaired by adding a degree of freedom for redirection of the foveal point. The result is similar to (13.15), with the nonlinearities $A^{ij}[\phi]$ and $B[\phi]$ replaced by functions $A^{ij}(\mathbf{x})$ and $B(\mathbf{x})$, respectively.

The question may arise of how to choose the formal transformation. A strong constraint is obtained by requiring the spatial transformation to leave equation (13.2) locally *form invariant*. It is a remarkably strong constraint, for suppose that $A^{ij}(\mathbf{x})$ equals the contravariant metric $\eta^{ij}(\mathbf{x})$, then we have (cf. (13.16))

$$\sqrt{\frac{g}{\eta}}\, g^{ij} = \eta^{ij} . \tag{13.45}$$

Taking the determinant of this equation yields

$$\sqrt{\frac{g}{\eta}}^{\,d} = \frac{g}{\eta} , \tag{13.46}$$

so that by necessity we either have $d = 2$ or $g = \eta$. Furthermore we see from (13.45) that the mapping $\eta_{ij} \to g_{ij}$ must be conformal. This, in turn, shows that the case $g = \eta$ is trivial (it corresponds to $g_{ij} = \eta_{ij}$).

So let us concentrate on the 2-dimensional case, and let us furthermore assume that the conformal metric is given by (13.40) with a rotationally invariant function $\tau = \tau(r)$. The integrability requirement $R^i_{jkl}(\mathbf{x}) = 0$ has been shown to correspond to (13.44), which in this case reduces to the homogeneous Laplace equation

$$\eta^{ij} D_i D_j \tau = 0 . \tag{13.47}$$

In polar coordinates it reads (using a prime to denote a derivative)

$$r\,\tau'' + \tau' = 0 , \tag{13.48}$$

the general solution of which is

$$\tau(r) = p \ln \frac{r}{r_0} , \tag{13.49}$$

with integration constants $p \in \mathbb{R}$ and $r_0 \in \mathbb{R}^+$. It is not so difficult to solve for the active spatial transformation that corresponds to this conformal mapping. For $p = 1$ (the other cases are left to the reader) one obtains a particularly interesting case: starting from the usual polar coordinates (r, θ) one finds the "log-polar" transform [371]:

$$\begin{cases} \tilde{x} &= r_0 \theta \\ \tilde{y} &= r_0 \ln \frac{r}{\rho_0} \end{cases} , \tag{13.50}$$

in which r_0 and ρ_0 are constants with the dimension of length. In the $(\tilde{x}, \tilde{y})$-coordinate system the line element corresponding to the conformal metric g_{ij} becomes

$$d\ell^2 = \left(\frac{r_0}{r}\right)^2 (dr^2 + r^2 d\theta^2) = d\tilde{x}^2 + d\tilde{y}^2 . \tag{13.51}$$

Subjecting the warped image $\tilde{\phi}$, defined by $\tilde{\phi}(\tilde{x}, \tilde{y}) = \phi(r, \theta)$, to isotropic scaling:

$$\tilde{\phi}_{\tilde{x}\tilde{x}} + \tilde{\phi}_{\tilde{y}\tilde{y}} = \tilde{\phi}_{\tilde{s}} , \tag{13.52}$$

boils down to plain isotropic diffusion of ϕ:

$$\phi_{rr} + \frac{1}{r}\phi_r + \frac{1}{r^2}\phi_{\theta\theta} = \phi_s . \tag{13.53}$$

Here, the relation between the parameters $\tilde{s}$ and s is given by

$$\tilde{s} = \left(\frac{r_0}{r}\right)^2 s . \tag{13.54}$$

In other words, for a given "warped scale" $\tilde{\sigma}$ (defined as usual: $2\tilde{s} = \tilde{\sigma}^2$, etc.) the ordinary scale σ increases linearly with eccentricity r. Note the periodicity of the $\tilde{x}$-coordinate, and the singularity at the foveal centre $r = 0$, which indicates that the transformation is topologically nontrivial. The so-called "log-polar coordinates" $(\tilde{x}, \tilde{y})$ are the canonical coordinates for the planar similarity group, that is, the local coordinates in terms of which spatial scalings and rotations look like plain translations [335].

4. Conclusion and Discussion

We have proposed a generalised scale-space theory based on an active spatial transformation of conventional scale-space. By construction, a generalised scale-space can be regarded as a conventional scale-space in disguise. The transformation reflects an a priori bias, and may or may not depend on the image data. It is important to appreciate that, despite such a bias, there is essentially no loss of information; it is all a matter of assigning *a priori weights* to a common set of data. The bias, operationally described by the choice of transformation as a functional of the data, is the paradigm. It is the degree of freedom that can be utilised to facilitate the scale-space representation prior to any *specific* visual routine.

We have illustrated the theory by three examples. The first example elaborates on the crucial difference between passive transformations, or mere coordinate reparametrisations, and active transformations, that is, physically significant mappings of points within the image plane. The latter type of transformation induces a special kind of metric transform, one that leaves the intrinsic local properties of the base space unaffected (singularities apart). One may well imagine that such transformations lie at the basis of a data transfer mechanism between a general, linear scale-space representation and more specific data-interpreting, generally nonlinear routines.

In the second example we have shown that the nonlinear diffusion equation as implemented by Perona and Malik corresponds to a conformal metric transform, but cannot be regarded as an active spatial mapping of linear scale-space.

A final example shows that there are nontrivial manifestations of scale-space that preserve linearity, producing the well-known "log-polar transform". It is obtained by selecting a preferred location (the *foveal point*) and a conformal metric which is rotationally invariant with respect to that location. It is globally well-defined outside the foveal point, which itself corresponds to a singularity. At a given point, ordinary scale and warped "log-polar" scale are linearly related, with a proportionality constant that increases linearly with eccentricity.

It should be clear that many problems related to our definition of generalised scale-spaces are all but trivial. For example, we have been deliberately sloppy concerning important issues such as singularities, topology, etc.

Another problem yet to be solved concerns the *operational construction* of a generalised, notably nonlinear scale-space. The straightforward Euler method used in the example has a serious drawback, at least in the context of generalised scale-space theory: it does not exploit the very idea of transforming an existing linear scale-space representation at all! Although it does yield the solution of the nonlinear diffusion equation underlying the representation of interest, it obscures a transparent, modular connection with the "universal" linear scale-space representation. A possible solution may be to recast the diffusion equation into an *integral equation*, and to solve this by iteration. Such an iteration scheme may reflect the continuous spatial transformation that actually defines the nonlinear representation in terms of the underlying linear representation. This, however, remains to be investigated.

Our generalised scale-space theory can be generalised even further by relaxing the integrability condition underlying the active spatial transformation, or by admitting invertible scalar greyvalue transformations, or both. The first option entails the possibility of warping the spatial image domain by a general metric transform (thereby inducing a nonzero Riemann tensor). The second option seems a rather natural one to take into account, since, were it not for resolution limitations, it is merely a conventional matter whether one chooses to consider a given scalar field ϕ or any monotonic function of this. As an example of this, consider $\tilde{\phi} = T(\phi)$, with $\tilde{\phi}$ subject to isotropic diffusion:

$$\tilde{\phi}_s = \Delta\tilde{\phi} .$$

Of course we must have $T'(\phi) \neq 0$ for all ϕ. It is easily verified that if we take

$$T(\phi) = \frac{1}{\lambda}e^{\lambda\phi} ,$$

for some $\lambda \neq 0$, then we have in terms of the field ϕ (cf. [111]):

$$\phi_s = \Delta\phi + \lambda\|\nabla\phi\|^2 .$$

The limit $\lambda \to 0$ takes us back to linear scale-space, while the "morphological" limits $\lambda \to \pm\infty$, combined with a redefinition $\rho = \pm\lambda s$ (depending on whether λ is positive or negative), result in

$$\phi_\rho = \pm\|\nabla\phi\|^2 .$$

According to [369], these are the equations governing the morphological *dilation* and *erosion* of a given initial image, ϕ_0 say, by the quadratic structuring function

$$q_\rho(\mathbf{x}) = -\frac{\|\mathbf{x}\|^2}{4\rho}.$$

(Cf. [180].) Using $\oplus$ and $\ominus$ to denote dilation and erosion, we have

$$\phi^+(\mathbf{x};\rho) = (\phi_0 \oplus q_\rho)(\mathbf{x}) \overset{\text{def}}{=} \sup_{\mathbf{y} \in \mathbb{R}^d} [\phi_0(\mathbf{x}-\mathbf{y}) + q_\rho(\mathbf{y})] \qquad (\rho > 0),$$

$$\phi^-(\mathbf{x};\rho) = (\phi_0 \ominus q_\rho)(\mathbf{x}) \overset{\text{def}}{=} \inf_{\mathbf{y} \in \mathbb{R}^d} [\phi_0(\mathbf{x}-\mathbf{y}) - q_\rho(\mathbf{y})] \qquad (\rho < 0).$$

For more details on morphology in this context the reader is referred to [369, 87]. This simple example demonstrates the potential ability of a generalised scale-space theory to unify various models of interest in computer vision.

Acknowledgment

We are indebted to Andrea van Doorn for reviewing preliminary versions of the manuscript, Ruud Geraets for clarifying discussions on covariance principles, and Wiro Niessen for his implementation of nonlinear diffusion equations. This work has been carried out as part of the national priority research programme "3D Computer Vision", supported by the Netherlands Ministries of Economic Affairs and Education & Science through a SPIN grant. The support from the participating industrial companies is gratefully acknowledged.

Appendix

1. Appendix: The Covariant Formalism

1.1. Cartesian Coordinate Transformations

Taking a Cartesian coordinate system we may represent (13.1) as:

$$\partial_i \partial_i \phi = \partial_s \phi . \tag{13.55}$$

The spatial indices in (13.55) relate to the group of *Cartesian coordinate transformations* (CCT henceforth), comprising orthogonal transformations and translations:

Definition 1 (Cartesian Coordinate Transformations)
A Cartesian coordinate transformation is an orthogonal transformation combined with a translation:

$$x_i = r_{ij}\tilde{x}_j + \xi_i ,$$

with the orthogonality conditions $\det r_{ij} = \pm 1$ *and* $r_{ik}r_{jk} = r_{ki}r_{kj} = \delta_{ij}$, *in which* δ_{ij} *is the Kronecker symbol defined by* $\delta_{ij} = 1$ *whenever* $i = j$ *and* $\delta_{ij} = 0$ *otherwise.*

Upon such a Cartesian coordinate transformation the components of a Cartesian tensor $T_{i_1 \ldots i_n}$ transform according to the following law:

$$T_{i_1 \ldots i_n} = r_{i_1 j_1} \ldots r_{i_n j_n} \tilde{T}_{j_1 \ldots j_n} . \tag{13.56}$$

The following proposition explains why it is convenient to consider tensors in the first place:

Proposition 1 (Covariance for Cartesian Coordinates)
A Cartesian tensor equation that holds in a given Cartesian coordinate system holds in any other Cartesian coordinate system, i.e. if $T_{i_1 \ldots i_n} = 0$ *in a given Cartesian coordinate system* **x**, *then also* $\tilde{T}_{i_1 \ldots i_n} = 0$ *in any other Cartesian coordinate system* **x̃**.

The proof of this proposition is a triviality because of the homogeneous transformation law (13.56).

Example 4 (Cartesian Tensors)
Using the chain rule for differentiation the reader may verify that the n-th order partial derivative $\partial_{i_1 \ldots i_n}$ formally behaves as a Cartesian n-tensor under the CCT group. Another important example of a Cartesian n-tensor is the n-th order monomial $\delta x_{i_1} \ldots \delta x_{i_n}$ consisting of n factors of a displacement vector δx_i.

By the orthogonality property it follows that a full contraction of Cartesian indices yields a Cartesian invariant. In particular, equation (13.55) is indeed

Cartesian invariant. It can be considered as a coordinate representation of the isotropic diffusion equation (13.1) after imposing a *gauge condition*, i.e. a restriction on the set of all possible coordinate systems. In this case, the gauge is partially fixed to admit Cartesian frames only. It leaves us with a residual gauge freedom of Cartesian coordinate transformations that link all possible Cartesian frames. This residual CCT-invariance can be used to impose an additional gauge condition in order to single out a unique coordinate system. This is the usual procedure when doing actual calculations on digital images. The sampling grid usually serves to define the unique gauge system.

However, coordinates are introduced merely to label the points of the image domain in a unique way; they have no significance of themselves. Therefore, if we aim for a genuinely coordinate independent description of image structure, *any* coordinate transformation should be allowed in principle. For this reason the Cartesian representation (13.55) is unsatisfactory. The next subsection shows how to generalise it to the more general class of rectilinear coordinates. Although this is not our ultimate goal, it serves to illustrate the notions of *covariance* and *contravariance* as well as their duality through the *metric* in the simplest possible context.

1.2. Affine Coordinate Transformations

As an intermediate step towards an invariant coordinatisation of (13.1), consider the following generalisation of (13.55):

$$\partial_i \partial^i \phi = \partial_s \phi \, . \tag{13.57}$$

Here we have made a distinction between *covariant* and *contravariant* indices and adopted the convention that subscripts label covariant and superscripts label contravariant components. Their transformation behaviour will be considered below, as well as some examples. Contractions are now to be performed on pairs consisting of an upper and a lower index only. The dual interpretation of a single geometrical object (such as the gradient operator ∂ above) is implicit in the choice of a *metric*, which allows one to map a contravector to a unique covector and vice versa, formally indicated by *lowering* or *raising* the corresponding index. We will henceforth use η_{ij} to denote a *flat* covariant metric tensor, appropriate for our image domain, in whatever coordinate representation. It is by virtue of the metric that we can measure scalar properties of, say, a vector $\mathbf{v}$, the components of which are v^i, such as its length. To this end, simply contract the vector "onto itself", or rather, onto its dual covector $\tilde{\mathbf{v}}$, the components of which are given by $v_i = \eta_{ij}v^j$: $v^2 \stackrel{\text{def}}{=} v^i \eta_{ij} v^j \stackrel{\text{def}}{=} v^i v_i$. Similarly we may write v^i instead of $\eta^{ij} v_j$, in which η^{ij} is the inverse of η_{ij}, hence the metric acts like a "mirror", mapping a contravectorspace to its dual covectorspace in a one-to-one way [190]. One can likewise "raise" or "lower" indices in an arbitrary mixed tensor through the metric. This explains the notation of (13.57).

In the Cartesian case, η_{ij} by definition equals the Kronecker tensor δ_{ij}. In any rectilinear coordinate system the metric tensor is a constant all over the image domain, and in the most general case of curvilinear coordinates (to be discussed in the next section) it generally has coordinate dependent components. The common classical way of introducing the metric tensor is by relating a displacement vector to its squared length (that is, the Pythagorean rule in its most general form):

Definition 2 (Flat Metric)
The covariant flat metric tensor η_{ij} on a flat domain is defined by the squared arc length $d\ell^2$ of the infinitesimal arc connecting $\mathbf{x}$ to $\mathbf{x} + d\mathbf{x}$:

$$d\ell^2 = \eta_{ij}(\mathbf{x}) dx^i dx^j ,$$

with $\eta_{ij} = \eta_{ji}$ and $\eta \stackrel{\text{def}}{=} \det \eta_{ij} \neq 0$. The dual, contravariant metric tensor η^{ij} is given by its inverse:

$$\eta^{ik} \eta_{kj} = \delta^i_j .$$

In a Cartesian system, the metric tensor η_{ij} reduces to δ_{ij}.

As we will see below, the dual index convention manifestly captures the invariance under *affine coordinate transformations* (ACT). These are transformations of the following type:

Definition 3 (Affine Coordinate Transformations)
An affine coordinate transformation is a regular, linear transformation combined with a translation:

$$x^i = a^i_j \tilde{x}^j + \xi^i ,$$

with $\det a^i_j \neq 0$.

So ACT is slightly more general than its CCT subgroup. The effect of this kind of transformations on the covariant and contravariant components of a tensor $T^{i_1 \ldots i_m}_{j_1 \ldots j_n}$ is given by:

$$T^{i_1 \ldots i_m}_{j_1 \ldots j_n} = a^{i_1}_{k_1} \ldots a^{i_m}_{k_m} b^{l_1}_{j_1} \ldots b^{l_n}_{j_n} \tilde{T}^{k_1 \ldots k_m}_{l_1 \ldots l_n} , \tag{13.58}$$

in which $\mathbf{b} \stackrel{\text{def}}{=} \mathbf{a}^{\text{inv}}$. Since $a^i_k b^k_j = b^i_k a^k_j = \delta^i_j$ it follows again that a full contraction of upper and lower indices yields an affine invariant. Note that latin indices now relate to the affine group rather than the Cartesian group. Again we have:

Proposition 2 (Covariance for Rectilinear Coordinates)
An affine tensor equation that holds in a given rectilinear coordinate system holds in any other rectilinear coordinate system, i.e. if $T^{i_1...i_m}_{j_1...j_n} = 0$ in a given rectilinear coordinate system $\mathbf{x}$, then also $\tilde{T}^{k_1...k_m}_{l_1...l_n} = 0$ in any other rectilinear coordinate system $\tilde{\mathbf{x}}$.

The proof is a direct consequence of (13.58).

Example 5 (Covariant and Contravariant Affine Tensors)
A typical example of a covariant affine n-tensor is given by an n-th order partial derivative: $\partial_{i_1...i_n}$ formally behaves as a cotensor under the ACT group (which, as before, becomes clear by simply doing the transformation and using the chain rule). The n-th order monomial $\delta x^{i_1}...\delta x^{i_n}$ may serve to illustrate the dual case: it is a contratensor under ACT.

The reason why there is no need for a distinction between covariant and contravariant indices for the CCT group is the orthogonality condition that holds for a Cartesian transformation matrix, cf. (13.58): if we take $\mathbf{a}$ to be an orthogonal matrix, then we have $\mathbf{b} = \mathbf{a}^T$. Consequently, we can rewrite (13.58) into a subscript form like (13.56) without ambiguity and forget about the duality.

In order to make the geometrical insignificance of coordinates obvious, the manifest invariant incorporation of affine coordinates is of course still insufficient ((13.57) is only valid in rectilinear coordinates). We would like to represent the image and everything derived from it in a genuinely coordinate independent way. This is the purpose of the next subsection.

1.3. General Coordinate Transformations

It turns out to be possible to rewrite equation (13.57) in such a way that it is manifest invariant against *general coordinate transformations* (GCT henceforth):

Definition 4 (General Coordinate Transformations)
A general coordinate transformation is a smooth, invertible transformation:

$$x^i = x^i(\tilde{\mathbf{x}}) \,,$$

with $\det \frac{\partial x^i}{\partial \tilde{x}^j} \neq 0$.

This means that one does not have to impose any a priori restrictions on the choice of admissible coordinates for which the equation holds: the only admissibility requirement is that the coordinates are in a one-to-one correspondence with the points of the image domain and are sufficiently smooth (meaning that the transformation in Definition (4) is smooth and has a smooth inverse). This includes the possibility of taking *curvilinear coordinates*. The requirement of invariance under the GCT group is familiar from the theory of general relativity and from differential geometry. It is generally a good principle if one wants to adopt an *intrinsic* point of view, in which the irrelevance of coordinates is made explicit through GCT invariance.

The tensor transformation law for the GCT group is given by:

$$T^{i_1...i_m}_{j_1...j_n} = \frac{\partial x^{i_1}}{\partial \tilde{x}^{k_1}} \cdots \frac{\partial x^{i_m}}{\partial \tilde{x}^{k_m}} \frac{\partial \tilde{x}^{l_1}}{\partial x^{j_1}} \cdots \frac{\partial \tilde{x}^{l_n}}{\partial x^{j_n}} \tilde{T}^{k_1...k_m}_{l_1...l_n} . \tag{13.59}$$

The principle of invariance for this general group, once again a triviality by (13.59), is an extremely powerful result:

Proposition 3 (Covariance for General Coordinates)
A general tensor equation that holds in a given coordinate system holds in any other coordinate system, i.e. if $T^{i_1...i_m}_{j_1...j_n} = 0$ in a given coordinate system **x**, *then also $\tilde{T}^{k_1...k_m}_{l_1...l_n} = 0$ in any other coordinate system* $\tilde{\mathbf{x}}$.

Example 6 (Covariant and Contravariant General Tensors)
It seems pretty straightforward to come up with examples as we did before. Indeed, $\delta x^{i_1} \ldots \delta x^{i_n}$ is easily recognized as a contratensor under GCT, but an n-th order partial derivative does *not* behave as a cotensor anymore! To see this, use the chain rule to write down the relation between the n-th order derivatives in x and $\tilde{x}$ coordinates:

$$\partial_{i_1} \ldots \partial_{i_n} = \frac{\partial \tilde{x}^{j_1}}{\partial x^{i_1}} \tilde{\partial}_{j_1} \ldots \frac{\partial \tilde{x}^{j_n}}{\partial x^{i_n}} \tilde{\partial}_{j_n} . \tag{13.60}$$

Since the transformation matrices $\frac{\partial \tilde{x}^i}{\partial x^j}$ are functions of the $\tilde{x}^k$ (contrary to the linear case), we cannot simply shuffle them to the left; commuting them with the differential operators $\tilde{\partial}_k$ generally induces extra terms that violate the tensorial nature (cf. (13.59)). This implies that ordinary partial derivatives of a scalar image (except for its first order derivative) cannot be the components of some coordinate independent tensor.

Because the partial differential operator ∂_i (∂^i) does not formally behave as a covariant (contravariant) vector under GCT, we have reached the important conclusion that equation (13.57) as it stands is *not* invariant under GCT (it is *not* a general tensor equation).

The cause of this is the intrinsically multilocal nature underlying the definition of a derivative, which requires you to subtract values of its operand taken at two *distinct* locations. Since the local coordinates for these two neighboring points generally have no clear relation to one another, coordinate dependent components of quantities defined at these distinct points, such as the components of a vector field v^i, can no longer be compared, at least not in the naive way. Put differently, if $v^i(\mathbf{P}) = v^i(\mathbf{Q})$ for two neighboring points $\mathbf{P}$ and $\mathbf{Q}$, and all components $i = 1, \ldots, d$, this does not at all guarantee that the vectors are parallel. Clearly, this problem does not arise when we restrict ourselves to rectilinear coordinates, for which the variation of components after "parallel transport"—rigorously defined below—vanishes identically. Also, there is no problem when comparing *scalar* fields, which do not have components relative to some local frame.

When defining a meaningful differential operator (and more generally, any multilocal operator), one intuitively would like to compare the components of quantities defined at two distinct points only after a *parallel transport* of all quantities involved to a single base point, so that their components are given with respect to the same basis and hence become comparable. This can be made precise as follows:

Definition 5 (Parallel Transport of a Scalar)
Let $A(\mathbf{x})$ be a scalar, then the parallel transport of $A(\mathbf{x}+d\mathbf{x})$ from base point $\mathbf{x} + d\mathbf{x}$ to base point $\mathbf{x}$ induces a variation given by:

$$\delta A(\mathbf{x}) = 0 .$$

Definition 6 (Parallel Transport of a Covector)
Let $A_i(\mathbf{x})$ be a covector, then the parallel transport of $A_i(\mathbf{x} + d\mathbf{x})$ from base point $\mathbf{x} + d\mathbf{x}$ to base point $\mathbf{x}$ induces a variation given by:

$$\delta A_i(\mathbf{x}) = \Gamma_{ij}^{k}(\mathbf{x}) A_k(\mathbf{x}) dx^j .$$

The motivation for these two definitions is that a scalar is independent of the coordinate system, whereas the components of a covector are to be taken relative to the coordinate system at the base point of interest. By carrying the covector over an infinitesimal displacement dx^i (while keeping it parallel) its components with respect to the locally defined frame will change by an amount proportional to dx^i. Of course, the variation is also linear with respect to the covector itself and so the latter definition shows the most general form for this variation. The coefficients Γ_{ij}^{k} introduced in this way define a so-called *affine connection* or *affinity*. Despite its notation one should realize that this is *not* a tensor. Rather, its transformation law is determined by the following requirement:

Definition 7 (Covariant Derivative of a Covector)
Let A_i be a covector, then its covariant derivative $D_j A_i$ is the cotensor defined by:

$$D_j A_i = \partial_j A_i - \delta_j A_i ,$$

in which the affinity term $\delta_j A_i$ is given by:

$$\delta_j A_i dx^j = \delta A_i .$$

In other words, $D_j A_i = \partial_j A_i - \Gamma_{ij}^k A_k$. A straightforward transformation of left and right hand side in this equation shows that the affinity has to transform as follows:

Proposition 4 (Affinity Transformation Law)

If $\tilde{\mathbf{x}} = \tilde{\mathbf{x}}(\mathbf{x})$ is a general coordinate transformation and Γ_{ij}^k is an affinity given in the $\mathbf{x}$-representation, then in the $\tilde{\mathbf{x}}$ coordinate system we have:

$$\tilde{\Gamma}_{ij}^k = \frac{\partial \tilde{x}^k}{\partial x^n} \frac{\partial x^l}{\partial \tilde{x}^i} \frac{\partial x^m}{\partial \tilde{x}^j} \Gamma_{lm}^n + \frac{\partial \tilde{x}^k}{\partial x^n} \frac{\partial^2 x^n}{\partial \tilde{x}^i \partial \tilde{x}^j} .$$

The appearance of the inhomogeneous term on the r.h.s. accounts for the non-tensorial, hence gauge dependent nature of the affinity. For this reason Γ_{ij}^k is sometimes called a *gauge field*. The reader may verify that the difference of two affinities does transform as a tensor, and that the sum of an affinity and a tensor A_{ij}^k is again an affinity by the above rule.

The consistent generalisation of Definition 6 to arbitrary mixed tensors follows by repeated application of Definitions 5 and 6, writing out $0 = \delta(X_i A^i) = \delta X_i A^i + X_i \delta A^i$ for arbitrary X_i and then eliminating the X_i, etc:

Definition 8 (Parallel Transport of a Mixed Tensor)

Let $A_{j_1 \ldots j_q}^{i_1 \ldots i_p}(\mathbf{x})$ be a mixed tensor, then the parallel transport of $A_{j_1 \ldots j_q}^{i_1 \ldots i_p}(\mathbf{x} + d\mathbf{x})$ from base point $\mathbf{x} + d\mathbf{x}$ to base point $\mathbf{x}$ induces a variation given by:

$$\delta A_{j_1 \ldots j_q}^{i_1 \ldots i_p} = \Big(-\Gamma_{l_1 k}^{i_1} A_{j_1 \ldots j_q}^{l_1 i_2 \ldots i_p} - \ldots - \Gamma_{l_p k}^{i_p} A_{j_1 \ldots j_q}^{i_1 \ldots i_{p-1} l_p} +$$
$$\Gamma_{j_1 k}^{l_1} A_{l_1 j_2 \ldots j_q}^{i_1 \ldots i_p} + \ldots + \Gamma_{j_q k}^{l_q} A_{j_1 \ldots j_{q-1} l_q}^{i_1 \ldots i_p} \Big) dx^k .$$

We conclude that, once we know the affinity in one coordinate system (say $\Gamma_{ij}^k = 0$ as in rectilinear coordinates), we can calculate it for arbitrary reparametrisations of the coordinates, immediately yielding a covariant derivative for tensor fields:

Definition 9 (Covariant Derivative of a Mixed Tensor)

Let $A_{j_1 \ldots j_q}^{i_1 \ldots i_p}$ be a mixed tensor, then its covariant derivative $D_k A_{j_1 \ldots j_q}^{i_1 \ldots i_p}$ is the mixed tensor given by:

$$D_k A_{j_1 \ldots j_q}^{i_1 \ldots i_p} = \partial_k A_{j_1 \ldots j_q}^{i_1 \ldots i_p} - \delta_k A_{j_1 \ldots j_q}^{i_1 \ldots i_p} ,$$

in which the affinity terms $\delta_k A_{j_1 \ldots j_q}^{i_1 \ldots i_p}$ are given by:

$$\delta_k A_{j_1 \ldots j_q}^{i_1 \ldots i_p} dx^k = \delta A_{j_1 \ldots j_q}^{i_1 \ldots i_p} .$$

We will henceforth consider the image domain to be a *Riemannian space* with a general *Riemannian metric*:

Definition 10 (Riemannian Metric)
The covariant Riemannian metric tensor g_{ij} is defined by the squared arc length $d\ell^2$ of the infinitesimal arc connecting $\mathbf{x}$ to $\mathbf{x} + d\mathbf{x}$:

$$d\ell^2 = g_{ij}(\mathbf{x})dx^i dx^j \ ,$$

with $g_{ij} = g_{ji}$ and $g \stackrel{\text{def}}{=} \det g_{ij} \neq 0$. The dual, contravariant metric tensor g^{ij} is given by its inverse:

$$g^{ik}g_{kj} = \delta^i_j \ .$$

Note that we now omitted the additional restriction characteristic of a flat metric, cf. Definition 2: although it is generally possible to transform the metric tensor to one appropriate for a Cartesian frame in each given point $\mathbf{x}$, one cannot generally (i.e. given a general Riemannian metric) expect this to be possible on a full neighborhood of $\mathbf{x}$.

The index-raising and index-lowering manipulations can be consistently carried over to hold for covariant derivatives, provided we have $D_k g_{ij} = 0$ and $D_k g^{ij} = 0$. This is achieved by using a *metric affinity*:

Definition 11 (Metric Affinity)
In a Riemannian space with metric $d\ell^2 = g_{ij}(\mathbf{x})dx^i dx^j$ the metric affinity Γ^k_{ij} is defined by:

$$\Gamma^k_{ij} = \left\{ \begin{matrix} k \\ i\,j \end{matrix} \right\} \ .$$

Here, we have defined the Christoffel symbol of the second kind $\left\{ \begin{matrix} k \\ i\,j \end{matrix} \right\}$ by:

$$\left\{ \begin{matrix} k \\ i\,j \end{matrix} \right\} = g^{kl}\,[\,i\,j, l\,] \quad \text{(Christoffel Symbol of the Second Kind)} \ ,$$

i.e. the symbol obtained by raising the third index of the Christoffel symbol of the first kind $[\,i\,j, k\,]$:

$$[\,i\,j, k\,] = \frac{1}{2}\left(\partial_i g_{jk} + \partial_j g_{ki} - \partial_k g_{ij} \right) \quad \text{(Christoffel Symbol of the First Kind)} \ .$$

The reader may verify that this is a good definition with regard to Proposition 4.

Using a metric affinity the metric tensor indeed behaves as a constant under covariant differentiation. More generally, we have:

Proposition 5 (Uniqueness of Metric Affinity)
$\Gamma^k_{ij} = \left\{ \begin{matrix} k \\ i\,j \end{matrix} \right\}$ *if and only if $D_k g_{ij} = D_k g^{ij} = D_k \delta^j_i = 0$.*

So a given metric generates a unique metric affinity (the reverse is not true). The proof of Proposition 5 is left to the reader (see also [206, 55] and [348, vol. II]).

Henceforth we will silently assume that we have a metric affinity. In that case it is not difficult to show that there exist "geodesic coordinates" in a neighborhood of any given point **P**, such that the affinity vanishes in **P** (but generally not in its immediate neighborhood). This means that any Riemannian space locally looks flat, but generally is not flat; its flat appearance is only a matter of *scale*.

Covariant derivatives generally do not commute. If their commutator is nonvanishing, this is an indication of an intrinsic local curvature property of the Riemannian space. This property is captured by the *Riemann curvature tensor*, which can be defined as follows:

Definition 12 (Riemann Curvature Tensor)
The Riemann curvature tensor is defined by the following commutator:

$$R^i_{jkl} A_i = [D_l, D_k] A_j \, ,$$

or, in terms of the metric affinity:

$$R^i_{jkl} = \Gamma^i_{rk}\Gamma^r_{jl} - \Gamma^i_{rl}\Gamma^r_{jk} + \partial_k \Gamma^i_{jl} - \partial_l \Gamma^i_{jk} \, .$$

Related to this are the *Ricci curvature tensor* $R_{ij} = R^k_{ijk}$, the *Ricci curvature scalar* $R = R^i_i$, and the *Einstein curvature tensor* $E^i_j = R^i_j - \frac{1}{2}\delta^i_j R$ ($D_i E^i_j = 0$).

Let us consider a few simple examples.

Example 7 (The Role of the Metric Affinity in GCT Invariance)
An example in which we have a Euclidean space, but the coordinates are curvilinear (i.c. polar coordinates):

$$(x, y) = (r \cos \phi, r \sin \phi) \, .$$

In this coordinate system, all Γ^k_{ij} vanish identically except $\Gamma^\phi_{r\phi} = \Gamma^\phi_{\phi r} = r^{-1}$, and $\Gamma^r_{\phi\phi} = -r$. Therefore we have an extra, first order term in the GCT invariant Laplacian $D_i D^i \psi$, viz. $\Gamma^\phi_{r\phi}\partial_r \psi = r^{-1}\partial_r \psi$:

$$D_i D^i \psi = (\partial_{rr} + r^{-2}\partial_{\phi\phi} + r^{-1}\partial_r) \, \psi \, .$$

Clearly, this is the appropriate expression for $\Delta \psi$ in polar coordinates. In this example the Riemann curvature tensor vanishes identically, $R^l_{ijk} = 0$ (in whatever coordinate system), reflecting the flat nature of the spatial domain. Indeed, we can make the Γ^k_{ij} vanish globally simply by switching back to a Cartesian (or otherwise rectilinear) coordinate system.

Example 8 (An Intrinsically Curved Space)

One of the simplest examples in which we have a non-Euclidean metric is provided by the surface of a sphere: we can use the polar angles θ and ϕ in which we have $d\ell^2 = d\theta^2 + \sin^2\theta d\phi^2$, so: $g_{\theta\theta} = g^{\theta\theta} = 1$, $g_{\phi\phi} = \sin^2\theta$, $g^{\phi\phi} = \sin^{-2}\theta$ and $g_{\theta\phi} = g_{\phi\theta} = g^{\theta\phi} = g^{\phi\theta} = 0$. By a straightforward calculation we then find:

Affinity

All $\Gamma^k_{ij} = 0$ except $\Gamma^\theta_{\phi\phi} = -\sin\theta\cos\theta$, $\Gamma^\phi_{\theta\phi} = \Gamma^\phi_{\phi\theta} = \cot\theta$.

Riemann Tensor

All $R^i_{jkl} = 0$ except $R^\theta_{\phi\theta\phi} = -R^\theta_{\phi\phi\theta} = \sin^2\theta$, $R^\phi_{\theta\phi\theta} = -R^\phi_{\theta\theta\phi} = 1$.

Ricci Tensor

All $R_{ij} = 0$ except $R_{\theta\theta} = -1$ and $R_{\phi\phi} = -\sin^2\theta$. Alternatively: $R^i_j = \text{diag}\{-1, -1\}$. Note that the eigenvalues correspond to the principal curvatures of the umbilical surface, and that the Gaussian curvature is an overall constant: $\det R^i_j = 1$.

Ricci Scalar

$R = -2$ (i.e. the sum of principal curvatures).

Einstein Tensor

$E^i_j = R^i_j - \frac{1}{2}\delta^i_j R = 0$.

Since there exists no coordinate system in which the Riemann tensor vanishes identically, it is apparently impossible to find flat coordinates for the above Riemann space.

A DIFFERENTIAL GEOMETRIC APPROACH TO ANISOTROPIC DIFFUSION

David Eberly

Computer Science Department
University of North Carolina
Chapel Hill, NC 27599-3175, USA

1. Introduction

The necessity of using a multiscale analysis of images has clearly been established in the literature. The introduction of a continuous scale-space can be found in [187], [386], and [396]. The fundamental constraint on a continuous scale-space is that it be *causal*; that is, no spurious detail should be generated with increasing scale. Additional constraints involving linearity and symmetry lead to the fact that the Gaussian kernel is the unique scale-space filter. A detailed investigation of scale-space, including its natural differential operators and differential invariants is found in [358]. The issues of discretization of the operators are found in [213].

The essential foundations of a scale-space are that an image $I(\vec{x})$ is a physical observable with an inner scale σ_0, determined by the resolution of the sampling device, and an outer scale σ_1, limited by the field of view. A front–end vision system which allows for superposition of input stimuli (linearity), which samples and preprocesses its input in a symmetric way (rotational and translational invariance) and which has no preferred scale of measurement (scale invariance), can be modeled by the diffusion equation $B_\sigma(\vec{x}, \sigma) = \sigma \nabla^2 B(\vec{x}, \sigma)$ for $\vec{x} \in \mathbb{R}^n$ and $\sigma \in [\sigma_0, \sigma_1]$, with initial conditions $B(\vec{x}, \sigma_0) = I(\vec{x})$ for $\vec{x} \in \mathbb{R}^n$. The information derived at scale σ is $B(\vec{x}, \sigma) = K(\vec{x}, \sigma) \oplus I(\vec{x})$, which is the convolution of a radially symmetric Gaussian kernel of standard deviation σ with the input data. Researchers have investigated nonlinear diffusion processes as models of a front–end vision system [67, 293, 380], where the assumption of linearity is not made so that object interactions within the image can be accounted for.

A key idea in [358] is the invoking of dimensional analysis: A function relating physical observables must be independent of dimensional units. A set of *natural spatial coordinates* is proposed, namely $\vec{y} = \vec{x}/\sigma$. The natural distance between two points $\vec{x}_1$ and $\vec{x}_2$ at scale σ is $\|\vec{x}_1 - \vec{x}_2\|/\sigma$. Moreover, a *natural scale parameter* is proposed, namely $\tau = \ln(\sigma/\epsilon)$, where ϵ is a hidden scale whose unit is a dimension of length and which is dependent on the image acquisition process. The main consequence of their development is that differentiation at some selected scale σ can be made a well–posed operation if kernels constructed as derivatives of a Gaussian at the same scale are used. At a fixed scale σ, first–order dimensionless derivatives are σB_{x_i} for each spatial component x_i. Second–order dimensionless derivatives are $\sigma^2 B_{x_i x_j}$. In general, the dimensionless spatial derivatives are obtained from the usual partial derivatives by multiplying by the appropriate power of scale.

A definition is proposed for scale-space that is similar, but fundamentally different, from that described in [358]. It is desired that the front–end vision system shows rotational invariance, translational invariance, and zoom invariance (invariance with respect to changes in the units of measurement, which is equivalent to a uniform magnification in the space and scale variables). The change of variables to obtain the natural spatial coordinates $\vec{y} = \vec{x}/\sigma$ preserves translational invariance *if scale σ is assumed to be a constant*. It does not preserve translational invariance *through varying scale*; that is, the natural coordinates place an unnatural emphasis on the spatial origin $\vec{0}$. A definition for scale space should be developed which has all the desired invariances for all scales, not just for a fixed scale.

To obtain the desired invariances, note that a measured *spatial difference* is meaningful only in the context of the scale at which it is measured. Similarly, when making multiscale measurements, a measured *scale difference* is meaningful only in the context of the scale at which it is measured. These assumptions suggest specifying differential forms as the measurement tools. The dimensionless 1–forms to be used for scale-space measurements are

$$\frac{d\vec{x}}{\sigma} \quad \text{and} \quad \frac{d\sigma}{\sigma}.$$

In contrast, the differential forms induced by the change of variables $\vec{y} = \vec{x}/\sigma$ and $\tau = \ln(\sigma/\epsilon)$ proposed in [358] are

$$d\vec{y} = \frac{d\vec{x}}{\sigma} - \frac{\vec{x}}{\sigma}\frac{d\sigma}{\sigma} \quad \text{and} \quad d\tau = \frac{d\sigma}{\sigma}.$$

Note that for a *fixed scale* σ_0, the induced forms for natural coordinates are $d\vec{y} = d\vec{x}/\sigma_0$, which agree with the proposed forms. But for non–constant scale, the spatial forms are fundamentally different. In fact, one major consequence of using the proposed forms is that the geometry of scale-space is non–Euclidean, whereas the geometry of scale space using $d\bar{y}$ and $d\tau$ is still Euclidean.

This chapter provides the mathematical formalism of scale-space as a geometric entity. A concise coverage of the mathematics used in this chapter can be found in the previous chapter of this book (L. Florack et al.) and in [175]. In particular these references cover tensor calculus, Riemannian geometry, Christoffel symbols, and covariant derivatives. Section 2 gives the definition for the metric tensor of scale-space and shows how differentiation must be defined. Scale-space is shown to be Riemannian with constant negative curvature. The isometries of the space verify that scale space has the desired invariance under rotation, translation, and zoom. Section 3 describes how to compute the gradient and Hessian of a real–valued function defined on this scale-space metric. The theory of curves, geodesics, distance, integration, and curvature of surfaces is discussed. Section 4 introduces a more general metric for scale-space which depends on the image data itself. The selection of the metric automatically determines which anisotropic diffusion process must be used to generate multiscale data. The conductance term and the density term which occur in the more general model for heat transfer show up as parameters in the metric. Finally, Section 5 briefly mentions the applicability of the ideas to construction of cores of objects in gray scale images and to subsequent applications.

2. The Structure of Scale Space

Let scale-space be denoted by $\mathcal{S} = \mathbb{R}^n \times (0, \infty)$ with typical element denoted by $\vec{\xi} = (\vec{x}, \sigma)$. The vector $\vec{x} \in \mathbb{R}^n$ represents the spatial information of the point and $\sigma > 0$ represents the scale information of the point. For indexing purposes, we have $\xi_i = x_i$ for $1 \leq i \leq n$ and $\xi_{n+1} = \sigma$.

2.1. Metric Tensor

The measuring tool used for component ξ_i is the 1–form $d\xi_i/\sigma$. The motivation is that a measured spatial or scale difference $d\xi_i$ is meaningful only in the context of the scale at which it is measured. That is, we need only be concerned with the relative measurement $d\xi_i/\sigma$. Note that the 1–forms are dimensionless quantities. The metric of the space is therefore determined by

$$ds^2 = \sum_{i=1}^{n} \frac{dx_i^2}{\sigma^2} + \frac{1}{\rho^2} \frac{d\sigma^2}{\sigma^2}.$$

The parameter $\rho \geq 0$ relates the importance of scale measurements to spatial measurements. The $(n+1) \times (n+1)$ metric tensor (as a 2×2 block diagonal matrix) is

$$G = [g_{ij}] = \frac{1}{\sigma^2} \text{diag}\left(I, \frac{1}{\rho^2}\right).$$

Given two vectors $\vec{V}_i = (\vec{W}_i, \gamma_i) \in \mathbb{R}^n \times \mathbb{R}$, $i = 1, 2$, with initial point at $(\vec{x}, \sigma) \in \mathcal{S}$, their dot product with respect to the metric G is defined as

$$\vec{V}_1 \odot \vec{V}_2 = \vec{V}_1{}^t G \vec{V}_2 = \frac{1}{\sigma^2}\left(\vec{W}_1 \cdot \vec{W}_2 + \frac{\gamma_1\gamma_2}{\rho^2}\right),$$

where the single dot symbol on the right–hand side of the definition represents the Euclidean dot product. Note that, unlike the Euclidean dot product, the scale-space dot product does depend on the initial point of the vectors. The length of a vector $\vec{V} = (\vec{W}, \gamma)$ with initial point at $(\vec{x}, \sigma) \in \mathcal{S}$ is defined by

$$\|\vec{V}\| = \sqrt{\vec{V} \odot \vec{V}} = \frac{1}{\sigma}\sqrt{|\vec{W}|^2 + \frac{\gamma^2}{\rho^2}},$$

where the single bars represent Euclidean length. The angle between the two vectors is determined by

$$\cos\theta = \frac{\vec{V}_1 \odot \vec{V}_2}{\|\vec{V}_1\| \, \|\vec{V}_2\|}, \quad \theta \in [0, \pi].$$

The angle is well–defined since the right–hand side can be shown to be no larger than 1 in magnitude; i.e. the Cauchy–Schwartz inequality holds in this space. The condition for orthogonality of vectors is $\vec{V}_1 \odot \vec{V}_2 = 0$.

2.2. Christoffel Symbols and Covariant Derivatives

The analysis involves tensor quantities which need to be differentiated. In general, the partial derivative of a tensor is not necessarily a tensor. Tensor differentiation requires the use of some nontensorial objects called Christoffel symbols.

Let $\vec{e}_k \in \mathbb{R}^{n+1}$ denote the $(n + 1) \times 1$ unit length vector whose components are all 0 except for the kth component which is 1. The Christoffel symbols of the second–kind are defined by

$$\Gamma_{ij}^k = \frac{1}{2}\sum_{\ell=1}^{n+1} g^{k\ell}\left(\frac{\partial g_{j\ell}}{\partial x_i} + \frac{\partial g_{i\ell}}{\partial x_j} - \frac{\partial g_{ij}}{\partial x_\ell}\right)$$

where the g_{ij} are the components of G and the g^{ij} are the components of the inverse matrix G^{-1}. For the scale-space metric G, define the matrix

$\Gamma^k = [\Gamma^k_{ij}]$; then

$$\Gamma^k = -\frac{1}{\sigma}\left\{\begin{array}{ll} \vec{e}_k\vec{e}^t_{n+1} + \vec{e}_{n+1}\vec{e}^t_k, & 1 \le k \le n \\ \vec{e}_{n+1}\vec{e}^t_{n+1} - \rho^2\sum_{\ell=1}^n \vec{e}_\ell\vec{e}^t_\ell, & k = n+1 \end{array}\right\}.$$

For example, if $n = 1$,

$$\Gamma^1 = -\frac{1}{\sigma}\begin{bmatrix} 0 & 1 \\ 1 & 0 \end{bmatrix}, \quad \Gamma^2 = -\frac{1}{\sigma}\begin{bmatrix} -\rho^2 & 0 \\ 0 & 1 \end{bmatrix}$$

and if $n = 2$,

$$\Gamma^1 = -\frac{1}{\sigma}\begin{bmatrix} 0 & 0 & 1 \\ 0 & 0 & 0 \\ 1 & 0 & 0 \end{bmatrix}, \Gamma^2 = -\frac{1}{\sigma}\begin{bmatrix} 0 & 0 & 0 \\ 0 & 0 & 1 \\ 0 & 1 & 0 \end{bmatrix}, \Gamma^3 = -\frac{1}{\sigma}\begin{bmatrix} -\rho^2 & 0 & 0 \\ 0 & -\rho^2 & 0 \\ 0 & 0 & 1 \end{bmatrix}$$

Note that each Γ^k is a symmetric matrix.

Let $\vec{T} : \mathbb{R}^{n+1} \to \mathbb{R}^{n+1}$ be a vector field, say $\vec{T}(\vec{x}) = [T_i(\vec{x})]$. As a vector field defined on Euclidean space, the *derivative of* $\vec{T}$ is the $(n+1) \times (n+1)$ matrix of partial derivatives given by $d\vec{T}/d\vec{\xi} = [\partial T_i/\partial \xi_j]$, where i is the row index and j is the column index. In a space whose curvature is not identically zero, the matrix of partial derivatives is not a tensor. The derivative of a tensor should again be a tensor. Intuitively, the measured differences are in a curved space, so the the usual derivative measurements need to be "corrected" to be consistent with the curvature of the space. *Covariant differentiation* is the correct generalization of ordinary differentiation. The application of covariant differentiation to vectors requires knowing whether the vector is *covariant* or *contravariant*. A covariant vector is analogous to the gradient of a function, which is normal to level surfaces. A contravariant vector is analogous to tangent vectors of the level surfaces. Let $\vec{U} = [U_i] = (\vec{W}, \gamma)$ be a covariant vector defined on scale-space. The *covariant derivative of* $\vec{U}$ is the second–order tensor defined by

$$\frac{d\vec{U}}{d\vec{\xi}} = \left[\frac{\partial U_i}{\partial \xi_j} - \sum_{k=1}^{n+1} \Gamma^k_{ij}U_k\right] = \frac{d\vec{U}}{d\vec{\xi}} + \frac{1}{\sigma}\left[\begin{array}{c|c} -\gamma\rho^2 I & \vec{W} \\ \hline \vec{W}^t & \gamma \end{array}\right].$$

Let $\vec{V} = [V^i] = (\vec{W}, \gamma)$ be a contravariant vector. The *covariant derivative of* $\vec{V}$ is the second–order tensor defined by

$$\frac{d\vec{V}}{d\vec{\xi}} = \left[\frac{\partial V^i}{\partial \xi_j} + \sum_{k=1}^{n+1} \Gamma^i_{jk}V^k\right] = \frac{d\vec{V}}{d\vec{\xi}} + \frac{1}{\sigma}\left[\begin{array}{c|c} -\gamma I & -\vec{W} \\ \hline \rho^2\vec{W}^t & -\gamma \end{array}\right].$$

In either case, the covariant derivative definitions include not only the usual partial derivatives but also correction terms dependent on the Christoffel symbols. A contravariant vector $\vec{V}$ can be converted to a covariant vector by $\vec{U} = G\vec{V}$. The two covariant derivative tensors are related by

$$\frac{\mathrm{d}\vec{U}}{\mathrm{d}\vec{\xi}} = \frac{\mathrm{d}(G\vec{V})}{\mathrm{d}\vec{\xi}} = G\frac{\mathrm{d}\vec{V}}{\mathrm{d}\vec{\xi}} \; ;$$

that is, the metric tensor is treated as a constant by covariant differentiation.

2.3. Riemann Tensor and Curvature

The implication of using the 1–forms $d\xi_i/\sigma$ for relative measurements is that the geometry of scale-space is Riemannian. The Riemann tensor of the second kind is defined by

$$R^i_{jk\ell} = \frac{\partial \Gamma^i_{j\ell}}{\partial x_k} - \frac{\partial \Gamma^i_{jk}}{\partial x_\ell} + \sum_{m=1}^{n+1} \Gamma^m_{j\ell}\Gamma^i_{mk} - \sum_{m=1}^{n+1} \Gamma^m_{jk}\Gamma^i_{m\ell}.$$

The Riemann curvature relative to the metric G is defined for each pair of contravariant vectors $\vec{U} = [U^i]$ and $\vec{V} = [V^i]$ as

$$K(\vec{x}; \vec{U}, \vec{V}) = \frac{\sum_{i,j,k,\ell} R_{ijk\ell} U^i V^j U^k V^\ell}{\sum_{i,j,k,\ell} G_{ijk\ell} U^i V^j U^k V^\ell}$$

where $R_{ijk\ell} = \sum_{m=1}^{n+1} g_{im} R^m_{jk\ell}$ and $G_{ijk\ell} = g_{ik}g_{j\ell} - g_{i\ell}g_{jk}$. It can be shown that the only nonzero independent components are

$$R_{ijij} = -\frac{\rho^2}{\sigma^4} \quad \text{and} \quad G_{ijij} = \frac{1}{\sigma^4},$$

for $1 \leq i < j \leq n+1$. The other components are either 0 or are determined by the values of the terms mentioned above. Consequently, the Riemann curvature of the space is identically a constant, $K = -\rho^2$. When $\rho = 1$, this particular Riemannian geometry is called *hyperbolic geometry* and has been studied extensively in the differential geometry literature. A basic description of hyperbolic geometry is found in [102]. An advanced treatise on the subject is [30].

2.4. Isometries

When analyzing an image using multiscale techniques, the measurements should be invariant with respect to certain transformations. In particular, they should be invariant under spatial rotations, spatial translations, and spatial reflections. Object shape information produced by an algorithm should not depend on the orientation of the object being measured.

Invariance with respect to *zoom* is also required. That is, if scale measurements are made for a given object, the scale measurements for a magnified version of the object should be the magnified measurements for the original object.

An *isometry* for scale-space is a function $\vec{\psi} : S \to S$ whose differential preserves the dot product of vectors. Specifically, if $\vec{\psi} = (\psi_1, \ldots, \psi_{n+1})$ and $d\vec{\psi}/d\vec{\xi}$ is the matrix whose $(i,j)^{\text{th}}$ entry is $\partial\psi_i/\partial\xi_j$, then $\vec{\psi}$ is an isometry iff

$$\left(\frac{d\vec{\psi}}{d\vec{\xi}}\vec{V_1}\right) \odot \left(\frac{d\vec{\psi}}{d\vec{\xi}}\vec{V_1}\right) = \vec{V_1} \odot \vec{V_2},$$

where the vectors $\vec{V_i}$ are positioned at $\vec{\xi}$ and the vectors $(d\vec{\psi}/d\vec{\xi})\vec{V_i}$ are positioned at $\vec{\psi}(\vec{\xi})$. Scale-space has the following isometries which provide the invariances described previously. It has an additional isometry which is mentioned for completeness.

$$
\begin{aligned}
\vec{\psi}(\vec{x},\sigma) &= (\vec{x}+\vec{a},\sigma), && \text{translation by constant vector } \vec{a} \\
\vec{\psi}(\vec{x},\sigma) &= (R\vec{x},\sigma), && R \text{ is a rotation or reflection matrix} \\
\vec{\psi}(\vec{x},\sigma) &= (\mu\vec{x},\mu\sigma), && \text{for any } \mu > 0 \text{ (zoom with magnification } \mu) \\
\vec{\psi}(\vec{x},\sigma) &= \frac{(\vec{x},\sigma)}{\|(\vec{x},\sigma)\|}, && \text{inversion with respect to a hypersphere}
\end{aligned}
$$

Intuitively, the first three isometries indicate that if measurements involving angles or lengths are made at a given point, the measurements will be the same if you translate or rotate the spatial coordinates or if you change the units of measurement by zooming both space and scale. A more detailed discussion of the isometries is found in [363].

3. Measurements in Scale-Space

In this section some basic formulas are derived that are needed in multiscale algorithms. Since the space is Riemannian, the measurements must take into account the curvature of the space.

3.1. Gradient and Hessian

Let $f : S \to \mathbb{R}$ be a twice–differentiable function. Let f_{ξ_i} denote the partial derivative of f with respect to ξ_i and let $f_{\xi_i\xi_j}$ denote the second partial derivative of f with respect to ξ_i and ξ_j. In Euclidean space the gradient of f is the vector $\nabla f = [f_{\xi_i}]$, and the Hessian of f is the matrix $Hf = [f_{\xi_i\xi_j}]$.

The differential of f is given by

$$df = \underbrace{\sum_{i=1}^{n+1} f_{\xi_i} d\xi_i}_{\text{Euclidean}} = \underbrace{\sum_{i=1}^{n} \sigma f_{x_i} \frac{dx_i}{\sigma} + \rho\sigma f_\sigma}_{\text{scale-space}}.$$

Therefore, the *scale-space gradient of* f is defined to be

$$\hat{\nabla} f = \sqrt{G^{-t}} \, \nabla f = (\sigma f_{x_1}, \dots, \sigma f_{x_n}, \rho\sigma f_\sigma).$$

The natural derivative operators are $\sigma \, \partial/\partial\xi_i$, as indicated in [218] and [358]. The relationship of scale-space gradient and Euclidean gradient is reminiscent of the equation obtained when making a change of variables $\vec{\zeta} = M\vec{\xi}$ and $g(\vec{\zeta}) = f(\vec{\xi})$, where $\nabla g = M^{-t}\nabla f$. (The notation M^{-t} is a concise representation of $(M^{-1})^t$.) Taking second derivatives yields

$$\mathrm{H}g = M^{-t} \frac{d\nabla f}{d\vec{\xi}} M^{-1} = M^{-t}\mathrm{H}f M^{-1}.$$

The similarity carries over to defining the *scale-space Hessian of* f, but covariant differentiation of the covariant vector ∇f must be used, denoted by d:

$$\begin{aligned}
\hat{H}f &= \sqrt{G^{-t}} \frac{\mathrm{d}\nabla F}{\mathrm{d}\vec{\xi}} \sqrt{G^{-1}} \\
&= \mathrm{diag}(I,\rho) \left(\sigma^2 \mathrm{H}f + \sigma \left[\begin{array}{c|c} -\rho^2 f_\sigma I & \nabla_{\vec{x}} f \\ \hline \nabla_{\vec{x}} f^t & f_\sigma \end{array} \right] \right) \mathrm{diag}(I,\rho),
\end{aligned}$$

where $\mathrm{H}f$ is the Hessian matrix of f in Euclidean space and where $\nabla f = (\nabla_{\vec{x}} f, f_\sigma)$. In [218] and [358] it is suggested that the natural second derivatives are just $\sigma^2 f_{\xi_i \xi_j}$. But in this Riemannian setting, the correction terms must be included. The derivatives $\sigma^2 f_{\xi_i \xi_j}$ are natural only if the scale is fixed and the differential calculations are made on the manifold $\sigma = \sigma_0$, which is embedded in scale-space.

3.2. Curves

Consider curves $\vec{\xi}(t) = (\vec{x}(t), \sigma(t))$ in scale-space. The speed of a particle traveling along the curve is $ds/dt = \|\vec{\xi}'(t)\|$, the length of the tangent vector $\vec{\xi}'(t)$. If $t \in [t_0, t_1]$, the arc length of the curve is $\int_{t_0}^{t_1} ds/dt$. A curve is parametrized by arc length if $ds/dt \equiv 1$. To obtain a unit length tangent vector $\vec{T}(t)$, make the usual adjustment $\vec{T}(t) = \vec{\xi}'(t)/\|\vec{\xi}'(t)\|$.

If the curve is parameterized by arc length s, the unit length tangent vector is just $\vec{T}(s) = \vec{\xi}'(s)$. In Euclidean geometry, the s– derivative of $\vec{T}(s)$ is taken to obtain a vector which is normal to the curve. In Riemannian geometry the ordinary derivative is not necessarily a tensor. The analog is to take the *absolute derivative* along the curve, which does yield a tensor quantity. For a contravariant vector $\vec{V}(\vec{\xi})$ defined on a curve $\vec{\xi}(t)$, its absolute derivative is defined by

$$\frac{\mathrm{d}\vec{V}}{\mathrm{d}t} = \frac{\mathrm{d}\vec{V}}{\mathrm{d}\vec{\xi}}\frac{\mathrm{d}\vec{\xi}}{\mathrm{d}t}.$$

For a curve parametrized by arc length s, the unit principal normal is the contravariant vector $\vec{N}(s)$ defined by $\mathrm{d}\vec{T}/\mathrm{d}s = \kappa(s)\vec{N}$, where $\kappa(s) = \|\mathrm{d}\vec{T}/\mathrm{d}s\| \geq 0$ is the curvature of the curve. The explicit components of the absolute derivative of the tangent vector are

$$\left(\frac{\mathrm{d}T}{\mathrm{d}s}\right)_i = \frac{d^2\xi_i}{ds^2} + \sum_{j=1}^{n+1}\sum_{k=1}^{n+1}\Gamma^i_{jk}\frac{d\xi_j}{ds}\frac{d\xi_k}{ds}$$

$$= \begin{cases} \frac{d^2x_i}{ds^2} - \frac{2}{\sigma}\frac{dx_i}{ds}\frac{d\sigma}{ds}, & 1 \leq i \leq n \\ \frac{d^2\sigma}{ds^2} - \frac{1}{\sigma}\left[\left(\frac{d\sigma}{ds}\right)^2 - \rho^2\sum_{j=1}^{n}\left(\frac{dx_j}{ds}\right)^2\right], & i = n+1 \end{cases}.$$

3.3. Geodesics

A curve is a *geodesic* if its curvature is identically zero, which means

$$(\mathrm{d}\vec{T}/\mathrm{d}s)_i = 0$$

for all i. The geodesics are then solutions to the system of ordinary differential equations

$$\frac{d^2\vec{x}}{ds^2} = \frac{2}{\sigma}\frac{d\sigma}{ds}\frac{d\vec{x}}{ds} \quad \text{and} \quad \frac{d^2\sigma}{ds^2} = \frac{1}{\sigma}\left[\left(\frac{d\sigma}{ds}\right)^2 - \left|\rho\frac{d\vec{x}}{ds}\right|^2\right].$$

These can be solved in closed form to obtain either

$$(\vec{x}(s), \sigma(s)) = (\vec{c}, r\exp(\rho s)) \tag{14.1}$$

for some constants $\vec{c}$ and $r > 0$, in which case the geodesic is a line in the direction of the scale axis, or

$$(\vec{x}(s), \sigma(s)) = (\vec{c} + r\tanh(\rho s)\rho^{-1}\vec{u}, r\operatorname{sech}(\rho s)) \tag{14.2}$$

for constants $\vec{c}, \vec{u} \in \mathbb{R}^n$ with $|\vec{u}| = 1$, and $r > 0$, in which case the geodesics are curves on half–ellipses of the form $|\rho(\vec{x} - \vec{c})|^2 + \sigma^2 = r^2$ with center on the hyperplane $\sigma = 0$.

Geodesics act as paths of minimum distance between points in scale-space; see the next subsection. They also can be used to illustrate why scale-space

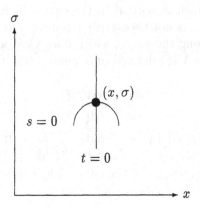

<figure>Figure 14.1. Geodesic coordinate axes</figure>

derivatives (of first and second order) are the natural derivatives to compute in this space. Define $\vec{\xi}(s,t) = (\vec{x} + \sigma e^{\rho t}\tanh(\rho s)\vec{e}_k, \sigma e^{\rho t}\text{sech}(\rho s))$. The curve $\vec{\xi}(s,0)$ is a half–ellipse geodesic whose north pole is at $(\vec{x}, \sigma)$, and the curve $\vec{\xi}(0,t)$ is a straight–line geodesic. The point $(\vec{x}, \sigma)$ can be thought of as the origin of a coordinate system (corresponding to $(s,t) = (0,0)$). Figure 14.1 shows typical (local) coordinate axes given by the curves when $s = 0$ and $t = 0$. Define $\phi(s,t) = f(\vec{\xi}(s,t))$. Some computations will show that $\phi_s(0,0) = \sigma f_{x_k}$, $\phi_t(0,0) = \rho \sigma f_\sigma$, $\phi_{ss}(0,0) = (\sigma^2 f_{x_k x_k} - \rho^2 \sigma f_\sigma)$, $\phi_{st}(0,0) = \rho(\sigma^2 f_{x_k \sigma} + \sigma f_{x_k})$, and $\phi_{tt}(0,0) = \rho^2(\sigma^2 f_{\sigma\sigma} + \sigma f_\sigma)$, where the derivatives of f are all evaluated at $(\vec{x}, \sigma)$. These relationships show that at the origin of the geodesic coordinates (s,t) the components of the first– and second–order covariant derivatives for f are the same as the first– and second–order partial derivatives of the function $\phi(s,t)$ at the origin. The analogy does not hold for third and higher order derivatives of ϕ since covariant differentiation of those orders is not necessarily commutative because of the effects of the curvature of the space.

3.4. Distance Between Points

Given two points $(\vec{x}_k, \sigma_k)$ for $k = 1,2$, the distance between them is measured along the unique geodesic path connecting the points. Let $(\vec{x}(s), \sigma(s))$ be a parameterization by scale–space arc length of a geodesic such that $(\vec{x}_k, \sigma_k) = (\vec{x}(s_k), \sigma(s_k))$ for some s_k, $k = 1, 2$. Without loss of generality assume that $\sigma_1 \leq \sigma_2$ and $s_1 \leq s_2$. The distance between the two points is $s_2 - s_1$.

If the geodesic is a line in the σ direction, then $\vec{x}_1 = \vec{x}_2$ and the distance

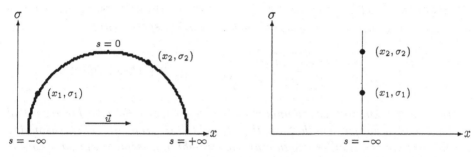

Figure 14.2. Geodesics as shortest paths

between the two points is

$$\text{dist}((\vec{x}_1, \sigma_1), (\vec{x}_2, \sigma_2)) = \frac{1}{\rho} \ln \left(\frac{\sigma_1}{\sigma_2} \right).$$

Otherwise, the geodesic connecting the two points is a half–ellipse, and the distance between the two points is derived as follows. Let $\vec{u} = (\vec{x}_2 - \vec{x}_1)/L$ where $L = |(\vec{x}_2 - \vec{x}_1)|$. The choice of $\vec{u}$ and the ordering of scales $\sigma_1 \leq \sigma_2$ implies that $s_1 < 0$ and $s_2 > s_1$ (see Figure 14.2). From the geodesic equations, $\text{sech}(\rho s_k) = \sigma_k/r$. Using the identity $\tanh^2(z) + \text{sech}^2(z) \equiv 1$, $\tanh(\rho s_1) = -[1 - \text{sech}^2(\rho s_1)]^{1/2} = -[1 - (\sigma_1/r)^2]^{1/2}$, where the minus sign is a result of $s_1 < 0$. Also, $(\vec{x}_2 - \vec{x}_1) = (r/\rho)[\tanh(\rho s_2) - \tanh(\rho s_1)]\vec{u}$, which leads to $\tanh(\rho s_2) = L/r - \sqrt{1 - (\sigma_1/r)^2}$. Finally, it can be shown that

$$\kappa = \frac{1}{r} = \frac{2L}{\sqrt{(\sigma_1^2 - \sigma_2^2)^2 + L^2[L^2 + 2(\sigma_1^2 + \sigma_2^2)]}}.$$

The distance between the two points is

$$
\begin{aligned}
\text{dist}((\vec{x}_1, \sigma_1), (\vec{x}_2, \sigma_2)) &= \frac{1}{\rho} \left(\ln \left(\frac{\text{sech}(\rho s_2)}{1 - \tanh(\rho s_2)} \right) - \ln \left(\frac{\text{sech}(\rho s_1)}{1 - \tanh(\rho s_1)} \right) \right) \\
&= \lambda \ln \left(\frac{\sigma_2}{\sigma_1} \frac{1 + \sqrt{1 - (\kappa \sigma_1)^2}}{1 + \sqrt{1 - (\kappa \sigma_1)^2} - \kappa L} \right).
\end{aligned}
$$

Note that this distance formula also applies in the case $\vec{x}_1 = \vec{x}_2$ because $L = 0$. Figure 14.2 illustrates two geodesic curves and the relative position of the points on them.

Example   Let $\rho = 1$. Define $a = 1/\sqrt{2}$ and let $(x(t), \sigma(t)) = (t, a)$ for $|t| \leq a$. The scale-space arc length between $(-a, a)$ and (a, a) along the specified constant scale arc is

$$\ell = \int_{-a}^{a} \frac{\sqrt{\dot{x}^2(t) + \dot{\sigma}^2(t)}}{\sigma(t)} \, dt = \int_{-a}^{a} \frac{1}{a} \, dt = 2.$$

Along the geodesic circular arc $(x(s), \sigma(s)) = (\tanh(s), \text{sech}(s))$ between $(x_1, \sigma_1) = (-a, a)$ and $(x_2, \sigma_2) = (a, a)$, the scale-space distance is

$$\delta = \left| \ln\left(\frac{a}{a} \frac{1 + \sqrt{1 - a^2} - 2a}{1 + \sqrt{1 - a^2}} \right) \right| = 2\ln(\sqrt{2} + 1) \simeq 1.763 < \ell.$$

If the same lengths were computed in Euclidean space, the circular arc would have had a larger length than the straight line segment. The reason the circular arc has smaller length than the straight line segment in scale-space is that the spatial domain is in effect "less dense" at larger scales. The distance between spatial locations x_1 and x_2 at scale $\sigma = 2$ is half the distance between the same locations at scale $\sigma = 1$. Initially traversing the circular arc through increasing scale will accumulate a total distance which is smaller than that accumulated by traversing the line segment at a given scale. The north pole of the geodesic represents the point at which further increasing the scale of the path is no longer cost effective (as measured by total distance traveled).

3.5. Volume Integrals, Hyperspheres

Integration in scale-space must also take into account the relative 1–forms that define the space. If $f(\vec{x}, \sigma)$ is a real–valued function defined on a region V, the *scale-space integral of f over V* is given by

$$\int_V f(\vec{x}, \sigma) \frac{d\sigma \prod_{i=1}^{n} dx_i}{\rho \sigma^{n+1}}.$$

For example, let $\rho = 1$ and consider the scale-space hypersphere of radius R centered at $(\vec{x}_1, \sigma_1)$. The set consists of all points $(\vec{x}, \sigma)$ which are R units of scale-space distance from the central point. The defining equation is

$$\left| \ln\left(\frac{\text{sech}(s)}{1 - \tanh(s)} \frac{1 - \tanh(s_1)}{\text{sech}(s_1)} \right) \right| = R,$$

where s and s_1 are the arc length parameter values for the points $(\vec{x}, \sigma)$ and $(\vec{x}_1, \sigma_1)$, respectively, along the unique geodesic containing the points. Some algebraic computation will show that this equation is equivalent to

$$|\vec{x} - \vec{x}_1|^2 + [\sigma - \sigma_1 \cosh(R)]^2 = [\sigma_1 \sinh(R)]^2,$$

which is the equation of a Euclidean hypersphere centered at $(\vec{x}_1, \sigma_1 \cosh(R))$ and whose radius is $\sigma_1 \sinh(R)$. Note that the center of the scale-space hypersphere is not the same as the center of the Euclidean hypersphere. Let S_{n+1} be the set of points $(\vec{x}, \sigma)$ satisfying $|\vec{x} - \vec{x}_1|^2 + [\sigma -$

$\sigma_1 \cosh(R)]^2 \le [\sigma_1 \sinh(R)]^2$. In scale-space the volume of the hypersphere is

$$V_{\mathrm{sc}}^{(n+1)}(R) = \int_{S_{n+1}} \frac{dx_1 \cdots dx_n \, d\sigma}{\sigma^n \, \sigma} = \frac{2\pi^{\frac{n+1}{2}}}{\Gamma\left(\frac{n+1}{2}\right)} \int_0^R \sinh^n(r) \, dr.$$

Note that the volume is independent of the center of the hypersphere, even though the hypersphere appears to be "larger" as the scale component of the center is increased. In comparison, in Euclidean space the volume of the hypersphere is

$$V_{\mathrm{eu}}^{(n+1)}(R) = \int_{S_{n+1}} dx_1 \cdots dx_n d\sigma = \frac{2\pi^{\frac{n+1}{2}}}{\Gamma\left(\frac{n+1}{2}\right)} \int_0^R r^n \, dr.$$

For the special cases $n = 1, 2, 3$, the scale-space volumes are $2\pi[\cosh(R) - 1]$, $\pi[\sinh(2R) - 2R]$, and $2\pi^2[\cosh^3(R) - 3\cosh(R) + 2]/3$, respectively. For the same special cases, the Euclidean volumes are πR^2, $4\pi R^3/3$, and $2\pi^2 R^4$, respectively.

3.6. Curvature of Surfaces

Given an n–dimensional surface embedded in $(n+1)$–dimensional Euclidean space, *principal curvatures* and *principal directions* can be constructed at each point on the surface. The analogous quantities are derived for an n–dimensional surface embedded in $(n + 1)$–dimensional Riemannian space. For a self–contained discussion, the constructions for Euclidean space are given first.

3.6.1. *Graphs in Euclidean Space*

Let $f : \mathbb{R}^n \to \mathbb{R}$ be a function, say $\sigma = f(\vec{x})$, whose graph lives in Euclidean space $\mathbb{R}^n \times \mathbb{R}$. If $(\vec{x}(s), \sigma(s))$ is a curve on the graph which is parametrized by arc length, then $\sigma(s) = f(\vec{x}(s))$, $\sigma' = \vec{x}' \cdot \nabla f$ and $\sigma'' = \vec{x}'' \cdot \nabla f + \vec{x}' \cdot (\mathrm{H}f)\vec{x}'$. Unit tangent vectors to the curve are $\vec{T}(s) = (\vec{x}'(s), \sigma'(s))$. A unit normal to the graph of f is $\vec{N}(s) = (-\nabla f, 1)/\ell_N$, where $\ell_N^2 = 1 + |\nabla f|^2$. For an arbitrary vector $\vec{V} \in \mathbb{R}^n$, the unit tangents to the graph are $\vec{T} = (\vec{V}, \vec{V} \cdot \nabla f)/\ell_T$, where $\ell_T^2 = \vec{V}^t(I + \nabla f \nabla f^t)\vec{V}$. The curvature of the curve in the normal section determined by $\vec{V}$ and $\vec{N}$ is given by

$$\kappa = \vec{N} \cdot \frac{d\vec{T}}{ds} = \frac{\vec{V}^t\left[\frac{1}{\ell_N}\mathrm{H}(f)\right]\vec{V}}{\vec{V}^t(I + \nabla f \nabla f^t)\vec{V}}.$$

The principal curvatures κ and principal directions $\vec{V}$ are the eigenvalues and eigenvectors solving the general eigensystem

$$\left(\frac{H(f)}{\sqrt{1+|\nabla f|^2}} - \kappa\left(I + \nabla f \nabla f^{\,t}\right)\right)\vec{V} = \vec{0}.$$

The matrices representing the first and second fundamental forms for the graph of f are $I + \nabla f \nabla f^{\,t}$ and $H(f)/\sqrt{1+|\nabla f|^2}$, respectively. The shape operator is the matrix

$$W = (I + \nabla f \nabla f^{\,t})^{-1}\frac{H(f)}{\sqrt{1+|\nabla f|^2}},$$

so the principal curvatures and principal directions are also solutions to the regular eigensystem $W\vec{V} = \kappa\vec{V}$.

3.6.2. Graphs in Scale-Space

Let $f : \mathbb{R}^n \to \mathbb{R}$ be a function, say $\sigma = f(\vec{x})$, whose graph lives in scale-space $\mathbb{R}^n \times (0,\infty)$ with metric $G = \sigma^{-2}\mathrm{diag}(I,\rho^{-1})$. As in the Euclidean case, a curve $(\vec{x}(s),\sigma(s))$ on the graph and which is parameterized by arc length satisfies $\sigma(s) = f(\vec{x}(s))$, $\sigma' = \vec{x}'\cdot\nabla f$ and $\sigma'' = \vec{x}''\cdot\nabla f + \vec{x}'\cdot(Hf)\vec{x}'$. The unit tangent vectors are still $\vec{T}(s) = (\vec{x}'(s),\sigma'(s))$, but unit normals to the graph of f are now $\vec{N}(s) = (-\rho^{-2}\nabla f, 1)/\ell_N$, where $\ell_N^2 = \sigma^{-2}(1+|\rho^{-1}\nabla f|^2)$. Select a constant vector $\vec{V} \in \mathbb{R}^n$ such that the unit tangents are $\vec{T} = (\vec{V}, \vec{V}\cdot\nabla f)/\ell_T$, where $\ell_T^2 = (\rho\sigma)^{-2}\vec{V}^{\,t}(\rho^2 I + \nabla f \nabla f^{\,t})\vec{V}$. The curvature of the curve in the normal section determined by $\vec{V}$ and $\vec{N}$ is given by

$$\kappa = \vec{N}\odot\frac{\mathrm{d}\vec{T}}{\mathrm{d}s} = \frac{\vec{V}^{\,t}\left[\frac{1}{\ell_N}H\left(\frac{|\rho\vec{x}|^2+f^2}{2}\right)\right]\vec{V}}{\vec{V}^{\,t}(\rho^2 I + \nabla f \nabla f^{\,t})\vec{V}}.$$

The principal curvatures κ and principal directions $\vec{V}$ are the eigenvalues and eigenvectors solving the general eigensystem

$$\left(\frac{H\left(\frac{|\rho\vec{x}|^2+f^2}{2}\right)}{\sqrt{1+|\rho^{-1}\nabla f|^2}} - \kappa(\rho^2 I + \nabla f \nabla f^{\,t})\right)\vec{V} = \vec{0}.$$

The matrices representing the first and second fundamental forms for the graph of f, as a surface embedded in scale-space, are $\rho^2 I + \nabla f \nabla f^{\,t}$ and $H((|\rho\vec{x}|^2 + f^2)/2)/\sqrt{1+|\rho^{-1}\nabla f|^2}$, respectively. The shape operator is the matrix

$$W = (\rho^2 I + \nabla f \nabla f^{\,t})^{-1}\frac{H\left(\frac{|\rho\vec{x}|^2+f^2}{2}\right)}{\sqrt{1+|\rho^{-1}\nabla f|^2}},$$

so the principal curvatures and principal directions are also solutions to the regular eigensystem $W\vec{V} = \kappa\vec{V}$.

3.6.3. Implicitly Defined Surfaces in Euclidean Space

Let $F : \mathbb{R}^n \times (0,\infty) \to \mathbb{R}$ be a function whose level surfaces implicitly defined by $F(\vec{\xi}) = c$ live in Euclidean space $\mathbb{R}^n \times (0,\infty)$. If $\vec{\xi}(s)$ is a curve parametrized by arc length and lives on the level surface defined by $F \equiv 0$, then $F(\vec{\xi}(s)) \equiv 0$. The unit tangent vectors are $\vec{T}(s) = \vec{\xi}'(s)$. Unit scale-space length normals are given by $\vec{N}(s) = \widehat{\nabla}F(\vec{\xi}(s))/|\widehat{\nabla}F(\vec{\xi}(s))|$. Assume that the curve is such that $\vec{T}$ and $\vec{N}$ determine a normal section. The curvature of the curve is

$$\kappa = \vec{N} \cdot \frac{d\vec{T}}{ds} = -\vec{T} \cdot \frac{d\vec{N}}{ds} = -\vec{T} \cdot \frac{d\vec{N}}{d\vec{\xi}}\vec{T} = -\frac{\vec{T}^{\,t}\frac{d\vec{N}}{d\vec{\xi}}\vec{T}}{\vec{T}^{\,t}\vec{T}},$$

where the identities $\vec{T} \cdot \vec{N} \equiv 0$ and $\vec{T} \cdot \vec{T} \equiv 1$ were used. The principal curvatures κ and principal directions $\vec{T}$ are the eigenvalues and tangential eigenvectors solving

$$-\frac{d\vec{N}}{d\vec{\xi}}\vec{T} = \kappa\vec{T}.$$

The matrix $-d\vec{N}/d\vec{\xi}$ represents the shape operator as an operation applied to the tangent spaces in the ambient $(n+1)$–dimensional space, as compared to the usual representation as an operator on the n–dimensional tangent spaces.

3.6.4. Implicitly Defined Surfaces in Scale Space

Let $F : \mathbb{R}^n \times (0,\infty) \to \mathbb{R}$ be a function whose level surfaces implicitly defined by $F(\vec{\xi}) = c$ live in scale-space $\mathbb{R}^n \times (0,\infty)$ with metric $G = \sigma^{-2}\mathrm{diag}(I,\rho^{-1})$. As in the Euclidean case, a curve $\vec{\xi}(s)$ parametrized by arc length and living on the level surface defined by $F \equiv 0$ satisfies $F(\vec{\xi}(s)) \equiv 0$. The unit tangent vectors are still $\vec{T}(s) = \vec{\xi}'(s)$. Unit normals are now given by $\vec{N}(s) = \nabla F(\vec{\xi}(s))/\|\nabla F(\vec{\xi}(s))\|$, where the length calculation is with respect to the metric. Assume that the curve is such that $\vec{T}$ and $\vec{N}$ determine a normal section. The curvature of the curve is

$$\kappa = \vec{N} \odot \frac{d\vec{T}}{ds} = -\vec{T} \odot \frac{d\vec{N}}{ds} = -\vec{T} \odot \frac{d\vec{N}}{d\vec{\xi}}\vec{T} = -\frac{\vec{T}^{\,t}\, d\vec{N}/d\vec{\xi}\,\vec{T}}{\vec{T}^{\,t}\vec{T}},$$

where the identities $\vec{T} \odot \vec{N} \equiv 0$ and $\vec{T} \odot \vec{T} \equiv 1$ were used. The principal curvatures κ and principal directions $\vec{T}$ are the eigenvalues and tangential eigenvectors solving

$$-\frac{\mathrm{d}\vec{N}}{\mathrm{d}\vec{\xi}}\,\vec{T} = \kappa\vec{T}.$$

Note that the development for scale-space is identical to that for Euclidean space, except that differentiation of tangents and normals is replaced by covariant differentiation.

4. Anisotropic Diffusion as a Consequence of the Metric

The purpose of this section is to show that the generation of multiscale data via anisotrophic diffusion is intimately related to the geometry of the underlying scale-space. In particular, anisotropic diffusion is in a loose sense "linear" diffusion in non–Euclidean space. This is in contrast to the view that anisotropic diffusion is a nonlinear diffusion in Euclidean space. Moreover, the selection of the metric for scale-space based on invariance requirements for a front–end vision system is a more natural approach to solving vision problems. Once the metric is selected, the anisotropic diffusion process for generating the multiscale data is automatically determined.

4.1. Linear Diffusion in Euclidean Space

The simplest diffusion process is given by

$$u_\sigma = \nabla \cdot (\nabla u), \; x \in \mathbb{R}^n, \; \sigma > 0.$$

As a heat transfer process, the conductance and density of the material are constants. No preference is given for direction of transfer (rotational invariance) or for location of origin (translational invariance). The units of measurement are significant here (no scale invariance). For example, if space and scale are transformed by $(x, \sigma) \rightarrow \lambda(x, \sigma)$ and if $v(x, \sigma) = u(\lambda x, \lambda \sigma)$, then $v_\sigma = \lambda^{-1}\nabla \cdot (\nabla v)$. To retain invariance, the temperatures must also be transformed by $u \rightarrow \lambda v$.

The standard analysis of the linear diffusion equation assumes that the underlying space is Euclidean. That is, the metric of the space is determined by

$$ds^2 = dx \cdot dx + d\sigma^2,$$

and the total derivative is given by

$$du = \nabla u \cdot dx + u_\sigma d\sigma.$$

Any measurements in the (x, σ) space are independent of the multiscale data $u(x, \sigma)$. For example, distance between points in scale-space are measured using the standard formula for Euclidean distance. Objects in the initial image $I(x)$ are analyzed essentially independently of the image intensities and of the proximity of other objects.

The standard finite difference scheme (forward difference in scale, central difference in space) in solving the linear diffusion equation uses a grid consisting of a rectangular lattice of points. In one spatial dimension, the approximations

$$u_\sigma \simeq \frac{u(x, \sigma + k) - u(x, \sigma)}{k}$$

and

$$u_{xx} \simeq \frac{u(x + h, \sigma) - 2u(x, \sigma) + u(x - h, \sigma)}{h^2}$$

lead to the difference scheme

$$u(x, \sigma + k) = u(x, \sigma) + \frac{k}{h^2} \left[u(x + h, \sigma) - 2u(x, \sigma) + u(x - h, \sigma) \right].$$

For n spatial variables with constant sample spacing h across dimensions, the scheme is stable if $2nk < h^2$. For a given scale, a spatial grid point is always a constant (Euclidean) distance h from its nearest neighbors at the same scale. For a given spatial location, a scale grid point is always a constant (Euclidean) distance k from its nearest neighbors at adjacent scales.

4.2. Linear Diffusion in Non–Euclidean Space

In addition to requiring rotational and translational invariance in space, a front–end vision system might be required to have invariance with respect to units of measurement. A metric which has all three invariances is determined by

$$ds^2 = \frac{dx \cdot dx}{\sigma^2} + \frac{d\sigma^2}{\sigma^2}.$$

Consequently scale-space becomes non–Euclidean (the geometry is hyperbolic). The total derivative is given by

$$du = \sigma \nabla u \cdot \frac{dx}{\sigma} + \sigma u_\sigma \frac{d\sigma}{\sigma}.$$

Any measurements in the (x, σ) space are still independent of the multiscale data $u(x, \sigma)$, but there is interaction between the space and scale variables. Geodesics are now semicircles with centers at $\sigma = 0$ as compared to lines, which are the geodesics in Euclidean space. Objects in the initial image $I(x)$ are still analyzed independently of the image intensities and of the proximity of other objects, but the dependence on units of measurement has been removed.

The linear diffusion process corresponding to this metric is given by

$$\sigma u_\sigma = \sigma \nabla \cdot (\sigma \nabla u), \quad x \in \mathbb{R}^n, \quad \sigma > 0.$$

In Euclidean space, if the diffusion scale is t ($u_t = u_{xx}$), then the relationship to the scale in this non–Euclidean space is $t = \sigma^2/2$. However, now the derivatives are in a unitless form due to the zoom invariance of the metric. Note that the left–hand side of the diffusion equation is a single application of the scale-space σ derivative (as specified in the total derivative formula for du) and the right–hand side of the equation is two applications of the scale-space x gradient operator: $(\sigma \partial / \partial \sigma) u = (\sigma \nabla)^2 u$.

A finite difference scheme for the one spatial dimension case may be used to solve the diffusion equation in the current setting. The extension to more spatial dimensions is apparent. The following approximations are used,

$$\sigma u_\sigma \simeq \frac{u(x, b\sigma) - u(x, \sigma)}{\ln b},$$

and

$$\sigma^2 u_{xx} \simeq \frac{u(x + h\sigma, \sigma) - 2u(x, \sigma) + u(x - h\sigma, \sigma)}{h^2},$$

where $b > 1$. The difference scheme is

$$u(x, b\sigma) = u(x, \sigma) + \frac{\ln b}{h^2} \left[u(x + h\sigma, \sigma) - 2u(x, \sigma)) + u(x - h\sigma, \sigma) \right].$$

For n spatial variables with the same pixel spacing $h > 0$, the scheme is stable as long as $2n \ln b < h^2$. Note that this finite difference scheme will get you to a larger scale more quickly than the one using the Euclidean metric, since the scale parameter increases as a geometric sequence rather than as an arithmetic sequence.

Now notice that the implied grid of points is no longer a "rectangular" grid as in the Euclidean case. The implied sampling in scale requires us to use a geometric sequence of scales. As scale increases, the implied spatial samples are sparsely placed as compared to the placement at small scale. A closer look shows that in fact the implied grid points are a "constant" distance apart, but now distance is measured with respect to the metric. For example, the distance between $(x, b\sigma)$ and (x, σ) is $\ln b$, a constant distance in this non–Euclidean space. For a given scale σ, the distance between $(x + h\sigma, \sigma)$ and (x, σ) is h units, again a constant.

4.3. Anisotropic Diffusion

The last two sections have a common theme. The linear diffusion equation is related to the metric assigned to the space. In both cases, the left–hand side of the diffusion is one application to u of the scale derivative which is natural to the metric. The right–hand side is two applications to u of the spatial gradient which is natural to the metric. In the Euclidean metric case, the diffusion is

$$\left(\frac{\partial}{\partial\sigma}\right)u = (\nabla)^2\, u$$

and in the non–Euclidean metric case, it is

$$\left(\sigma\frac{\partial}{\partial\sigma}\right)u = (\sigma\nabla)^2\, u.$$

More generally, anisotropic diffusion can be viewed as the multiscale process one must apply given an appropriate metric for scale-space. Let c denote the conductance function for anisotropic diffusion. The conductance may be a function of space, scale, or image data (and its derivatives). Let ρ denote the density function which also appears in the more general model of heat transfer:

$$\frac{\partial u}{\partial\sigma} = \frac{1}{\rho}\nabla\cdot(c\nabla u)\,.$$

In the Perona–Malik model [293], no density term is included, so $\rho \equiv 1$. Some candidates for the conductance are

$$c(|\nabla u|) = \exp(-|\nabla u|^2/k^2) \text{ or } c(|\nabla u|) = (1 + |\nabla u|^2/k^2)^{-1}$$

for some parameter $k > 0$. Of course this makes the diffusion process nonlinear. The idea is that "homogeneous regions" (where gradient intensity is small) have large conductance and are diffused significantly, but "edge regions" (where gradient intensity is large) have small conductance and are not diffused much. In this way the image is smoothed to preserve noticeable boundaries.

The conductance function has been viewed as a "stretching" of space. This notion is made formal by introducing a metric on scale-space which contains both the conductance and density functions. Specifically, let the metric be determined by the arc length form

$$ds^2 = \frac{dx\cdot dx}{c^2} + \frac{d\sigma^2}{\rho^2 c^2}.$$

Both c and ρ are allowed to depend on space, scale, and image data. It makes sense that a front–end vision system will make measurements (via a metric) that are data–dependent. Object interference is a natural phenomena, so the metric should reflect this and depend on functionals of the intensity. The total derivative in this metric is therefore

$$du = c\nabla u\cdot\frac{dx}{c} + \rho c u_\sigma\frac{d\sigma}{\rho c}.$$

The natural diffusion equation for this metric has left–hand side given by one application of the scale derivative and right–hand side given by two applications of the spatial gradient, namely

$$\left(\rho c \frac{\partial}{\partial \sigma}\right) u = (c\nabla)^2 u.$$

This simplifies to the anisotropic diffusion equation mentioned earlier: $u_\sigma = \rho^{-1}\nabla \cdot (c\nabla u)$. As special cases, for the linear diffusion equation in Euclidean space, the parameters are $c \equiv 1$ and $\rho \equiv 1$. For the linear diffusion in a non–Euclidean space, the parameters are $c \equiv \sigma$ and $\rho \equiv 1$. Finally, a number of researchers have been studying

$$u_\sigma = |\nabla u|\nabla \cdot \left(\frac{\nabla u}{|\nabla u|}\right)$$

either in the context of embedded curve evolution or for edge and corner detection [9, 104, 181]. In this case, $c = 1/|\nabla u| = \rho$.

Now that the metric is quantified in scale-space, appropriate choices can be made for c and ρ so that the metric has the desired invariances for the given application. The construction of multiscale data is then a *consequence* of metric selection. Any object analysis in the image is performed with respect to the metric, for example, measuring distances between objects. Such distances now naturally depend on the multiscale image data, so phenomena such as object interference can be accounted for.

Moreover, the metric could be used in constructing finite difference schemes for solving the anisotropic diffusion equation. In the linear diffusion case, the grid points are positioned at constant distances from each other, distance being measured according to the metric. Grid selection for the general anisotropic case can be done similarly, although shown here is a grid selection for which spacing is not necessarily uniform. The selection is adaptive depending on what the values of conductance and density are. For example, if conductance is $c(x, \sigma)$ and density is $\rho(x, \sigma)$ (no image dependence for now), then the first derivative approximations are

$$cu_x \simeq \frac{u(x + \frac{h}{2}c(x, \sigma), \sigma) - u(x - \frac{h}{2}c(x, \sigma), \sigma)}{h}$$

and

$$\rho c u_\sigma \simeq \frac{u(x, \sigma + k\rho(x, \sigma)c(x, \sigma)) - u(x, \sigma)}{k}$$

for some small positive constants h and k. The second derivative spatial approximation is somewhat more complicated, but it just involves the first–order difference operator applied twice to the function. In the following equation, the dependence on σ is dropped just for notational simplicity. The second derivative is approximated by

$$c(cu_x)_x \simeq \frac{1}{h^2}\Big[u(x + \tfrac{h}{2}c(x) + \tfrac{h}{2}c(x + \tfrac{h}{2}c(x))) - u(x + \tfrac{h}{2}c(x)$$
$$- \tfrac{h}{2}c(x + \tfrac{h}{2}c(x))) - u(x - \tfrac{h}{2}c(x) + \tfrac{h}{2}c(x - \tfrac{h}{2}c(x)))$$
$$+ u(x - \tfrac{h}{2}c(x) - \tfrac{h}{2}c(x - \tfrac{h}{2}c(x)))\Big]$$

It can be shown that the right–hand side converges to $c(cu_x)_x$ as $h \to 0$. Notice that when $c \equiv 1$ and $\rho \equiv 1$, the finite difference method is exactly the one given earlier for $u_\sigma = u_{xx}$. When $c = \sigma$ and $\rho \equiv 1$, the scale derivative approximation is slightly different. The scales are sampled geometrically, but in place of the term $\ln b$ appearing in the approximation we have $b - 1$. The replacement by $\ln b$ can be viewed as a minor refinement which has some effect on the approximation error and on the range of b which provides a stable algorithm.

If the conductance and/or density depend on image values, now the difference schemes become implicit. They may be more difficult to implement, but hopefully, like many implicit schemes, they will exhibit unconditional stability.

5. Discussion

The development of the differential properties of scale-space is not to be viewed as simply an exercise in mathematics. The consequences of scale-space measurements are far–reaching in applications to image analysis. Objects in an image impose their own "geometry" in the image. Any measurements made should depend on position, scale of measurement, and interference by other objects.

A major application has been in the construction of *cores* of objects in gray scale images [294]. Cores are essentially indicators of medial locations and corresponding widths for gray scale objects. They are formally defined as ridges of medialness functions derived from multiimage data. A discussion of ridge definitions for both Euclidean and scale-space can be found in [92]. Specific approaches to constructing cores from gray scale images using scale-space ideas can be found in [93], [120], and [261]. Applications dependent on the core include portal image registration [120], volume registration [233], object–based interpolation [302], and fusion of CT image data taken at different orientations.

Figure 14.3.   The Starry Night from Vincent van Gogh. The original figure is shown and the result after applying affine shortening flow. We see how regions become smooth, while the distinction between different regions remains sharp. Implementation described in next chapter (Niessen et al.).

NUMERICAL ANALYSIS OF GEOMETRY-DRIVEN DIFFUSION EQUATIONS

Wiro J. Niessen

and

Bart M. ter Haar Romeny

and

Max A. Viergever

Utrecht University, Computer Vision Research Group,
Utrecht University Hospital, Room E.02.222,
Heidelberglaan 100, NL-3584 CX Utrecht, The Netherlands

1. Introduction

The various equations described in the previous chapters are of the parabolic and hyperbolic types. They arise when evolutionary processes are modeled: the solution of the equations can be interpreted as evolving from an initial state as time is increasing. In our analysis the initial state will be an image acquired at a certain resolution, and the evolution will generate a multi-scale representation. The nonlinear partial differential equations in the previous chapters cannot be solved analytically, so we have to resort to methods of numerical analysis. Some approximation schemes have already been mentioned by the various authors. This chapter provides a general framework for the numerical approximation of evolution equations. The organization of the chapter is as follows: In section 2.1 we treat the numerical approximation of the linear diffusion equation to introduce the concepts of explicit and implicit methods and discuss their stability criteria. We then (section 2.2) deal with the implementation of the hyperbolic advective equation, and introduce the concept of upwind schemes. In section 3 we will show how these methods can be used to find finite difference schemes for the nonlinear equations that were proposed. Since some of the equations are very anisotropic, their implementation on a fixed grid is problematic. We therefore introduce a scheme in which the solution is approximated using

Gaussian derivative operators which are rotationally invariant. This scheme is applicable to all diffusion (parabolic type) equations. For hyperbolic terms we need to use an upwind scheme as proposed by Osher & Sethian [285]. In section 5 we show several results on test images and medical images.

2. Classical Difference Schemes

2.1. Parabolic Equations

2.1.1. *Explicit Schemes*
The usual way to discretize parabolic type evolution equations is to choose equally spaced points both in the spatial direction and in the scale direction. In practice, the spacing in the spatial domain is determined by the distance between pixels in the input image (initial value), while the time-step Δt has to be set. Partial derivatives are discretized in the following way:

$$\frac{\partial}{\partial t} I_j \simeq \frac{I_j^{n+1} - I_j^n}{\Delta t}$$

$$\frac{\partial}{\partial x} I_j \simeq \frac{I_{j+1}^n - I_{j-1}^n}{2\Delta x} \tag{15.1}$$

$$\frac{\partial^2}{\partial x^2} I_j \simeq \frac{I_{j+1}^n - 2I_j^n + I_{j-1}^n}{(\Delta x)^2} \tag{15.2}$$

where $I_{j,l}^n$ denotes the luminance value $I(t_n, x_j, y_l)$. For the spatial derivatives we use centered differences while for the time we use forward differences. The linear diffusion equation with constant conductivity can be approximated in $2D$ as:

$$\frac{I_{j,l}^{n+1} - I_{j,l}^n}{\Delta t} = \frac{I_{j+1,l}^n - 2I_{j,l}^n + I_{j-1,l}^n}{(\Delta x)^2} + \frac{I_{j,l+1}^n - 2I_{j,l}^n + I_{j,l-1}^n}{(\Delta y)^2} \tag{15.3}$$

This difference equation approximates the solution of the analytical equation *running forward in time*. All difference equations in which the value at time $n + 1$ ($I_{j,l}^{n+1}$) depends on values that are already known are referred to as explicit schemes. Stability of the difference equation can be tested using the Von Neumann stability analysis [14]. This analysis is local and effectively ignores boundary conditions. In the remainder of the chapter we will assume that all our schemes are pure initial value problems, so we will neglect the influence of boundary conditions on the stability requirement. A difference scheme is said to be unstable if there exist initial values for which the solution $I_{j,l}^n$ blows up for $n \to \infty$. Since every initial value can be expressed as:

$$I_{jl}^0 = \sum_p \sum_q A_{pq} e^{ik_{pq}(j\Delta x + l\Delta y)} \tag{15.4}$$

(here k denotes a real spatial wavenumber and the A_{pq} are (complex) Fourier coefficients) we can look at the individual terms in this expansion separately. These evolve according to:

$$I_{j,l}^n = \xi^n e^{ik(j\Delta x + l\Delta y)} \tag{15.5}$$

where ξ is a complex number which depends on k and is called the amplification factor. If $|\xi(k)| < 1$ for all k the equation is stable. Upon inserting (15.5) in (15.3) we find:

$$\xi = 1 - \frac{\Delta t}{(\Delta x)^2}(4\sin^2(\frac{k\Delta x}{2})) - \frac{\Delta t}{(\Delta y)^2}(4\sin^2(\frac{k\Delta y}{2})) \tag{15.6}$$

If we assume $\Delta x = \Delta y$ we need to impose:

$$\frac{\Delta t}{(\Delta x)^2} \le \frac{1}{4} \tag{15.7}$$

in order to assure stability. More general in D dimensions we have:

$$\frac{\Delta t}{(\Delta x)^2} \le \frac{1}{2D} \tag{15.8}$$

2.1.2. Implicit schemes

We will now write the linear diffusion equation in the following way:

$$\frac{I_{j,l}^{n+1} - I_{j,l}^n}{\Delta t} = \frac{I_{j+1,l}^{n+1} - 2I_{j,l}^{n+1} + I_{j-1,l}^{n+1}}{(\Delta x)^2} + \frac{I_{j,l+1}^{n+1} - 2I_{j,l}^{n+1} + I_{j,l-1}^{n+1}}{(\Delta y)^2} \tag{15.9}$$

The derivatives at the right hand side of the equation are now evaluated at the time-step $n+1$ rather than at time-step n. The consequence is that we now have to solve a set of linear equations at each time-step. In the $1D$-case this boils down to the tridiagonal set of equations:

$$(1 + \frac{2\Delta t}{(\Delta x)^2})I_j^{n+1} - \frac{\Delta t}{(\Delta x)^2}I_{j+1}^{n+1} - \frac{\Delta t}{(\Delta x)^2}I_{j-1}^{n+1} = I_j^n \tag{15.10}$$

In the $2D$-case the situation gets a little more complex. We have to solve a $3D$-matrix, which is fortunately still sparse (tridiagonal by blocks) and can be solved using standard sparse matrix techniques. However, we can also use the alternating direction implicit (ADI) method, see for example [15, 298] in which each time-step is divided in sub-steps in which a different dimension is treated implicitly:

$$\begin{cases} I_{j,l}^{n+\frac{1}{2}} &= D_1(I_{j,l}^{n+\frac{1}{2}}) + D_2(I_{j,l}^n) \\ I_{j,l}^{n+1} &= D_1(I_{j,l}^{n+\frac{1}{2}}) + D_2(I_{j,l}^{n+1}) \end{cases} \tag{15.11}$$

where D_1 and D_2 denote the difference schemes in the x and y directions respectively. Now we only have to solve a tridiagonal system at each time-substep.

The amplification factor of the implicit formula can be calculated inserting equation (15.5) in (15.9):

$$\xi = \frac{1}{1 + 4\frac{\Delta t}{(\Delta x)^2}\sin^2(\frac{k\Delta x}{2}) + 4\frac{\Delta t}{(\Delta y)^2}\sin^2(\frac{k\Delta y}{2})} \tag{15.12}$$

which is smaller than 1 for all k. The scheme therefore is unconditionally stable (this result holds for arbitrary dimensions). The popular *Crank Nicholson* scheme [74, 135] which is stable and second-order accurate in both space and time is simply the average of the implicit and explicit method where the terms at both sides are centered at time-step $n + \frac{1}{2}$.

2.2. Hyperbolic Equations

Prototypical examples of hyperbolic equations are the wave equation:

$$\frac{\partial^2 u}{\partial t^2} = c^2 \frac{\partial^2 u}{\partial^2 x} \tag{15.13}$$

and the advective equation:

$$\frac{\partial I}{\partial t} = v\frac{\partial I}{\partial x} \tag{15.14}$$

In the latter case we used I rather than u to denote the evolving function, already suggesting that it is interesting to investigate this equation in the evolution of images. This is exactly the kind of equation in which the level curves of the image move in the direction of the gradient with speed v. In mathematical morphology this is similar to the dilation or (depending on the sign of v) the erosion with a disk with radius $\|v\|$ as structuring element, see also the chapter by Alvarez and Morel in this book. If we apply the transformation $\xi = x - vt$ the equation reduces to:

$$\frac{\partial I}{\partial \xi} = 0 \tag{15.15}$$

Thus along the lines $\xi = (x - vt)$ in the x, t-plane, called the characteristics of the equation, the solution does not change. For the problem with initial value $I(x, t = 0) = I_0$ the solution of (15.14) is given as:

$$I(x, t) = I_0(x - vt) \tag{15.16}$$

Therefore, it makes no sense to use centered spatial differences to approximate the solution. We have to be aware of this effect when trying to finite difference the equation. For example, the naive way to difference (15.14) is:

$$\frac{I_j^{n+1} - I_j^n}{\Delta t} = v_j^n \frac{I_{j+1}^n - I_{j-1}^n}{2\Delta x} \tag{15.17}$$

Not only is this scheme unstable (an easy calculation shows that the amplification factor in the Von Neumann stability analysis is always larger than 1) but information also propagates to mesh points which should not be affected according to the analytic equation. The straightforward way to solve this problem is to rewrite the equation in a form depending on the direction of the velocity (upwind schemes):

$$\begin{cases} \frac{I_j^{n+1} - I_j^n}{\Delta t} = v_j^n \frac{I_{j+1}^n - I_j^n}{\Delta x}, & v_j^n > 0 \\ \frac{I_j^{n+1} - I_j^n}{\Delta t} = v_j^n \frac{I_j^n - I_{j-1}^n}{\Delta x}, & v_j^n < 0 \end{cases} \tag{15.18}$$

This formulation can easily be extended to the $2D$-case, either by splitting the normal velocity into a velocity component in the x and y directions, or by a transformation into the (v, w)-gauge described in the chapter by Lindeberg and ter Haar Romeny. In this gauge the difference scheme can be written as:

$$\frac{I_j^{n+1} - I_j^n}{\Delta t} = v_{\hat{w}} \frac{I_{j+1}^n - I_j^n}{\Delta x} \tag{15.19}$$

where I_{j+1}^n is some estimate of the value of I a distance Δx in the direction $\hat{w}$ from the position j. The latter approach is more expensive from the computational point of view than splitting the velocity into its x and y components, but it has no preference for a certain grid direction. The upwind (15.18) scheme is stable for sufficiently small time-steps; inserting (15.5) into (15.18) we find the Courant-Friedrichs-Lewy stability criterion [15]:

$$\|v\| \frac{\Delta t}{\Delta x} \leq 1 \tag{15.20}$$

This upwind scheme is only first order accurate in the spatial derivatives. Osher and Sethian studied higher order upwind schemes [285], which are generalizations of Godunov's scheme. This method has been described in the chapter by Kimia et al. in this book.

3. Nonlinear Diffusion Equations

In this section we discuss the numerical implementation of the Perona and Malik equation [291], the Euclidean [181] and affine [321, 323] shortening flow and the reaction-diffusion equation [177, 180] which contains a (hyperbolic) constant motion term. One problem which often occurs is caused by the anisotropic character of these schemes: it is difficult, if possible at all, to find rotationally invariant implicit schemes on a fixed grid. Fortunately, (heuristic) stability criteria can be obtained for the Euler forward scheme in the case of the (parabolic) diffusion-type cases. This motivates the search for a rotational invariant difference equation for these equations which will be the topic of the next section. Hyperbolic terms require special care. We will discuss some of the problems that one can encounter. Kimia et al. in this book already reviewed the Osher and Sethian scheme [285] which can fruitfully be applied in many circumstances.

3.1. Anisotropic Diffusion

In chapter 3 the Perona and Malik equation with variable conductance $c(x, y, t)$ was introduced:

$$\begin{cases} I_t &= \vec{\nabla} \cdot (c(x, y, t) \nabla I) \\ I(\vec{x}, 0) &= I_0(\vec{x}) \end{cases} \tag{15.21}$$

Two choices of $c(x, y, t) = g(\|\nabla I(x, y, t)\|)$ are

$$g_1(\nabla I) = e^{-(\frac{\|\nabla I\|}{K})^2} \tag{15.22}$$

$$g_2(\nabla I) = \frac{1}{1 + (\frac{\|\nabla I\|}{K})^2} \tag{15.23}$$

Note that the Taylor expansion up to first order is identical for both choices. One possible problem is that in the case that cI_x decreases with respect to I_x we have locally an inverse heat equation. This happens in the particular cases that $I_x > K$ and $I_x > \frac{1}{2}\sqrt{2}K$. For more discussion on this topic we refer the reader to the discussion at the end of chapter 3.

We will now look at the implementation of a regularized version of the Perona & Malik equation. Nitzberg & Shiota remark [271] that an implementation on a fixed grid always imposes implicit smoothing which regularizes the equation (differentiation is impossible without increasing scale). They however argue that it is essential to regularize the equation itself in order to reduce the effect of the grid size and control the behavior of discrete approximations. A regularization of (15.21) with each of the conduction functions g_1 and g_2 (15.22) (15.23) is due to Catté et al. [57]. They evaluate the gradient at a certain scale $DG_\sigma * I$ rather than using the gradient ∇I at the scale of the original image (which is physically not possible anyway, since differentiation implies an increase of scale).

This makes the problem well-posed and consequently the results shown in chapter 3 are experimentally stable.

We will discuss both the explicit and the implicit difference schemes for the Perona & Malik equation. A straightforward forward Euler discretization (which is almost similar to the discretization in the chapter by Perona, Shiota and Malik) gives:

$$\frac{I_{j,l}^{n+1} - I_{j,l}^{n}}{\Delta t} = \frac{(c_{j+1,l}+c_{j,l})(I_{j+1,l}^{n}-I_{j,l}^{n})-(c_{j,l}+c_{j-1,l})(I_{j,l}^{n}-I_{j-1,l}^{n})}{2(\Delta x)^2} +$$
$$\frac{(c_{j,l+1}+c_{j,l})(I_{j,l+1}^{n}-I_{j,l}^{n})-(c_{j,l}+c_{j,l-1})(I_{j,l}^{n}-I_{j,l-1}^{n})}{2(\Delta y)^2} \qquad (15.24)$$

From considerations in the previous section we find that the stability criterion is:

$$\frac{\Delta t}{(\Delta x)^2} \leq min_{j,l}\frac{1}{4c_{j,l}} \qquad (15.25)$$

Since the functions c that Perona & Malik have an upper bound of 1 we still can use the stability condition $\Delta t < 0.25(\Delta x)^2$. It is important to note that the function c may be evaluated at any scale. This is in keeping with the idea of scale as a free parameter. The limitation to four points implies that the method violates rotational invariance except for angles of $\frac{\pi}{2}$ (Using an eight point neighborhood would improve this to angles of $\frac{\pi}{4}$). In the implicit scheme we evaluate c as in equation (15.24), but the image values are evaluated at the time-step $n + 1$. We therefore can write the scheme in the form:

$$(\mathbf{Id} + \Delta t\mathbf{A})I^{n+1} = I^n \qquad (15.26)$$

where $\mathbf{Id}$ denotes the identity matrix and $\mathbf{A}$ is tridiagonal by blocks. As described in the previous section this matrix can be inverted using suitable sparse matrix techniques. However, the ADI-method (15.11) discussed in the same section is also fruitful. The implicit method is unconditionally stable, but still not rotationally invariant.

3.2. Geometrical Heat Flows

We recall that one approach to obtain a multi-scale representation of images has been inspired by the mathematical treatment of the evolution of curves as a function of their geometrical properties [125, 124, 95, 17, 18, 139]. This method has been adopted by computer vision researchers and was described in the chapters by Olver, Sapiro, Tannenbaum and Kimia, Zucker and Tannenbaum in this book. In the 2D-planar case a curve is fully described by its curvature and we can model the general evolution process by:

$$C_t = F(\kappa) \cdot \vec{\mathcal{N}} \qquad (15.27)$$

where F is an arbitrary function and $\vec{N}$ denotes the normal direction pointing outwards. In order to numerically approximate these equations Osher & Sethian [285] introduced a Lipschitz function $\phi(x, y, t)$ which is smaller than 1 inside the curve and larger than 1 outside the curve and equal to 1 on the curve itself. This scheme was reviewed by Kimia in chapter 12. Since we are primarily interested in grey-value images we take a somewhat different approach and view the image as an embedded set of isophotes which evolve according to equation (15.27). If we consider a displacement both in the spatial and scale directions of a point on an isophote, a straightforward calculation gives that the evolution of a $2D$-image is given by:

$$\frac{\partial I}{\partial t} = F(\kappa)\|\vec{\nabla}I\| \qquad (15.28)$$

which gives up to 1st order in κ:

$$\frac{\partial I}{\partial t} = (\beta_0 + \gamma_0\kappa)\|\vec{\nabla}I\| \qquad (15.29)$$
$$= \beta_0 L_w + \gamma_0 L_{vv} \qquad (15.30)$$

Note that β_0 and γ_0 have different dimensions, the value $\frac{\beta_0}{\gamma_0}$ is not invariant under a spatial rescaling $\vec{x} \to \lambda\vec{x}$. We either have to work in natural coordinates, or multiply nth order derivatives with the nth power of the scale at which the derivative is evaluated. Depending on the values of β_0 and γ_0 we have constant motion, curvature motion or a reaction-diffusion type equation.

3.2.1. Euclidean and Affine Curve Shortening

In this section we will look at the numerical implementation of the Euclidean shortening flow and its affine analogue. The first equation is obtained setting β_0 to 0 and γ_0 to 1. The flow only diffuses in the direction of the gradient and therefore is very anisotropic! We will describe the Euclidean evolution of $2D$-images in a convenient form using the gauge invariant coordinates (see definition in the chapter by Lindeberg and ter Haar Romeny):

$$\frac{\partial I}{\partial t} = I_{vv} \qquad (15.31)$$

The affine curve shortening is the affine analogue of the equation and can be written as:

$$\frac{\partial I}{\partial t} = I_{vv}{}^{\frac{1}{3}} I_w{}^{\frac{2}{3}} \qquad (15.32)$$

In the Euclidean case an implicit scheme for the numerical implementation was introduced by Alvarez [9]. He introduces an eight pixel neighborhood, looks for the direction η which is closest to the gradient direction and assumes that I_{vv} is equal to $I_{\eta\eta}$. He then defines an implicit scheme for the resulting (linear diffusion like) equation. As noted by Alvarez himself, the scheme is only rotationally invariant for angles of $\frac{\pi}{4}$. A disk will evolve to a polygon after a certain evolution time.

Since we would like our equation to be rotationally invariant and since moreover implicit schemes seem not feasible for equation (15.32) we will look at the explicit schemes for both equations in the remainder of this section:

$$\frac{I_{j,l}^{n+1} - I_{j,l}^{n}}{\Delta t} = (I_{vv})_{j,l}^{n} \tag{15.33}$$

$$\frac{I_{j,l}^{n+1} - I_{j,l}^{n}}{\Delta t} = (I_{vv}^{\frac{1}{3}} I_{w}^{\frac{2}{3}})_{j,l}^{n} \tag{15.34}$$

Using a similar argument as for the regularized Perona & Malik equation we can see that the stability criterion $\Delta < 0.25(\Delta x)^2$ still holds for equation (15.33). Using the Von Neumann stability analysis one can not come up with a stability requirement for equation (15.34).

We have to find a rotationally invariant difference scheme for the righthandside of equations (15.33) and (15.34). In [70, 68] Cohignac describes a scheme proposed by Alvarez & Guichard in which the Euclidean curvature is calculated as rotationally invariant as possible on a 8 pixel neighborhood. We have developed another approach [268] which is in keeping with the role of scale as a free parameter. The partial derivatives are not approximated using finite differences, but using the scaled Gaussian derivative operators which were discussed in the first chapter of this book. This approach is rotationally invariant and we describe it in section 4.

3.2.2. Constant motion

If β_0 is not equal to zero, we have a hyperbolic term in our equation. A classical solution may fail to exist, since discontinuities may occur. Kimia shows in chapter 12 that an integral formulation of the equation leads to the concept of weak solutions (the space of admissible solutions is extended to cover those functions that are measurable and bounded). To pick out the right solution, an entropy condition needs to be imposed. The entropy condition for shape was formulated by Kimia [177] and states that the process of merging or splitting can not be reversed. This implies that when a disconti- nuity in the evolution occurs, the characteristics have to flow into this discontinuity. The desired entropy condition is found both in the case that a grey-value image is considered as an embedded set of isophotes, and in the case that we use the Osher & Sethian scheme [285].

4. Scheme using scale-space operators

4.1. Implementation of general evolution equations

In a finite difference scheme the distance between the pixels arbitrarily determines the scale at which a partial derivative is evaluated. If we try to model a continuous equation one naturally wants a grid size as small as possible, and if in the limit that the grid size tends to zero the difference equation converges to the differential equation, the choice of this "smallest possible" derivative operator is not a drawback. However, the basic idea of nonlinear scale-spaces is to make the evolution of the image dependent on geometrical properties of the image, which is likely to manifest itself at multiple scales. In a finite difference scheme there is always a preference for geometrical structures present at the scale of the grid size. We therefore introduce a scheme in which the scale parameter is allowed to vary.

The natural scaled derivative operators for physically measured data are the Gaussian derivative operators which were discussed in chapter 2. After having established a representation using these operators or, equivalently, having established the *geometry*, we can construct multi-scale representations using this information. The equations which have been mentioned in the preceding chapters and which we discussed in the previous sections, are a few examples. However, *any* evolution equation can be described in this general framework. We will write the general equation as:

$$\frac{\partial I}{\partial t} = F(I_i, I_{ij}, I_{ijk}, ...) \tag{15.35}$$

where the indices i, j, k represent the spatial dimension of the image. The equations for constant motion, Euclidean shortening flow, reaction-diffusion and affine shortening flow can be written in this form:

$$\frac{\partial I}{\partial t} = I_w = (I_i I_i)^{\frac{1}{2}} \tag{15.36}$$

$$\frac{\partial I}{\partial t} = \kappa I_w = I_{vv} = \frac{I_{ii} I_j I_j - I_i I_j I_{ij}}{I_k I_k} \tag{15.37}$$

$$\frac{\partial I}{\partial t} = \beta_0 I_w + \gamma_0 I_{vv}$$

$$= \beta_0 (I_i I_i)^{\frac{1}{2}} + \gamma_0 \frac{I_{ii} I_j I_j - I_i I_j I_{ij}}{I_k I_k} \tag{15.38}$$

$$\frac{\partial I}{\partial t} = I_{vv}^{\frac{1}{3}} I_w^{\frac{2}{3}} = (I_{ii} I_j I_j - I_i I_j I_{ij})^{\frac{1}{3}} \tag{15.39}$$

We can write the image at time $t_0 + \Delta t$ as:

$$I(\vec{x}, t_0 + \Delta t) = \sum_{n=0}^{\infty} \frac{(\Delta t)^n}{n!} \partial t^n I(\vec{x}, t_0) \tag{15.40}$$

If we include terms up to $N = 1$ we obtain the Euler forward scheme which we mentioned before:

$$I(\vec{x}, t_0 + \triangle t) - I(\vec{x}, t_0) = \triangle t \ F(I_i, I_{ij}, I_{ijk}, ...) \qquad (15.41)$$

We can include higher order terms to improve the accuracy. We have to be aware, though, whether these higher order operators are still accurate at the scale at which they are evaluated.

All partial derivative terms at scale s can be obtained by a convolution of the original image with the equivalent partial derivative of the Gaussian at scale s (because the operations of convolution and differentiation commute):

$$\partial_{i_1...i_n} I_0(\vec{x}) * G(\vec{x}, s) = I_0(\vec{x}) * \partial_{i_1...i_n} G(\vec{x}, s) \qquad (15.42)$$

Convolutions are most conveniently calculated in the Fourier domain because the Fourier transform $\mathcal{F}$ of a convolution in the spatial domain is equal to the product of the individual Fourier transforms in the Fourier domain. We need to compute both the Fourier transform of the primal image and of the desired (combination of) derivatives of the Gaussian kernel. The Fourier transform of the n-th order derivative of a Gaussian kernel is obtained by an n-times multiplication with $-i\vec{\omega}$:

$$\mathcal{F}\{\frac{\partial^n}{\partial \vec{x}^n} G(\vec{x}; t)\} = (-i\vec{\omega})^n \mathcal{F}\{G(\vec{x}; t)\} \qquad (15.43)$$

After both terms have been multiplied the Fourier back transform $\mathcal{F}^{-1}$ yields the desired derivative. So e.g.

$$I_{xxy} = I * G_{xxy} = \mathcal{F}^{-1} \left\{ \mathcal{F}\{I(x, y; 0)\} \cdot (-\omega_x^2)(-i\omega_y)\mathcal{F}\{G(x, y, t)\} \right\} \quad (15.44)$$

An evolution equation which depends on local geometry (terms in the local jet) can now be numerically approximated. We first rewrite the differential equations in Cartesian coordinates. We therefore designed a *parser*, which is able to parse any expression in manifest invariant form or gauge coordinates, no matter its complexity, into its Cartesian form, given the number of dimensions. Examples are given in table 2.1 in the chapter by Lindeberg and ter Haar Romeny. This parser performs a formal expression tree evaluation and automatically generates C-code with pointers to the pre-calculated Gaussian partial derivative images of the input image (e.g. I_x, I_{yy}) to calculate the invariant Cartesian expression (e.g. ($I_{vv} = I_{xx}I_y^2 - 2I_{xy}I_xI_y + I_{yy}I_x^2)/(I_x^2 + I_y^2)$) in each pixel or voxel. The iteration is subsequently done according to (15.41). Free parameters are not only the time-step $\triangle t$ but also the scale of the differential operator. In the chapter by Florack an example of this implementation is given. There, on page 357 in figure 13.4 we see how a 256x256 MR-image of the brain is evolved using the Euclidean shortening flow.

Although the above scheme seems general at a first glance, we have to recall the discussion in section 2 when trying to implement the advective equation (15.14). We saw that the information to calculate the image at a next time-step can only depend on a part of the image domain, depending on the direction of the velocity. Since Gaussian derivatives are centered derivative operators, we will create spurious boundary conditions when implementing advective equations. A workaround is cumbersome. We first have to evaluate the direction of the velocity, and then calculate an approximation of the spatial derivative at a point nearby (15.19). The advantage of this method is that it is still rotationally invariant.

4.2. Stability

4.2.1. Time-step

We showed that the Euclidean shortening flow and the Perona & Malik equation satisfy the stability criterion for $\Delta t < 0.25(\Delta x)^2$ in 2D. We will now look whether the scheme is experimentally stable, using different time-steps to obtain an image at a certain evolution time. We evolved the images with the Euclidean shortening flow. The images are compared using Pearson's correlation coefficient which is given as:

$$r = \frac{\sum_{jl}(I_{jl} - \overline{I})(I_{jl}^* - \overline{I^*})}{\sqrt{\sum_{jl}(I_{jl} - \overline{I})^2}\sqrt{\sum_{jl}(I_{jl}^* - \overline{I^*})^2}} \qquad (15.45)$$

Here I and I^* denote the two images and $\overline{I}$ denotes the average luminance value of the image.

In figure 15.1 we plotted the correlation between an image obtained using a very small time-step (reference image) and several images using larger time-steps. The five lines represent a total evolution time of $t = 10, 20, 30, 40, 50$ square pixel units respectively. In the range where the heuristic stability criterion holds, the correlation coefficient is approximately 1. Perceptually the results are indistinguishable.

4.2.2. Ringing

Although most implementations show experimentally very robust results, we can in some circumstances see the effect of the grid size. Because differentiation is not possible without increasing scale, and the increase in scale is necessarily of the order of the grid size, it is not possible to extract differential information at the scale of the original image (s_0), but only at scales $s = s_0 + \mathcal{O}(\Delta\S)$. However, all evolution schemes are based on the evolution of the original image. The actual implementation in practice smoothes out the differential information of the image. From (15.40) we see that in the fortunate circumstance that we are able to calculate derivatives up to any order, we in fact exactly calculate these derivatives on the original image, smoothed with a Gaussian with a σ equal to the σ of the derivative operators. Even in this ideal situation we are not able to construct the multi-scale representation of the original image. In practice, only the first few terms can be calculated which results in an enlargement of this problem.

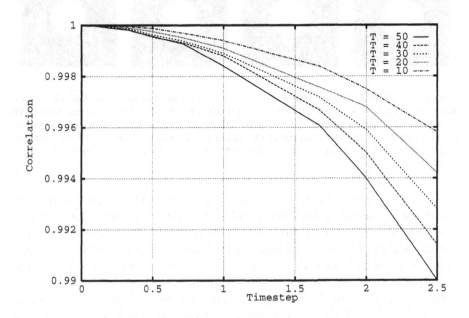

Figure 15.1.   Correlation coefficient between the best approximated image ($\Delta t = 0.025$) and an image evaluated using larger time-steps for five different evolution times. The five lines correspond to a total evolution time of $10, 20, 30, 40$ and 50 pixel square units respectively. The time-step Δt was chosen between 0.05 and 2.50 pixel square units.

In figure 15.2 we show the effect on the image of a circle which looks like a binary image. Because of the grid, curvature is not only measured at the location of the boundary, but also in a range surrounding the boundary. This creates spurious boundary conditions. After hundreds of iterations we see how many spurious edges have built up (ringing). The effect is considerably smaller if we choose a smaller scale for the Gaussian derivative operators. In noisy grey value images we hardly notice this effect and the scheme is robust for sufficiently small evolution times.

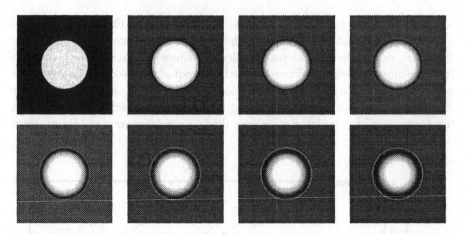

Figure 15.2. Illustration of the effect of the grid-size. We have applied the affine shortening flow on a white circle and deliberately evaluated the Gaussian derivative operators at a high scale of 3 pixel units. The effect is that we measure curvature in a region surrounding the contour of the (binary like) circle. This results in spurious edges. Iterating the equation many times we see how the number of spurious edges increases.

A scheme proposed by Koenderink & van Doorn [195] and adopted by Merriman, Bence & Osher [251] avoids the effect of ringing. They define a (normalized) characteristic function which measures some physical quantity. Setting a threshold on the characteristic function we can divide the image in figure & ground (we create a binary image). We proceed to calculate the evolution of this image (using a finite difference or a Gaussian-like approach) which results in a new characteristic function to which we can apply the same sequence of steps. One problem, noted by Cohignac [68] is the fact that the evolution will stop at low curvature points.

4.3. Measure of performance

We should clearly distinguish the stability of the process (no dependence on time-step) from the actual precision with which the exact solution is approximated. Linear scale-space hands us an example to actually measure the accuracy. We compare an image at scale t after a convolution with a Gaussian filter or using the approximation scheme:

$$I^{n+1} = I^n + \Delta t I_{ii}^n$$

In figure 15.3 we see a 256 x 256 MR image of the brain evaluated at different scales using both methods. Perceptually the results are hardly, if at all, distinguishable.

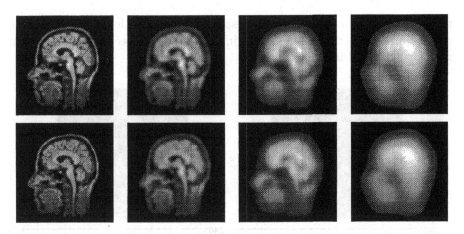

Figure 15.3. 4 levels in scale-space of a 256×256 MR image of the brain. The four upper images are obtained using a convolution of the primal image with a Gaussian kernel. The four lower images show the results obtained when we evolved the image using the Euler forward scheme until an evolution time which corresponds to the scale of the Gaussian kernel. We chose σ equal to $2, 8, 32, 128$.

5. Applications & Discussion

In many computer vision tasks we require information of the image at several levels of resolution. Examples range from low level vision tasks as noise removal, to high level visual tasks as pattern recognition. In many of these cases a geometry dependent multi-scale representation can be useful. The particular dependency is determined by the application at hand. However, the choice of nonlinearities is not necessarily ad hoc. The evolution equations which were described by Alvarez et al., Kimia et al. and Sapiro et al. have some very interesting global properties which have paved the way for a multi-scale shape-description. Koenderink already noted [195] that shape description only makes sense in a dynamic setting. In the future we can foresee a further axiomatic restriction on the shape evolution schemes, like conservation laws with regard to global properties of the objects like area or topological entities. This issue is in the limelight of present-day research in computer vision. Throughout this book one can find several results of the implementation of the nonlinear schemes. We present below three more examples, one on an artificial image and two on medical images. The images shown here are obtained using the Euler forward model, which is stable for small time-steps. Thanks to the strong regularizing power of the used differential operators we were able to apply this scheme onto a variety of nonlinear schemes. The obtained results are robust, even for noisy images and a large number of iterations. The method turned out to be rather efficient, both with respect to computational resources and ease of implemen- tation. The following pages show three more examples of nonlinear diffused images.

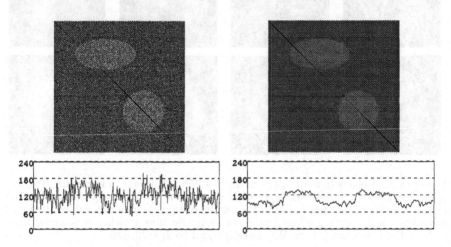

Figure 15.4. Left: 256 x 256 image generated using octree anti-aliasing (depth=5 [372]). Gaussian noise has been added. The original signal to noise ratio is $\frac{4}{3}$. The graph under the image shows the luminance values along the plotted diagonal. Segmenting the image using a thresholding is not feasible. Right: Image after 100 iterations with time-step 0.25 square pixel units using Euclidean curvature motion. The Gaussian derivative operators are evaluated at $s = 0.8$. The distinction between object and background using thresholding is now feasible.

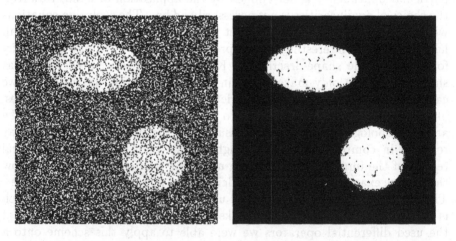

Figure 15.5. Original and evolved image thresholded to obtain a crude segmentation. The threshold was set to 120.

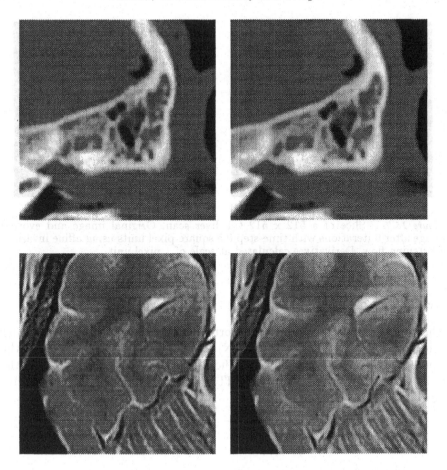

Figure 15.6.   Upper figure: a part of a CT-scan which shows the petrous bone and the semi-circular channels. Lower figure: part of an MR-scan which shows the cerebellum and cerebrum. Both figures are taken from the Elsevier's interactive anatomy atlas on CD-I/CD-ROM edited by Hillen [153]. Due to the limiting resolution of the acquisition, especially in the longitudinal direction and the small size of the region studied (inner ear) the data had to be resampled for display (the atlas provides all three viewing directions). The figures to the left show the images after they have been resampled using a cubic convolution algorithm. To the right we see the effect (we chose a larger evolution than in the data on CDI to improve clarity) after we applied the evolution equation $\frac{\partial L}{\partial t} = L_{vv}^{\frac{1}{3}}$ which evolves the isophotes as a function of their curvature with a thresholding depending on the norm of the gradient. The process improves the visual impression of the images on the display.

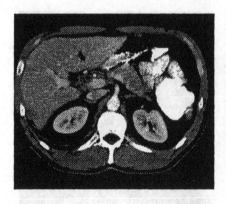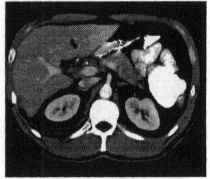

Figure 15.7.   Slice of a 512 x 512 CT liver scan. Original image and evolved image after 6 iterations with time-step 0.5 square pixel units using affine invariant shortening flow. Derivatives calculated at scale of 1 pixel unit.

Acknowledgement

The investigations were partly supported by the Foundation for Computer Science in the Netherlands (SION) with financial support from the Netherlands Organization for Scientific Research (NWO) and partly by the Real World Computing Program MITI, Japan through the project Active Perception & Recognition.

Bibliography

1. R. Abraham, J. E. Marsden, and T. Ratiu. *Manifolds, Tensor Analysis, and Applications*. 75. Springer-Verlag, New York, Berlin, Heidelberg, London, Paris, Tokyo, 2nd edition edition, 1988. Applied Mathematical Sciences.
2. M. Abramowitz and I. A. Stegun, editors. *Handbook of Mathematical Functions with Formulas, Graphs, and Mathematical Tables*. Applied Mathematics Series. National Bureau of Standards, 55 edition, 1964.
3. R. A. Adams. *Sobolev Spaces*, volume 65 of *Pure and Applied Mathematics*. Academic Press, New York, 1975.
4. J. Aisbett. Optical flow with an intensity-weighted smoothing. *IEEE Transactions on Pattern Analysis and Machine Intelligence*, 11(5):512–522, 1989.
5. L. Alvarez, F. Guichard, P. L. Lions, and J. M. Morel. Axiomatisation et nouveaux operateurs de la morphologie mathematique. *C. R. Acad. Sci. Paris*, 315:265–268, 1992.
6. L. Alvarez, F. Guichard, P. L. Lions, and J. M. Morel. Axiomes et equations fondamentales du traitement d'images. *C. R. Acad. Sci. Paris*, 315:135–138, 1992. Also: report 9216, CEREMADE, Université Paris Dauphine, 1992, presented at Arch. for Rat. Mech., July 1992.
7. L. Alvarez, F. Guichard, P. L. Lions, and J. M. Morel. Axioms and fundamental equations of image processing. *Arch. for Rational Mechanics*, 123(3):199–257, September 1993.
8. L. Alvarez and M. L. Signal and image restoration using shock filters and anisotropic diffusion. *SIAM J. Num. Anal.*, 31(1), January 1994.
9. L. Alvarez, P.-L. Lions, and J.-M. Morel. Image selective smoothing and edge detection by nonlinear diffusion. II. *SIAM J. Num. Anal.*, 29(3):845–866, June 1992.
10. L. Ambrosio. A compactness theorem for a special class of functions of bounded variation. *Boll. Un. Mat. It.*, 3-B:857–881, 1989.
11. L. Ambrosio. Variational problems on SBV. *Acta Applicandae Math.*, 17:1–40, 1989.
12. L. Ambrosio and V. Tortorelli. On the approximation of functionals depending on jumps by elliptic functionals. *Commun. Pure and Appl. Math.*, 43:999–1036, 1990.
13. L. Ambrosio and V. Tortorelli. Approximation of functionals depending on jumps by elliptic functionals via Γ–convergence. *Boll. Un. Mat. Ital.*, 7:105–123, 1992.
14. W. F. Ames. *Nonlinear Partial Differential Equations in Engineering*, volume 1–2. Academic Press, New York, San Francisco, London, 1972.
15. W. F. Ames. *Numerical Methods for Partial Differential Equations*. Academic Press, New York, San Francisco, 1977.
16. S. Angenent. On the formation of singularities in the curve shortening flow. *J. Differential Geometry*, 33:601–633, 1991.
17. S. Angenent. Parabolic equations for curves on surfaces, part I. curves with p-integrable curvature. *Annals of Mathematics*, 132:451–483, 1991.
18. S. Angenent. Parabolic equations for curves on surfaces, part II. intersections, blowup, and generalized solutions. *Annals of Mathematics*, 133:171–215, 1991.
19. S. Angenent, G. Sapiro, and A. Tannenbaum. On the affine heat equation for nonconvex curves. Technical report, MIT - LIDS, 1994. Submitted.
20. A. Arehart, L. Vincent, and B. B. Kimia. Mathematical morphology: The Hamilton-Jacobi connection. In *International Conf. on Computer Vision*, pages 215–219, 1993. Also: Brown University Technical Report LEMS 108.

411

21. V. I. Arnold. *Mathematical Methods in Classical Mechanics.* Springer-Verlag, New York, 1981.

22. H. Asada and M. Brady. The curvature primal sketch. *IEEE Trans. Pattern Analysis and Machine Intelligence,* 8(1):2–14, 1986.

23. H. Attouch. *Variational Convergence for Functions and Operators.* Pitman Publishing Inc., 1984.

24. J. Babaud, A. P. Witkin, M. Baudin, and R. O. Duda. Uniqueness of the Gaussian kernel for scale-space filtering. *IEEE Trans. Pattern Analysis and Machine Intelligence,* 8(1):26–33, 1986.

25. C. Ballester and M. Gonzalez. Affine invariant multiscale segmentation by variational methods. In *Eighth Workshop on Image and Multidimensional Image Processing,* pages 220–221, Cannes, September 8–10 1993. IEEE.

26. G. Barles. Remarks on a flame propagation model. Technical Report 464, INRIA, 1985.

27. G. Barles and G. Georgelin. A simple proof of convergence for an approximation scheme for computing motions by mean curvature. To appear, 1992.

28. G. Barles and P. E. Souganidis. Convergence of approximation schemes for fully nonlinear second order equations. *Asymp. Anal.,* 1993. to appear.

29. P. Belhumeur. A binocular stereo algorithm for reconstructing sloping, creased, and broken surfaces, in the presence of half-occlusion. In *International Conf. on Computer Vision,* Berlin, Germany, 1993.

30. R. Benedetti and C. Petronio. *Lectures on Hyperbolic Geometry.* Springer-Verlag, Berlin, 1987.

31. F. Bergholm. Edge focusing. *IEEE Trans. Pattern Analysis and Machine Intelligence,* 9(6):726–741, November 1987.

32. P. Bijl. *Aspects of Visual Contrast Detection.* PhD thesis, University of Utrecht, The Netherlands, May 1991.

33. G. L. Bilbro, W. E. Snyder, S. J. Garnier, , and J. W. Gault. Mean field annealing: A formalism for constructing GNC-like algorithms. *IEEE Trans. Neural Networks,* 3(1), January 1992.

34. A. Blake and A. Zisserman. *Visual Reconstruction.* MIT Press, Cambridge, Mass., 1987.

35. W. Blaschke. *Vorlesungen über Differentialgeometrie II.* Verlag Von Julius Springer, Berlin, 1923.

36. J. Blom. *Topological and Geometrical Aspects of Image Structure.* PhD thesis, Utrecht University, The Netherlands, 1992.

37. J. Blom, B. M. ter Haar Romeny, A. Bel, and J. J. Koenderink. Spatial derivatives and the propagation of noise in Gaussian scale-space. *J. of Vis. Comm. and Im. Repr.,* 4(1):1–13, March 1993.

38. J. Blom, B. M. ter Haar Romeny, and J. J. Koenderink. Affine invariant corner detection. Technical report, 3 D Computer Vison Research Group, Utrecht University NL, 1992.

39. D. Blostein and N. Ahuja. Representation and three-dimensional interpretation of image texture: An integrated approach. In *Proc. 1st Int. Conf. on Computer Vision,* pages 444–449, London, 1987. IEEE Computer Society Press.

40. H. Blum. Biological shape and visual science. *J. Theor. Biology,* 38:205–287, 1973.

41. G. W. Bluman and J. D. Cole. *Similarity Methods for Differential Equations.* Springer-Verlag, New York, Heidelberg, Berlin, 1974.

42. C. d. Boor. *A Practical Guide to Splines,* volume 27 of *Applied Mathematical Sciences.* Springer-Verlag, New York, 1978.

43. H. Brézis. *Opérateurs Maximaux Monotones et Semi-groupes de Contractions dans les Espaces de Hilbert,* volume 5 of *Mathematics Studies.* North-Holland, 1973.

44. R. W. Brockett and P. Maragos. Evolution equations for continuous-scale morphology. In *Proc. IEEE Int. Conf. on Acoust., Speech, Signal Processing,* 1992.

45. J. W. Bruce and P. J. Giblin. *Curves and Singularities.* Cambridge University

Press, Cambridge, 1984.

46. A. M. Bruckstein and A. N. Netravali. On differential invariants of planar curves and recognizing partially occluded planar shapes. In *Proc. of Visual Form Workshop*, Capri, May 1990. Plenum Press.

47. K. Brunnström, T. Lindeberg, and J.-O. Eklundh. Active detection and classification of junctions by foveation with a head-eye system guided by the scale-space primal sketch. In G. Sandini, editor, *Proc. Second European Conference on Computer Vision*, volume 588 of *Lecture Notes in Computer Science*, pages 701–709, Santa Margherita Ligure, Italy, May 1992. Springer-Verlag.

48. P. J. Burt. Fast filter transforms for image processing. *Computer Vision, Graphics, and Image Processing*, 16:20–51, 1981.

49. P. J. Burt and E. H. Adelson. The Laplacian pyramid as a compact image code. *IEEE Trans. Communications*, 9:4:532–540, 1983.

50. M. Campani and V. A. Computing optical flow from an overconstrained system of linear algebraic equations. In *Proc. Third Internatinal Conference on Computer Vision*, pages 22–26, Osaka, Japan, 1990.

51. J. Canny. A computational approach to edge detection. *IEEE Trans. Pattern Analysis and Machine Intelligence*, 8(6):679–698, 1986.

52. V. Cantoni and S. Levialdi, editors. *Pyramidal Systems for Computer Vision*. Springer-Verlag, Berlin, 1986.

53. M. Carriero and A. Leaci. S^k-valued maps minimizing the L^p norm of the gradient with free discontinuities. *Ann. Sc. Norm. Pisa, s.IV*, 18:321–352, 1991.

54. E. Cartan. *La Théorie des Groupes Finis et Continus et la Géometrie Différentielle traitées par la Méthode du Répère Mobile*. Gauthiers-Villars, 1937.

55. E. Cartan. *Leçons sur la Géométrie des Espaces de Riemann*. Gauthier-Villars, Paris, 2 edition, 1963.

56. V. Caselles, F. Catte, T. Coll, and F. Dibos. A geometric model for active contours in image processing. Technical Report 9210, CEREMADE, Université Paris Dauphine, 1992.

57. F. Catté, P.-L. Lions, J.-M. Morel, and T. Coll. Image selective smoothing and edge detection by nonlinear diffusion. *SIAM J. Num. Anal.*, 29(1):182–193, February 1992.

58. R. D. Chaney. Analytical representation of contours. A.I. memo 1392, Massachusetts Institute of Technology, October 1992.

59. A. Chehikian and J. L. Crowley. Fast computation of optimal semi-octave pyramids. In *Proc. 7th Scand. Conf. on Image Analysis*, pages 18–27, Aalborg, Denmark, August 1991.

60. Y.-G. Chen, Y. Giga, and S. Goto. Uniqueness and existence of viscosity solutions of generalized mean curvature flow equations. *J. Differential Geometry*, 33:749–786, 1991.

61. Y.-G. Chen, Y. Giga, T. Hitaka, and M. Honma. Numerical analysis of motion of a surface by its mean curvature. Preprint, Hokkaido University, 1989.

62. J. Clark. Singularity theory and phantom edges in scale-space. *IEEE Trans. Pattern Analysis and Machine Intelligence*, 10(5), 1988.

63. J. J. Clark. Authenticating edges produced by zero-crossing algorithms. *IEEE Trans. Pattern Analysis and Machine Intelligence*, 11:43–57, 1989.

64. F. Clarke. *Nonsmooth Analysis*. SIAM Publications, 1987.

65. J. Coggins and A. Jain. A spatial filtering approach to texture analysis. *Pattern Recognition Letters*, 3:195–203, 1985.

66. J. Cohen. Nonlinear variational method for optical flow computation. *Proc. 8th Scand. Conf. on Image Analysis*, 1:523–530, 1993.

67. M. A. Cohen and S. Grossberg. Neural dynamics of brightness perception: Features, boundaries, diffusion, and resonance. *Perception&Psychophysics*, 36(5):428–456, 1984.

68. T. Cohignac, F. Eve, F. Guichard, and C. Lopez. Affine morphological scale space:

Numerical analysis of its fundamental equation. Technical report, CEREMADE, Universite Paris Dauphine, 1993. unpublished.

69. T. Cohignac, F. Eve, F. Guichard, and J. M. Morel. Numerical analysis of the fundamental equation of image processing. Technical Report 9254, CEREMADE, Universite Paris Dauphine, 1992.

70. T. Cohignac, F. Guichard, and J. M. Morel. Multiscale analysis of shapes, images and textures. In *Eighth Workshop on Image and Multidimensional Image Processing*, pages 142–143, Cannes, September 8–10 1993. IEEE.

71. N. G. Cooper and G. B. West. *Scale and Dimension*, pages 4–21. Cambridge University Press, Los Alamos National Laboratory, 1988.

72. M. G. Crandall, H. Ishii, and P. L. Lions. User's guide to viscosity solutions of second order partial differential equations. *Bulletin of the American Mathematical Society*, 27:1–67, 1992.

73. M. G. Crandall and P. L. Lions. Viscosity solutions of Hamilton-Jacobi equations. *Trans. Amer. Math. Soc.*, 277:1–42, 1983.

74. J. Crank. *Mathematics of Diffusion*. Oxford University Press, London, 1956.

75. J. L. Crowley. *A Representation for Visual Information*. PhD thesis, Carnegie-Mellon University, Robotics Institute, Pittsburgh, Pennsylvania, 1981.

76. J. L. Crowley and A. C. Sanderson. Multiple resolution representation and probabilistic matching of 2-D gray-scale shape. *IEEE Trans. Pattern Analysis and Machine Intelligence*, 9(1):113–121, 1987.

77. J. L. Crowley and R. M. Stern. Fast computation of the Difference of Low Pass Transform. *IEEE Trans. Pattern Analysis and Machine Intelligence*, 6:212–222, 1984.

78. J. R. Crowley. A representation for shape based on peaks and ridges in the Difference of Low-Pass Transform. *IEEE Trans. Pattern Analysis and Machine Intelligence*, 6(2):156–170, 1984.

79. G. DalMaso, J. M. Morel, and S. Solimini. A variational method in image segmentation: Existence and approximation results. *Acta Math.*, 168:89–151, 1992.

80. I. Daubechies. *Ten Lectures on Wavelets*. SIAM, Philadelphia, 1992.

81. G. David and S. Semmes. On the singular sets of minimizers of the Mumford–Shah functional. preprint, IHES, 1993.

82. E. De Giorgi and L. Ambrosio. Un nuovo tipo di funzionale del calcolo delle variazioni. *Atti Accad. Naz. Lincei, s.8*, 82:199–210, 1988.

83. E. De Giorgi, M. Carriero, and A. Leaci. Existence theorem for a minimum problem with a free discontinuity set. *Arch. Rat. Mech. Anal.*, 108:195–218, 1989.

84. E. De Giorgi and T. Franzoni. Su un tipo di convergenza variazionale. *Ren. Sem. Mat. Brescia*, 3:63–101, 1979.

85. A. Degasperis, A. P. Fordy, and M. Lakshmanan, editors. *Nonlinear Evolution Equations: Integrability and Spectral Methods*, Oxford Road, Manchester M13 9PL, UK, 1990. Manchester University Press.

86. J. A. Dieudonné and J. B. Carrell. *Invariant Theory: Old and New*. Academic Press, New York, 1971.

87. L. Dorst and R. van den Boomgaard. Morphological signal processing and the slope transform. Submitted, 1993.

88. B. A. Dubrovin, A. T. Fomenko, and S. P. Novikov. *Modern Geometry—Methods and Applications*. Graduate Texts in Mathematics. Springer-Verlag, New York, 2 edition, 1992. Part I. The Geometry of Surfaces, Transformation Groups, and Fields.

89. R. O. Duda and P. E. Hart. *Pattern Classification and Scene Analysis*. John Wiley and Sons, New York, 1973.

90. G. Dudek and J. K. Tsotsos. Shape representation and recognition from curvature. In *IEEE Conference on Computer Vision and Pattern Recognition*, Hawaii, 1991.

91. P. Dupuis and J. Oliensis. An optimal control formulation and related numerical methods for a problem in shape reconstruction. Technical report, Department of

Computer Science, University of Massachusetts at Amherst, 1993.

92. D. Eberly. *Geometric Analysis of Ridges in N–Dimensional Images*. PhD thesis, University of North Carolina at Chapel Hill, Computer Science Department, 1994.

93. D. Eberly, D. Fritsch, and C. Kurak. Filtering with a normalized laplacian of a Gaussian filter. In *Proceedings of the SPIE International Symposium, Mathematical Methods in Medical Imaging*, San Diego, CA, 1992.

94. D. Eberly, R. Gardner, B. Morse, S. Pizer, and C. Scharlach. Ridges for image analysis. Submitted to Journal of Mathematical Imaging and Vision, July 1993, 1993.

95. C. L. Epstein and M. Gage. *Wave Motion: Theory, Modeling, and Computation*, chapter The Curve Shortening Flow. Springer Verlag, New York, 1987. A. Chorin, A. Majda, Editors.

96. ESPRIT - NSF. Geometry driven diffusion. Technical report, Collaboration between KUL Leuven, Utrecht Univ., Ceremade Paris, Las Palmas Univ., KTH Stockholm, ETH Zurich, Linkoping Univ., MIT Cambridge, Harvard Cambridge, CalTech, UCB Berkeley, 1993-1996. Technical Annex.

97. L. C. Evans and J. Spruck. Motion of level sets by mean curvature I. *J. Differential Geometry*, 33:635–681, 1991.

98. C. F., D. F., and K. G. A morphological approach to mean curvature motion. Technical Report 9310, CEREMADE, Universite Paris Dauphine, 1993.

99. O. Faugeras. Cartan's moving frame method and its applications to the geometry and evolution of curves in the Euclidean, affine and projective planes. Technical Report TR-2053, INRIA, 1993.

100. O. Faugeras. On the evolution of simple curves of the real projective plane. *Comptes rendus de l'Académie des Sciences de Paris*, 317:565–570, September 1993.

101. J. Favard. *Cours de Géométrie Différentielle Locale*. Gauthier-Villars, 1957.

102. W. Fenchel. *Elementary Geometry in Hyperbolic Space*. Number 11 in Studies in Mathematics. Walter de Gruyter, Berlin, 1989.

103. W. Fleming and H. Soner. *Controlled Markov Processes and Viscosity Solutions*. Springer-Verlag, New York, 1993.

104. L. M. J. Florack. *The Syntactical Structure of Scalar Images*. PhD thesis, University of Utrecht, Utrecht, The Netherlands, November 1993. Cum Laude.

105. L. M. J. Florack, A. H. Salden, B. M. ter Haar Romeny, J. J. Koenderink, and M. A. Viergever. Nonlinear scale-space. *Image and Vision Computing*, 1994. In press.

106. L. M. J. Florack, B. M. ter Haar Romeny, J. J. Koenderink, and M. A. Viergever. General intensity transformations. In P. Johansen and S. Olsen, editors, *Proc. 7th Scand. Conf. on Image Analysis*, pages 338–345, Aalborg, DK, August 1991.

107. L. M. J. Florack, B. M. ter Haar Romeny, J. J. Koenderink, and M. A. Viergever. Families of tuned scale-space kernels. In G. Sandini, editor, *Proceedings of the European Conference on Computer Vision*, pages 19–23, Santa Margherita Ligure, Italy, May 19–22 1992.

108. L. M. J. Florack, B. M. ter Haar Romeny, J. J. Koenderink, and M. A. Viergever. General intensity transformations and second order invariants. In P. Johansen and S. Olsen, editors, *Theory & Applications of Image Analysis*, volume 2 of *Series in Machine Perception and Artificial Intelligence*, pages 22–29. World Scientific, Singapore, 1992.

109. L. M. J. Florack, B. M. ter Haar Romeny, J. J. Koenderink, and M. A. Viergever. Scale and the differential structure of images. *Image and Vision Computing*, 10(6):376–388, July/August 1992.

110. L. M. J. Florack, B. M. ter Haar Romeny, J. J. Koenderink, and M. A. Viergever. Cartesian differential invariants in scale-space. *Journal of Mathematical Imaging and Vision*, 3(4):327–348, November 1993.

111. L. M. J. Florack, B. M. ter Haar Romeny, J. J. Koenderink, and M. A.

Viergever. General intensity transformations and differential invariants. *Journal of Mathematical Imaging and Vision*, 1993. In press.

112. L. M. J. Florack, B. M. ter Haar Romeny, J. J. Koenderink, and M. A. Viergever. The multiscale local jet. In M. A. Viergever, editor, *Proceedings of the VIP*, pages 21–24, Utrecht, The Netherlands, June 2–4 1993.

113. L. M. J. Florack, B. M. ter Haar Romeny, J. J. Koenderink, and M. A. Viergever. The multiscale local jet. Submitted, 1993.

114. L. M. J. Florack, B. M. ter Haar Romeny, J. J. Koenderink, and M. A. Viergever. Images: Regular tempered distributions. In Y.-L. O, A. Toet, H. J. A. M. Heijmans, D. H. Foster, and P. Meer, editors, *Proc. of the NATO Advanced Research Workshop Shape in Picture - Mathematical description of shape in greylevel images*, volume 126 of *NATO ASI Series F*, pages 651–660. Springer Verlag, Berlin, 1994.

115. L. M. J. Florack, B. M. ter Haar Romeny, J. J. Koenderink, and M. A. Viergever. Linear scale-space. *Journal of Mathematical Imaging and Vision*, 1994.

116. M. A. Förstner and E. Gülch. A fast operator for detection and precise location of distinct points, corners and centers of circular features. In *Proc. Intercommission Workshop of the Int. Soc. for Photogrammetry and Remote Sensing*, Interlaken, Switzerland, 1987.

117. D. Forsyth, J. L. Mundy, A. Zisserman, C. Coelho, A. Heller, and C. Rothwell. Invariant descriptors for 3-D object recognition and pose. *IEEE Trans. Pattern Analysis and Machine Intelligence*, 13(10):971–991, October 1991.

118. J. Fourier. *The Analytical Theory of Heat*. Dover Publications, Inc., New York, 1955. Replication of the English translation that first appeared in 1878 with previous corrigenda incorporated into the text, by Alexander Freeman, M.A. Original work: "Théorie Analytique de la Chaleur", Paris, 1822.

119. W. T. Freeman and E. H. Adelson. Steerable filters for early vision, image analysis and wavelet decomposition. In *Proc. 3rd Int. Conf. on Computer Vision*, Osaka, Japan, December 1990. IEEE Computer Society Press.

120. D. Fritsch. *Registration of Radiotherapy Images Using Multiscale Medial Descriptions of Image Structure*. PhD thesis, The University of North Carolina at Chapel Hill, Department of Biomedical Engineering, 1993.

121. D. Gabor. Theory of communications. *Journal IEEE London*, 93:429–457, 1946.

122. M. Gage. An isoperimetric inequality with applications to curve shortening. *Duke Mathematical Journal*, 50:1225–1229, 1983.

123. M. Gage. Curve shortening makes convex curves circular. *Invent. Math.*, 76:357–364, 1984.

124. M. Gage. On an area-preserving evolution equation for plane curves. *Contemporary Mathematics*, 51:51–62, 1986.

125. M. Gage and R. S. Hamilton. The heat equation shrinking convex plane curves. *J. Differential Geometry*, 23:69–96, 1986.

126. E. B. Gamble and T. Poggio. Visual integration and detection of discontinuities: The key role of intensity edges. Technical report, MIT A.I. Lab, 1987. A.I. Memo No. 970.

127. J. Gårding and T. Lindeberg. Direct estimation of local surface shape in a fixating binocular vision system. In J.-O. Eklundh, editor, *Proc. 3rd European Conference on Computer Vision*, volume 800 of *Lecture Notes in Computer Science*, pages 365–376, Stockholm, Sweden, May 1994. Springer-Verlag.

128. D. Geiger and F. Girosi. Parallel and deterministic algorithms for MRFs: Surface reconstruction and integration. Technical Report Memo No. 1114, MIT A.I.Lab., June 1989.

129. D. Geiger and A. Yuille. A common framework for image segmentation. Technical Report Tech. Rep. no. 89-7, Harvard Robotics Laboratory, 1989.

130. D. Geiger and A. Yuille. A common framework for image segmentation. *International Journal of Computer Vision*, 6(3):227–243, 1991.

131. I. M. Gelfand and S. V. Fomin. *Calculus of Variations*. Selected Russian

Publications in the Mathematical Sciences. Prentice-Hall, Englewood Cliffs, New York, 1963. Silverman, R. A., Ed.

132. S. Geman and D. Geman. Stochastic relaxation, gibbs distributions, and the bayesian restoration of images. *IEEE Trans. Pattern Analysis and Machine Intelligence*, 6:721–741, 1984.

133. G. Gerig, O. Kübler, R. Kikinis, and F. A. Jolesz. Nonlinear anisotropic filtering of MRI data. *IEEE Transactions on Medical Imaging*, 11(2):221–232, June 1992.

134. G. Gerig, J. Martin, R. Kikinis, O. Kübler, M. Shenton, and F. A. Jolesz. Unsupervised tissue type segmentation of 3D dual-echo MR head data. *Image and Vision Computing*, 10(6):349–360, July 1992. IPMI 1991 special issue.

135. I. Gladwell and R. Wait. *A Survey of Numerical Methods for Partial Differential Equations*. Clarendon Press, London, 1979.

136. J. H. Grace and A. Young. *Algebra of Invariants*. Chelsea Publishing Company, New York, 1965.

137. N. Graham. Does the brain perform a Fourier analysis of the visual scene? Tech. Report, 1979.

138. N. Graham. The visual system does a crude Fourier analysis of patterns. In S. Grossberg, editor, *SIAM–AMS Proceedings*, volume 13, pages 1–16, Hillsdale, New Jersey, 1981. American Mathematical Society, Lawrence Erlbaum Associates.

139. M. Grayson. The heat equation shrinks embedded plane curves to round points. *Journal of Differential geometry*, 26:285–314, 1987.

140. M. Grayson. Shortening embedded curves. *Annals of Mathematics*, 129:71–111, 1989.

141. M. L. Green. The moving frame, differential invariants and rigidity theorems for curves in homogeneous spaces. *Duke Mathematical Journal*, 45:735–779, 1978.

142. S. Grossberg and D. Todorovic. Neural dynamics of 1-D and 2-D brightness perception: A unified model of classical and recent phenomena. *Perception&Psychophysics*, 43:241–277, 1988.

143. H. W. Guggenheimer. *Differential Geometry*. McGraw-Hill Book Company, New York, 1963.

144. F. Guichard. Multiscale analysis of movies. In *Eighth Workshop on Image and Multidimensional Image Processing*, pages 236–237, Cannes, September 8–10 1993. IEEE.

145. W. Hackbush. *Multi-Grid Methods and Applications*. Springer-Verlag, New York, 1985.

146. A. R. Hanson and E. M. Riseman. Processing cones: A parallel computational structure for scene analysis. Technical Report 74C-7, Computer and Information Science, Univ. of Massachusetts, Amherst, Massachusetts, 1974.

147. D. N. Hào. *Inverse Heat Conduction Problems*. PhD thesis, Mathematics Department, Free University of Berlin, 1991.

148. R. Haralick and L. Shapiro. *Computer and Robot Vision, Volume 1*. Addison-Wesley, New York, 1992.

149. J. Harris, C. Koch, J. Luo, and J. Wyatt. *Analog VLSI Implementation of Neural Systems*, chapter Resistive Fuses: Analog Hardware for Detecting Discontinuities in Early Vision, pages 27–55. Kluwer Academic Publishers, Norwell, MA, 1989. Mead, C. and Ismail, M., Eds.

150. D. Hilbert. Ueber die vollen Invariantensystemen. *Math. Annalen*, 42:313–373, 1893.

151. E. Hildreth. *The Measurement of Visual Motion*. M. I. T. Press, Cambridge, Mass., 1983.

152. E. Hille and R. S. Phillips. *Functional Analysis and Semi-Groups*, volume XXXI. American Mathematical Society Colloquium Publications, 1957.

153. B. Hillen. Interactive anatomy of craniofacial structure. Technical report, Utrecht University, Faculty of Medicine, 1993. CD-ROM interactive.

154. I. I. Hirschmann and D. V. Widder. *The Convolution Transform*. Princeton

University Press, Princeton, New Jersey, 1955.

155. L. Hörmander. *Linear Partial Differential Operators*, volume 257 of *Grundlehren der mathematische Wissenshaften*. Springer-Verlag, 1963.

156. B. Horn and M. Brooks, editors. *Shape from Shading*. M. I. T. Press, Cambridge, Mass., 1989.

157. B. Horn and B. Schunck. Determining optical flow. *Artificial Intelligence*, 23:185–203, 1981.

158. B. K. P. Horn and E. J. Weldon. Filtering closed curves. *IEEE Trans. Pattern Analysis and Machine Intelligence*, 8:665–668, 1986.

159. D. H. Hubel. *Eye, Brain and Vision*, volume 22 of *Scientific American Library*. Scientific American Press, New York, 1988.

160. D. H. Hubel and T. N. Wiesel. Receptive fields, binocular interaction, and functional architecture in the cat's visual cortex. *Journal of Physiology*, 160:106–154, 1962.

161. D. H. Hubel and T. N. Wiesel. Brain mechanisms of vision. *Scientific American*, 241:45–53, 1979.

162. G. Huisken. Flow by mean curvature of convex surfaces into spheres. *Journal of Differential Geometry*, 20:237–266, 1984.

163. R. A. Hummel. Representations based on zero crossings in scale space. In *Proceedings of the IEEE Computer Vision and Pattern Recognition Conference*, pages 204–209, June 1986. Reproduced in: "Readings in Computer Vision: Issues, Problems, Principles and Paradigms", M. Fischler and O. Firschein (eds.), Morgan Kaufmann, 1987.

164. R. A. Hummel. The scale-space formulation of pyramid data structures. In L. Uhr, editor, *Parallel Computer Vision*, pages 187–123. Academic Press, New York, 1987.

165. R. A. Hummel, B. B. Kimia, and S. W. Zucker. Deblurring Gaussian blur. *Computer Vision, Graphics, and Image Processing*, 38:66–80, 1987.

166. A. K. Jain and R. C. Dubes. *Algorithms for Clustering Data*. Prentice-Hall, New Jersey, 1988.

167. P. Johansen. On the classification of toppoints in scale space. *Journal of Mathematical Imaging and Vision*, 1993. To appear.

168. J.-J. Jolion and A. Rozenfeld. *A Pyramid Framework for Early Vision*. Kluwer Academic Publishers, Dordrecht, Netherlands, 1994.

169. G. J. Jones and J. Malik. A computational framework for determining stereo correspondence from a set of linear spatial filters. In G. Sandini, editor, *Proceedings of the European Conference on Computer Vision*, pages 395–410, Santa Margherita Ligure, Italy, May 19–22 1992. Springer-Verlag.

170. B. Julesz and T. Caelli. On the limits of Fourier decompositions in visual texture perception. *Perception*, 8:69–73, 1979.

171. K. Kanatani. *Group-Theoretical Methods in Image Understanding*, volume 20 of *Series in Information Sciences*. Springer-Verlag, 1990.

172. G. Kanizsa. *Organization in Vision*. Praeger, New York, 1979.

173. S. Karlin. *Total Positivity*. Stanford Univ. Press, 1968.

174. M. Kass, A. Witkin, and D. Terzopoulos. Snakes: Active contour models. In *Proc. IEEE First Int. Comp. Vision Conf.*, 1987.

175. D. C. Kay. *Tensor Calculus*. Schaum's Outline Series. McGraw-Hill Book Company, New York, 1988.

176. R. Kikinis, M. . Shenton, G. Gerig, J. Martin, M. Anderson, D. Metcalf, C. R. G. Guttmann, R. W. McCarley, B. Lorensen, H. Cline, and F. A. Jolesz. Routine quantitative analysis of brain and cerebrospinal fluid spaces with MR imaging. *Journal of Magnetic Resonance Imaging*, 2(6):619–629, 1992.

177. B. B. Kimia. *Conservation Laws and a Theory of Shape*. PhD thesis, McGill University, 1990.

178. B. B. Kimia. *Towards a Computational Theory of Shape*. PhD thesis, Department of Electrical Enginering, McGill University, Montreal, Canada, August 1990.

179. B. B. Kimia. Entropy scale-space. In *Proc. of Visual Form Workshop*, Capri, Italy, May 1991. Plenum Press.

180. B. B. Kimia, A. Tannenbaum, and S. W. Zucker. Towards a computational theory of shape, an overview. In *Proc. First European Conference on Computer Vision*, volume 427, pages 402–407, New York, 1990. Springer-Verlag. Lecture Notes in Computer Science.

181. B. B. Kimia, A. Tannenbaum, and S. W. Zucker. On the evolution of curves via a function of curvature I: the classical case. *Journal of Mathematical Analysis and Applications*, 163:438–458, 1992.

182. B. B. Kimia, A. Tannenbaum, and S. W. Zucker. Shapes, shocks, and deformations, I. *International Journal of Computer Vision*, To appear, 1994.

183. R. Kimmel, K. Siddiqi, B. B. Kimia, and A. Bruckstein. Shape from shading via level sets. Technical report, Department of Computer Science, Technion, Haifa, Israel, 1992. Techn. Report, submitted to IJCV.

184. L. Kitchen and A. Rosenfeld. Gray-level corner detection. *Pattern Recognition Letters*, 1:95–102, 1982.

185. A. Klinger. Pattern and search statistics. In J. S. Rustagi, editor, *Optimizing Methods in Statistics*, New York, 1971. Academic Press.

186. C. B. Knudsen and H. I. Christensen. On methods for efficient pyramid generation. In *Proc. 7th Scand. Conf. on Image Analysis*, pages 28–39, Aalborg, Denmark, August 1991.

187. J. J. Koenderink. The structure of images. *Biol. Cybern.*, 50:363–370, 1984.

188. J. J. Koenderink. Scale-time. *Biol. Cybern.*, 58:159–162, 1988.

189. J. J. Koenderink. The brain a geometry engine. *Psychological Research*, 52:122–127, 1990.

190. J. J. Koenderink. *Solid Shape*. MIT Press, Cambridge, Mass., 1990.

191. J. J. Koenderink. Local image structure. In *Proc. Scand. Conf. on Image Analysis*, pages 1–7, Aalborg, DK, August 1991.

192. J. J. Koenderink, A. Kappers, and A. van Doorn. Local operations: The embodiment of geometry. In G. A. Orban and H. H. Nagel, editors, *Artificial and Biological Vision Systems*, ESPRIT: Basic Research Series, pages 1–23. DG XIII Commision of the European Communities, 1992.

193. J. J. Koenderink and W. Richards. Two-dimensional curvature operators. *Journal of the Optical Society of America-A*, 5(7):1136–1141, 1988.

194. J. J. Koenderink and A. J. van Doorn. Visual detection of spatial contrast; influence of location in the visual field, target extent and illuminance level. *Biol. Cybern.*, 30:157–167, 1978.

195. J. J. Koenderink and A. J. van Doorn. Dynamic shape. *Biol. Cybern.*, 53:383–396, 1986.

196. J. J. Koenderink and A. J. van Doorn. Representation of local geometry in the visual system. *Biol. Cybern.*, 55:367–375, 1987.

197. J. J. Koenderink and A. J. Van Doorn. Operational significance of receptive field assemblies. *Biol. Cybern.*, 58:163–171, 1988.

198. J. J. Koenderink and A. J. van Doorn. Receptive field families. *Biol. Cybern.*, 63:291–298, 1990.

199. J. J. Koenderink and A. J. van Doorn. Generic neighborhood operators. *IEEE Trans. Pattern Analysis and Machine Intelligence*, 14(6):597–605, June 1992.

200. J. J. Koenderink and A. J. van Doorn. Local features of smooth shapes: Ridges and courses. In *Proc. SPIE Geometric Methods in Computer Vision II*, volume 2031, pages 2–13, San Diego, CA, July, 12-13 1993. Proceedings SPIE.

201. J. J. Koenderink and A. J. van Doorn. Two-plus-one dimensional differential geometry. *Pattern Recognition Letters*, 21(15):439–443, May 1994.

202. G. Koepfler, C. Lopez, and J. M. Morel. A multiscale algorithm for image segmentation by variational method. *SIAM J. Num. Anal.*, 1994. to appear.

203. A. F. Korn. Toward a symbolic representation of intensity changes in images.

IEEE Trans. Pattern Analysis and Machine Intelligence, 10(5):610–625, 1988.

204. A. Kumar. *Visual Information in a Feedback Loop*. PhD thesis, University of Minnesota, 1994.

205. D. F. Lawden. *An Introduction to Tensor Calculus and Relativity*. Spottiswoode Ballantyne & Co Ltd, 1962.

206. D. F. Lawden. *An Introduction to Tensor Calculus and Relativity*. Spottiswoode Ballantyne & Co Ltd, 1962.

207. T.-S. Lee, D. Mumford, and A. Yuille. Texture segmentation by minimizing vector-valued energy functionals: The coupled-membrane model. In G. Sandidi, editor, *Proc. Second European Conference on Computer Vision*, pages 165–183, 1992.

208. R. J. LeVeque. *Numerical Methods for Conservation Laws*. Birkhäuser, Boston, 1992.

209. T. Levi-Civita. *Der Absolute Differentialkalkül*. Springer-Verlag, 1928.

210. S. Lie. Klassifikation und Integration von gewöhnlichen Differentialgleichungen zwischen x, y, die eine Gruppe von Transformationen gestatten I, II. *Math. Ann.*, 32:231–281, 1888. See also *Gesammelte Abhandlungen*, vol. 5, B. G. Teubner, Leipzig, 1924, pp. 240–310.

211. L. M. Lifshitz and S. M. Pizer. A multiresolution hierarchical approach to image segmentation based on intensity extrema. *IEEE Trans. Pattern Analysis and Machine Intelligence*, 12(6):529–541, 1990.

212. A. P. Lightman, W. H. Press, R. H. Price, and S. A. Teukolsky. *Problem Book in Relativity and Gravitation*. Princeton University Press, Princeton, New Jersey, 1975.

213. T. Lindeberg. Scale-space for discrete signals. *IEEE Trans. Pattern Analysis and Machine Intelligence*, 12(3):234–245, 1990.

214. T. Lindeberg. *Discrete Scale-Space Theory and the Scale-Space Primal Sketch*. PhD thesis, Royal Institute of Technology, Department of Numerical Analysis and Computing Science, Royal Institute of Technology, S-100 44 Stockholm, Sweden, May 1991.

215. T. Lindeberg. Scale-space behaviour of local extrema and blobs. *Journal of Mathematical Imaging and Vision*, 1(1):65–99, March 1992.

216. T. Lindeberg. Detecting salient blob-like image structures and their scales with a scale-space primal sketch — a method for focus-of-attention. *International Journal of Computer Vision*, 11(3):283–318, 1993.

217. T. Lindeberg. Discrete derivative approximations with scale-space properties: A basis for low-level feature extraction. *Journal of Mathematical Imaging and Vision*, 3(4):349–376, 1993.

218. T. Lindeberg. On scale selection for differential operators. In K. H. K. A. Høgdra, B. Braathen, editor, *Proc. 8th Scandinavian Conf. Image Analysis*, pages 857–866, Trømso, Norway, May 1993. Norwegian Society for Image Processing and Pattern Recognition.

219. T. Lindeberg. Scale selection for differential operators. Technical Report ISRN KTH/NA/P–9403-SE, Dept. of Numerical Analysis and Computing Science, Royal Institute of Technology, January 1994.

220. T. Lindeberg. Scale-space behaviour and invariance properties of differential singularities. In Y.-L. O, A. Toet, H. J. A. M. Heijmans, D. H. Foster, and P. Meer, editors, *Proc. of the NATO Advanced Research Workshop Shape in Picture — Mathematical Description of Shape in Greylevel Images*, volume 126 of *NATO ASI Series F*, pages 591–600. Springer Verlag, Berlin, 1994.

221. T. Lindeberg. Scale-space for N-dimensional discrete signals. In Y.-L. O, A. Toet, H. J. A. M. Heijmans, D. H. Foster, and P. Meer, editors, *Proc. of the NATO Advanced Research Workshop Shape in Picture - Mathematical Description of Shape in Greylevel Images*, volume 126 of *NATO ASI Series F*, pages 571–590. Springer Verlag, Berlin, 1994. (Also available in Tech. Rep. ISRN KTH/NA/P–92/26–SE from Royal Inst. of Technology).

222. T. Lindeberg. Scale-space theory: A basic tool for analysing structures at different scales. *Journal of Applied Statistics*, 21(2):223–261, 1994. Special issue on 'Statistics and Images' (In press).

223. T. Lindeberg. *Scale-Space Theory in Computer Vision*. The Kluwer International Series in Engineering and Computer Science. Kluwer Academic Publishers, Dordrecht, the Netherlands, 1994.

224. T. Lindeberg and J. O. Eklundh. On the computation of a scale-space primal sketch. *J. of Vis. Comm. and Im. Repr.*, 2(1):55–78, 1990.

225. T. Lindeberg and J. O. Eklundh. Scale detection and region extraction from a scale-space primal sketch. In *Proc. 3rd Int. Conf. on Computer Vision*, pages 416–426, Osaka, Japan, December 1990.

226. T. Lindeberg and J. O. Eklundh. On the computation of a scale-space primal sketch. *Journal of Visual Comm. and Image Rep.*, 2:55–78, 1991.

227. T. Lindeberg and J. O. Eklundh. The scale-space primal sketch: Construction and experiments. *Image and Vision Computing*, 10(1):3–18, January 1992.

228. T. Lindeberg and L. Florack. On the decrease of resolution as a function of eccentricity for a foveal vision system. Technical Report TRITA-NA-P9229, Dept. of Numerical Analysis and Computing Science, Royal Institute of Technology, November 1992. Submitted to Biological Cybernetics.

229. T. Lindeberg and J. Gårding. Shape from texture from a multi-scale perspective. In H. H. Nagel et al., editors, *Proceedings of the fourth ICCV*, pages 683–691, Berlin, Germany, 1993. IEEE Computer Society Press.

230. T. Lindeberg and J. Gårding. Shape-adapted smoothing in estimation of 3-D depth cues from affine distortions of local 2-D structure. In J.-O. Eklundh, editor, *Proc. 3rd European Conference on Computer Vision*, volume 800 of *Lecture Notes in Computer Science*, pages 389–400, Stockholm, Sweden, May 1994. Springer-Verlag.

231. T. P. Lindeberg. Effective scale: A natural unit for measuring scale-space lifetime. *IEEE Trans. Pattern Analysis and Machine Intelligence*, 15(10), October 1993.

232. P. L. Lions, E. Rouy, and A. Tourin. Shape from shading, viscosity and edges. Technical report, CEREMADE, University of Paris Dauphine, France, 1991.

233. A. Liu, S. M. Pizer, D. Eberly, B. Morse, J. Rosenman, and V. Carrasco. Volume registration using the 3D core. In *Proc. SPIE Medical Imaging VIII*, Newport Beach, CA, February 1994. to appear.

234. C. Lopez and J. M. Morel. Axiomatization of shape analysis and application to texture hyperdiscrimination. In *Surface Tension and Movement by Mean Curvature*, Trento, 1992. De Gruyter, Berlin.

235. D. G. Lowe. Organization of smooth image curves at multiple scales. *International Journal of Computer Vision*, 3:119–130, 1989.

236. Y. Lu and R. C. Jain. Behaviour of edges in scale space. *IEEE Trans. Pattern Analysis and Machine Intelligence*, 11(4):337–357, 1989.

237. M. J. M. and S. S. *Variational Methods in Image Segmentation*. Birkhäuser, 1993. to appear.

238. J. Malik and R. Rosenholtz. A differential method for computing local shape-from-texture for planar and curved surfaces. In *Proc. IEEE Comp. Soc. Conf. on Computer Vision and Pattern Recognition*, pages 267–273, 1993.

239. R. Malladi, J. Sethian, and B. Vemuri. Shape modeling with front propagation: a level set approach. To appear, 1994.

240. S. G. Mallat. Multifrequency channel decompositions of images and wavelet models. *IEEE Trans. Acoustics, Speech, and Signal Processing*, 37:2091–2110, 1989.

241. S. G. Mallat. Multiresolution approximations and wavelet orthonormal bases of $L^2(R)$. *Trans. Amer. Math. Soc.*, 315:69–87, 1989.

242. S. G. Mallat. A theory for multiresolution signal decomposition: The wavelet representation. *IEEE Trans. Pattern Analysis and Machine Intelligence*, 11(7):674–694, 1989.

243. S. G. Mallat and S. Zhong. Characterization of signals from multi-scale edges.

IEEE Trans. Pattern Analysis and Machine Intelligence, 14:710–723, 1992.

244. R. March. Computation of stereo disparity using regularization. *Patt. Recogn. Lett.*, 8(3):181–187, 1988.

245. R. March. A regularization model for stereo vision with controlled continuity. *Patt. Recogn. Lett.*, 10(4):259–263, 1989.

246. R. March. Visual reconstruction with discontinuities using variational methods. *Image and Vision Computing*, 10(1), Jan.-Febr. 1992.

247. D. Marr. *Vision*. W. H. Freeman & Co., 1882.

248. D. C. Marr and E. C. Hildreth. Theory of edge detection. *Proc. Roy. Soc. London B*, 207:187–217, 1980.

249. J. L. Marroquin. *Probabilistic Solution of Inverse Problems*. PhD thesis, Dept. of E.E.C.S., MIT, 1985.

250. G. Matheron. *Random Sets and Integral Geometry*. Wiley, New York, 1975.

251. B. Merriman, J. Bence, and S. Osher. Diffusion generated motion by mean curvature. Technical report, University of California, Department of Mathematics, Los Angeles, CA 90024-1555, 1992. CAM Report #92-18.

252. Y. Meyer. *Analysis at Urbana I*, chapter Wavelets and Operators. London Mathematical Society Lecture Notes Series. Cambridge University Press, 1989. E. R. Berkson, N. T. Peck, and J. Uhi Eds.

253. Y. Meyer. *Ondelettes et Algorithmes Concurrents*. Hermann, 1992.

254. C. W. Misner, K. S. Thorne, and J. A. Wheeler. *Gravitation*. Freeman, San Francisco, 1973.

255. S. K. Mitter. Markov random fields, stochastic quantization, and image analysis. In R. Spigler, editor, *Applied and Industrial Mathematics*. Kluwer Academic Publishers, 1991.

256. F. Mokhatarian and A. Mackworth. A theory of multi-scale, curvature-based shape representation for planar curves. *IEEE Trans. Pattern Analysis and Machine Intelligence*, 14:789–805, 1992.

257. F. Mokhtarian and A. Mackworth. Scale-based description of planar curves and two-dimensional shapes. *IEEE Trans. Pattern Analysis and Machine Intelligence*, 8:34–43, 1986.

258. J.-M. Morel. Segmentation d'images. Technical Report 9047, Ceremade, Université Paris-Dauphine, 1990.

259. J. M. Morel and S. Solimini. Segmentation of images by variational methods: A constructive approach. *Rev. Matematica de la Universidad Complutense*, 1(3):169–182, 1988.

260. P. Morrison. *Powers of Ten: About the Relative Size of Things in the Universe*. W. H. Freeman and Company, 1985.

261. B. S. Morse, S. M. Pizer, and A. Liu. Multiscale medial analysis of medical images. In H. Barrett and A. Gmitro, editors, *Information Processing in Medical Imaging (IPMI 14)*, Berlin, 1993. Springer-Verlag.

262. D. Mumford. On the computational architecture of the neocortex. i. the role of the thalamo-cortical loop. *Biol. Cybern.*, 65:135–145, 1991.

263. D. Mumford and J. Shah. Boundary detection by minimizing functionals. In *Proc. IEEE Conf. on Computer Vision and Pattern Recognition*, San Francisco, 1985.

264. D. Mumford and J. Shah. Optimal approximations by piecewise smooth functions and associated variational problems. *Communications on Pure and Applied Mathematics*, XLII:577–685, July 1989.

265. J. L. Mundy and A. Zisserman, editors. *Geometric Invariance in Computer Vision*. MIT Press, Cambridge, Massachusetts, 1992.

266. H. H. Nagel. Displacement vectors derived from second-order intensity variations in image sequences. *Comp. Graph. and Image Proc.*, 21:85–117, 1983.

267. H. H. Nagel and W. Enkelmann. An investigation of smoothness constraints for the estimation of displacement vector fields from image sequence. *IEEE Transactions on Pattern Analysis and Machine Intelligence*, 8(5):565–593, 1986.

268. W. J. Niessen, B. M. ter Haar Romeny, L. M. J. Florack, A. H. Salden, and M. A. Vergever. Nonlinear diffusion of scalar images using well-posed differential operators with applications in medical imaging. In *IEEE Conference on Computer Vision and Pattern Recognition, CVPR94*, Seattle, WA, June 19-23 1994. IEEE.

269. L. Nirenberg. A strong maximum principle for parabolic equations. *Communications on Pure and Applied Mathematics*, VI:167–177, 1953.

270. M. Nitzberg, D. Mumford, and T. Shiota. *Filtering, Segmentation and Depth*. Lecture Notes in Computer Science, 662. Springer-Verlag, 1993.

271. M. Nitzberg and T. Shiota. Nonlinear image filtering with edge and corner enhancement. *IEEE Trans. Pattern Analysis and Machine Intelligence*, 14(8):826–833, 1992.

272. K. Nomizu. *Lie groups and differential geometry*. Mathematical Society of Japan, Tokyo, 1956.

273. N. Nordström. Biased anisotropic diffusion — a unified regularization and diffusion approach to edge detection. *Image and Vision Computing*, 8(11):318–327, 1990. Also in: Proc. 1st European Conf. on Computer Vision, LNCS-Series Vol. 427, Springer-Verlag, pages 18–27.

274. N. Nordström. *Variational Edge Detection*. PhD thesis, University of California at Berkeley, May 1990.

275. J. Norton. Einstein, the hole argument and the reality of space. In J. Forge, editor, *Measurement, Realism and Objectivity: Essays on Measurements in the Social and Physical Sciences*, volume 5 of *Australian Studies in History and Philosophy of Science*, chapter 5, pages 153–188. D. Reidel Publishing Company, 1987.

276. J. Norton. Coordinates and covariance: Einstein's view of space-time and the modern view. *Foundations of Physics*, 19(10), 1989.

277. J. Oliensis. Local reproducible smoothing without shrinkage. *IEEE Trans. Pattern Analysis and Machine Intelligence*, 15:307–312, 1993.

278. P. Olver. Equivalence, invariants and symmetry. Preliminary version of book, 1994.

279. P. J. Olver. *Applications of Lie Groups to Differential Equations*, volume 107 of *Graduate Texts in Mathematics*. Springer-Verlag, 1986. Second Edition 1993.

280. P. J. Olver. Equivalence, invariants, and symmetry. Preliminary version of book, 1993.

281. P. J. Olver. Differential invariants. To appear in Acta Appl. Math., 1994.

282. P. J. Olver, G. Sapiro, and A. Tannenbaum. Invariant geometric evolutions of surfaces and volumetric smoothing. Technical report, MIT - LIDS, April 1994. Sbmitted for publication.

283. L. Onsager. Crystal statistics, I. a 2-dimensional model with order-disorder transitions. *Phys. Rev.*, 65:117–149, 1944.

284. S. Osher and L. I. Rudin. Feature-oriented image enhancement using shock filters. *SIAM J. Num. Anal.*, 27:919–940, 1990.

285. S. Osher and S. Sethian. Fronts propagating with curvature dependent speed: algorithms based on the Hamilton-Jacobi formalism. *J. Computational Physics*, 79:12–49, 1988.

286. K. Ottenberg. *Model Based Extraction of Geometric Structure from Digital Images*. PhD thesis, Utrecht University, The Netherlands, 1993.

287. L. V. Ovsiannikov. *Group Analysis of Differential Equations*. Academic Press, New York, 1982.

288. E. J. Pauwels, P. Fiddelaers, T. Moons, and L. J. van Gool. An extended class of scale-invariant and recursive scale-space filters. Technical Report KUL/ESAT/MI2/9316, Catholic University Leuven, 1993.

289. E. J. Pauwels, M. Proesmans, L. J. Van Gool, T. Moons, and A. Oosterlinck. Image enhancement using coupled anisotropic diffusion equations. In *Proc. on the 11th European Conference on Circuit Theory and Design*, volume 2, pages 1459–1464, 1993.

290. P. Perona.   Steerable-scalable kernels for edge detection and junction analysis.
      In *Proc. 2nd European Conf. on Computer Vision*, pages 3–18, Santa Margherita
      Ligure, Italy, May 1992.

291. P. Perona and J. Malik. Scale-space and edge detection using anisotropic diffusion.
      In *IEEE Computer Society Workshop on Computer Vision*, pages 16–22, Miami,
      FL, 1987.

292. P. Perona and J. Malik. A network for multiscale image segmentation. In *IEEE
      International Symposium on Circuits and Systems*, pages 2565–2568, Helsinki, June
      1988.

293. P. Perona and J. Malik. Scale-space and edge detection using anisotropic diffusion.
      IEEE Trans. Pattern Analysis and Machine Intelligence, 12(7):629–639, July 1990.

294. S. M. Pizer, C. A. Burbeck, J. M. Coggins, D. S. Fritsch, and B. S. Morse. Object
      shape before boundary shape: Scale-space medial axis. Technical Report TR92-
      023, The University of North Carolina, Chapel Hill, Department of Computer
      Science, 1992.

295. S. M. Pizer, J. M. Gauch, T. J. Cullip, and R. E. Fredericksen. Descriptions of
      intensity structure via scale and symmetry. In *Proceedings First Conference on
      Visualization in Biomedical Computing*, pages 94–101, 1990.

296. S. M. Pizer and B. M. ter Haar Romeny. Fundamental properties of medical image
      perception. *Journal of Digital Imaging*, 4(1):1–20, Febr. 1990.

297. T. Poston and I. Steward. *Catastrophe Theory and its Applications*. Pitman,
      London, 1978.

298. W. H. Press, B. P. Flannery, S. A. Teukolsky, and W. T. Vetterling. *Numerical
      Recipes in C; the Art of Scientific Computing*. Cambridge University Press,
      Cambridge, UK, 1988.

299. M. Proesmans, E. J. Pauwels, and L. J. Van Gool. Coupled geometry-driven
      diffusion equations for low-level vision. Technical Report Technical Report
      KUL/ESAT/MI2/9409, Catholic University Leuven, 1994.

300. M. H. Protter and H. F. Weinberger. *Maximum Principles in Differential
      Equations*. Springer-Verlag, New York, 1984.

301. E. Prugovečki. *Quantum Geometry. A Framework for Quantum General Relativity*.
      Fundamental Theories of Physics. Kluwer Academic Publishers, Dordrecht, NL,
      1992.

302. D. Puff, D. Eberly, and S. Pizer. Object-based interpolation via the multiscale
      medial axis. In *Proc. SPIE Medical Imaging VIII*, February 1994. to appear.

303. W. E. Reichardt and T. Poggio.   Figure-ground discrimination by relative
      movement in the visual system of the fly. part I: Experimental results. *Biol.
      Cybern.*, 35:81–100, 1980.

304. S. O. Rice. Mathematical analysis of random noise. *The Bell System Technical J.*,
      XXIV(1):46–156, 1945.

305. T. J. Richardson. *Scale Independent Piecewise Smooth Segmentation of Images via
      Variational Methods*. PhD thesis, Dept. of E.E. & C.S., M.I.T., 1990.

306. T. J. Richardson. Limit theorems for a variational problem arising in computer
      vision. *Annali della Scuola Normale*, xix, Fasc. 1, 1992.

307. T. L. Richardson and S. K. Mitter. A variational formulation based edge focusing
      algorithm. To appear, 1994.

308. A. Rosenfeld.   *Multiresolution Image Processing and Analysis*, volume 12 of
      Springer Series in Information Sciences. Springer-Verlag, 1984.

309. A. Rosenfeld and M. Thurston. Edge and curve detection for visual scene analysis.
      IEEE Trans. on Computers, C-20:562–569, May 1971.

310. P. L. Rosin, A. C. F. Colchester, and D. J. Hawkes. Early image representation
      using regions defined by maximum gradient profiles between singular points.
      Pattern Recognition, 25(7):695–711, 1992.

311. E. Rouy and A. Turin. A viscosity solutions approach to shape-from-shading.
      SIAM J. Num. Anal., 29:867–884, 1992.

312. J. Rubinstein, P. Sternberg, and J. B. Keller. Fast reaction, slow diffusion, and curve shortening. *SIAM J. Appl. Math.*, 49:116–133, 1989.

313. L. Rudin, S. Osher, and E. Fatemi. Nonlinear total variation based noise removal algorithms. In *Modelisations Matematiques pour le traitement d'images*, pages 149–179. INRIA, 1992.

314. W. B. Ruskai, G. Beylkin, R. Coifman, I. Daubechies, S. Mallat, Y. Meyer, and L. Raphael, editors. *Wavelets and Their Applications*. Jones and Barlett Publishers, Boston, Massachusetts, 1992.

315. D. Rutovitz. Datastructures for operations on digital images. In G. C. Cheng, D. K. Pollock, and A. Rosenfeld, editors, *Pictorial Pattern Recognition*, pages 105–133, Washington, DC, 1968. Thompson.

316. P. Saint-Marc, J. S. Chen, and G. Medioni. Adaptive smoothing: A general tool for early vision. *IEEE Trans. Pattern Analysis and Machine Intelligence*, 13(6):514–529, 1991.

317. A. H. Salden, L. M. J. Florack, and B. M. ter Haar Romeny. Differential geometric description of 3D scalar images. Technical Report 91-23, 3D Computer Vision, Utrecht, 1991.

318. A. H. Salden, B. M. ter Haar Romeny, and L. M. J. Florack. Algebraic invariants: a complete and irreducible set of local features of 2D scalar images. Technical report, 3D Computer Vision, 1991. Technical Report 3DCV no. 91-22.

319. A. H. Salden, B. M. ter Haar Romeny, L. M. J. Florack, J. J. Koenderink, and M. A. Viergever. A complete and irreducible set of local orthogonally invariant features of 2-dimensional images. In I. T. Young, editor, *Proceedings 11th IAPR Internat. Conf. on Pattern Recognition*, volume III: Image, Speech and Signal Analysis, pages 180–184, The Hague, the Netherlands, August 30–September 3 1992. IEEE Computer Society Press, Los Alamitos.

320. G. Sapiro, R. Kimmel, D. Shaked, B. B. Kimia, and A. M. Bruckstein. Implementing continuous-scale morphology via curve evolution. *Pattern Recognition*, 26(9), 1993.

321. G. Sapiro and A. Tannenbaum. Affine invariant scale-space. Technical report, Technion-Israel Institute of Technology, 1992.

322. G. Sapiro and A. Tannenbaum. Affine shortening of non-convex plane curves. Technical Report 845, Technion Israel Institute of Technology, Department of Electrical Engineering, Technion, IIT, Haifa 32000, Israel, July 1992. EE Publication.

323. G. Sapiro and A. Tannenbaum. Affine invariant scale-space. *International Journal of Computer Vision*, 11:25–44, 1993.

324. G. Sapiro and A. Tannenbaum. Area and length preserving geometric invariant scale-spaces. Technical Report LIDS-2200, MIT, 1993. Accepted for publication in IEEE–PAMI. Also in *Proc. ECCV '94*, Stockholm, May 1994.

325. G. Sapiro and A. Tannenbaum. Formulating invariant heat-type curve flows. In *SPIE Conference on Geometric Methods in Computer Vision II*, 1993.

326. G. Sapiro and A. Tannenbaum. Image smoothing based on an affine invariant flow. In *Conference on Information Sciences and Systems*, pages 196–201. Johns Hopkins University, 1993.

327. G. Sapiro and A. Tannenbaum. On invariant curve evolution and image analysis. *Indiana Journal of Mathematics*, 42(3):985–1009, 1993.

328. G. Sapiro and A. Tannenbaum. On affine plane curve evolution. *Journal of Functional Analysis*, 119(1):79–120, January 1994.

329. I. J. Schoenberg. Contributions to the problem of approximation of equidistant data by analytic functions. *Quarterly of Applied Mathematics*, 4:45–99, 1946.

330. I. J. Schoenberg. On Pòlya frequency functions. II. Variation-diminishing integral operators of the convolution type. *Acta Sci. Math. (Szeged)*, 12:97–106, 1950.

331. I. J. Schoenberg. On smoothing operations and their generating functions. *Bull. Amer. Math. Soc.*, 59:199–230, 1953.

332. J. A. Schouten. *Der Ricci-Kalkül.* Springer-Verlag, 1924.

333. L. Schwartz. *Théorie des Distributions,* volume I, II of *Actualités scientifiques et industrielles; 1091,1122.* Publications de l'Institut de Mathématique de l'Université de Strasbourg, Paris, 1950–1951.

334. R. Schwartz. The pentagram map. *Experimental Mathematics,* 1:71–81, 1992.

335. J. Segman and Y. Y. Zeevi. Image analysis by wavelet-type transforms: Group theoretic approach. *Journal of Mathematical Imaging and Vision,* 3:51–77, 1993.

336. J. Serra. *Image Analysis and Mathematical Morphology.* Academic Press, London, New York, Paris, San Diego, San Francisco, São Paulo, Sydney, Tokyo and Toronto, 1982.

337. J. A. Sethian. *An Analysis of Flow Propagation.* PhD thesis, University of California, 1982.

338. J. A. Sethian. Curvature and the evolution of fronts. *Comm. Math. Phys.,* 101:487–499, 1985.

339. J. A. Sethian. A review of recent numerical algorithms for hypersurfaces moving with curvature dependent speed. *J. Differential Geometry,* 31:131–161, 1989.

340. J. A. Sethian and J. Strain. Crystal growth and dendritic solidification. *Journal of Computational Physics,* 98, 1992.

341. J. Shah. Segmentation by nonlinear diffusion. *Conference on Computer Vision and Pattern Recognition,* pages 202–207, June 1991.

342. J. Shah. Segmentation by minimizing functionals: Smoothing properties. *SIAM J. Control and Optimization,* 30(1):99–111, January 1992.

343. J. Shah. A nonlinear model for discontinuous disparity and half-occlusions in stereo. In *Proc. Computer Vision and Pattern Recognition,* pages 34–40, 1993.

344. M. E. Shenton, R. Kikinis, R. W. McCarley, D. Metcalf, J. Tieman, and F. A. Jolesz. Application of automated MRI volumetric measurement techniques to the ventricular system in schizophrenics and normal controls. *Schizophrenia Research,* 5:103–113, 1991.

345. J. Smoller. *Shock Waves and Reaction-diffusion Equations.* Springer-Verlag, New York, 1983.

346. M. Snyder. On the mathematical foundations of smoothness constraints for the determination of optical flow and for surface reconstruction. *IEEE Trans. Pattern Analysis and Machine Intelligence,* 13:1105–1114, 1991.

347. G. A. Sod. *Numerical Methods in Fluid Dynamics.* Cambridge University Press, Cambridge, 1985.

348. M. Spivak. *A Comprehensive Introduction to Differential Geometry,* volume I–V. Publish or Perish, Inc., Houston, Texas, 1970.

349. M. Spivak. *Differential Geometry,* volume 1–5. Publish or Perish, Inc., Berkeley, California, USA, 1975.

350. G. Strang. *Introduction to Applied Mathematics.* Wellesley-Cambridge Press, Wellesley, MA, 1986.

351. J. O. Strömberg. A modified Franklin system and higher order splines as unconditional basis for Hardy spaces. In B. W. et al., editors, *Proc. Conf. in Harmonic Analysis in Honor of Antoni Zygmund,* volume II. Wadworth Mathematical Series, 1983.

352. S. Tanimoto, editor. *IEEE Trans. Pattern Analysis and Machine Intelligence,* volume 11:7. i, 1989.

353. S. Tanimoto and A. Klinger, editors. *Structured Computer Vision.* Academic Press, New York, 1980.

354. S. Tanimoto and T. Pavlidis. A hierarchical structure for picture processing. *Computer Vision, Graphics, and Image Processing,* 4:104–119, 1975.

355. B. M. ter Haar Romeny and L. M. J. Florack. A multiscale geometric model of human vision. In W. R. Hendee and P. N. T. Wells, editors, *Perception of Visual Information,* chapter 4, pages 73–114. Springer-Verlag, Berlin, 1993.

356. B. M. ter Haar Romeny, L. M. J. Florack, M. de Swart, J. Wilting, and M. A.

Viergever. Deblurring Gaussian blur. In *Proc. Mathematical Methods in Medical Imaging II*, volume 2299, San Diego, CA, July, 25-26 1994. SPIE.

357. B. M. ter Haar Romeny, L. M. J. Florack, J. J. Koenderink, and M. A. Viergever. Invariant third order properties of isophotes: T-junction detection. In P. Johansen and S. Olsen, editors, *Proc. 7th Scand. Conf. on Image Analysis*, pages 346–353, Aalborg, DK, August 1991.

358. B. M. ter Haar Romeny, L. M. J. Florack, J. J. Koenderink, and M. A. Viergever. Scale-space: Its natural operators and differential invariants. In A. C. F. Colchester and D. J. Hawkes, editors, *Information Processing in Medical Imaging*, volume 511 of *Lecture Notes in Computer Science*, pages 239–255, Berlin, July 1991. Springer-Verlag.

359. B. M. ter Haar Romeny, L. M. J. Florack, J. J. Koenderink, and M. A. Viergever. Invariant third order properties of isophotes: T-junction detection. In P. Johansen and S. Olsen, editors, *Theory & Applications of Image Analysis*, volume 2 of *Series in Machine Perception and Artificial Intelligence*, pages 30–37. World Scientific, Singapore, 1992.

360. B. M. ter Haar Romeny, L. M. J. Florack, A. H. Salden, and M. A. Viergever. Higher order geometrical image structure. In H. Barrett, editor, *Proc. Information Processing in Medical Imaging '93, Flagstaff AZ*, pages 77–93, Berlin, 1993. Springer-Verlag.

361. B. M. ter Haar Romeny, L. M. J. Florack, A. H. Salden, and M. A. Viergever. Higher order differential structure of images. *Image and Vision Computing*, July/August 1994.

362. D. Terzopoulos. Regularization of inverse visual problems involving discontinuities. *IEEE Transactions on Pattern Analysis and Machine Intelligence*, 8:413–424, 1981.

363. J. A. Thorpe. *Elementary Topics in Differential Geometry*. Springer-Verlag, New York, NY, 1985.

364. A. N. Tikhonov and V. Y. Arsenin. *Solution of Ill-Posed Problems*. Winston and Wiley, Washington DC, 1977.

365. M. Tistarelli and G. Sandini. Dynamic aspects in active vision. *CVGIP: Image Understanding*, 56(1):108–129, July 1992.

366. V. Torre and T. A. Poggio. On edge detection. *IEEE Trans. Pattern Analysis and Machine Intelligence*, 8(2):147–163, 1986.

367. L. Uhr. Layered 'recognition cone' networks that preprocess, classify and describe. *IEEE Trans. Computers*, pages 759–768, 1972.

368. W. A. van de Grind, J. J. Koenderink, and A. J. van Doorn. The distribution of human motion detector properties in the monocular visual field. *Vision Research*, 26(5):797–810, 1986.

369. R. van den Boomgaard. *Mathematical Morphology: Extensions towards Computer Vision*. PhD thesis, University of Amsterdam, March 1992.

370. P. A. Van den Elsen and M. A. Viergever. Fully automated ct and mr brain image registration by correlation of geometrical features. In H. Barrett, editor, *Proc. Information Processing in Medical Imaging '93, Flagstaff AZ*, Berlin, 1993. Springer Verlag.

371. A. J. van Doorn, J. J. Koenderink, and M. A. Bouman. The influence of the retinal inhomogeneity on the perception of spatial patterns. *Kybernetik*, 10:223–230, 1972.

372. K. L. Vincken and F. J. R. Appelman. Accurate conversion of geometrical objects to voxel-based images. Report 3DCV 91-20, Utrecht University, 1991.

373. K. L. Vincken, A. S. E. Koster, and M. A. Viergever. Probabilistic multiscale image segmentation — set-up and first results. In R. A. Robb, editor, *Visualization in Biomedical Computing 1992*, pages 63–77. Proceedings SPIE 1808, 1992.

374. R. M. Wald. *General Relativity*. The University of Chicago Press, Chicago, 1984.

375. J. Weber and J. Malik. Rigid body segmentation and shape description from optical flow. Technical report, University of California at Berkeley, 1993. EE/CS preprint.

376. J. Weber and J. Malik. Robust computation of optical flow in a multi-scale differential framework. In *Proc. Fourth International Conference on Computer Vision*, pages 12–20, 1993.

377. H. Wechsler. *Computational Vision*. Academic Press, San Diego, CA, 1990.

378. H. Weyl. *The Classical Groups, their Invariants and Representations*. Princeton University Press, Princeton, NJ, 1946.

379. R. T. Whitaker. *Geometry-Limited Diffusion*. PhD thesis, The University of North Carolina, Chapel Hill, North Carolina 27599-3175, 1993.

380. R. T. Whitaker. Geometry-limited diffusion in the characterization of geometric patches in images. *Computer Vision, Graphics, and Image Processing: Image Understanding*, 57(1):111–120, January 1993.

381. R. T. Whitaker and S. M. Pizer. A multi-scale approach to nonuniform diffusion. *Computer Vision, Graphics, and Image Processing: Image Understanding*, 57(1):99–110, January 1993.

382. R. T. Whitaker and S. M. Pizer. Geometry-based image segmentation using anisotropic diffusion. In Y.-L. O, A. Toet, H. J. A. M. Heijmans, D. H. Foster, and P. Meer, editors, *Proc. of the NATO Advanced Research Workshop Shape in Picture — Mathematical Description of Shape in Greylevel Images*, volume 126 of *NATO ASI Series F*, pages 641–650. Springer Verlag, Berlin, 1994.

383. D. V. Widder. *The Heat Equation*. Academic Press, New York, 1975.

384. E. J. Wilczynski. *Projective Differential Geometry of Curves and Ruled Surfaces*, volume 18. Teubner B. G., Leipzig, 1906.

385. R. Wilson and A. H. Bhalerao. Kernel design for efficient multiresolution edge detection and orientation estimation. *IEEE Trans. Pattern Analysis and Machine Intelligence*, 14(3):384–390, 1992.

386. A. P. Witkin. Scale space filtering. In *Proc. International Joint Conference on Artificial Intelligence*, pages 1019–1023, Karlsruhe, Germany, 1983.

387. S. Wolfram. *Mathematica: A System for doing Mathematics by Computer*. Addison-Wesley, 1994. Version 2.2.

388. T. S. Yoo and J. M. Coggins. Using statistical pattern recognition techniques to control variable conductance diffusion. In *Proc. Information Processing in Medical Imaging 1993*, pages 459–471, 1993.

389. R. A. Young. The Gaussian derivative theory of spatial vision: Analysis of cortical cell receptive field line-weighting profiles. Publication GMR-4920, General Motors Research Labs, Computer Science Dept., 30500 Mound Road, Box 9055, Warren, Michigan 48090-9055, May 28 1985.

390. R. A. Young. The Gaussian derivative model for machine vision: Visual cortex simulation. Publication GMR-5323, General Motors Research Labs, Computer Science Dept., 30500 Mound Road, Box 9055, Warren, Michigan 48090-9055, July 7 1986.

391. R. A. Young. The Gaussian derivative model for machine vision: Visual cortex simulation. *Journal of the Optical Society of America*, July 1986.

392. R. A. Young. Simulation of human retinal function with the Gaussian derivative model. In *Proc. IEEE CVPR CH2290-5*, pages 564–569, Miami, Fla., 1986.

393. R. A. Young. The Gaussian derivative model for machine vision: I. retinal mechanisms. *Spatial Vision*, 2(4):273–293, 1987.

394. A. L. Yuille. The creation of structure in dynamic shape. In *IEEE Second Conf. on Computer Vision*, pages 685–689, Tampa, 1988.

395. A. L. Yuille and T. Poggio. Fingerprint theorems for zero crossings. *JOSA, "A"*, 2:683–692, May 1985.

396. A. L. Yuille and T. A. Poggio. Scaling theorems for zero-crossings. *IEEE Trans. Pattern Analysis and Machine Intelligence*, 8:15–25, January 1986.

397. S. Zeki. *A Vision of the Brain*. Blackwell Scientific Publications, Oxford, 1993.

Index

Computational Imaging and Vision

Kluwer Academic Publishers – Dordrecht / Boston / London